Springer Series in Statistics

Advisors:
P. Bickel, P. Diggle, S. Fienberg, K. Krickeberg,
I. Olkin, N. Wermuth, S. Zeger

Springer
New York
Berlin
Heidelberg
Barcelona
Hong Kong
London
Milan
Paris
Singapore
Tokyo

Springer Series in Statistics

Andersen/Borgan/Gill/Keiding: Statistical Models Based on Counting Processes.
Andrews/Herzberg: Data: A Collection of Problems from Many Fields for the Student and Research Worker.
Anscombe: Computing in Statistical Science through APL.
Berger: Statistical Decision Theory and Bayesian Analysis, 2nd edition.
Bolfarine/Zacks: Prediction Theory for Finite Populations.
Borg/Groenen: Modern Multidimensional Scaling: Theory and Applications
Brémaud: Point Processes and Queues: Martingale Dynamics.
Brockwell/Davis: Time Series: Theory and Methods, 2nd edition.
Daley/Vere-Jones: An Introduction to the Theory of Point Processes.
Dzhaparidze: Parameter Estimation and Hypothesis Testing in Spectral Analysis of Stationary Time Series.
Fahrmeir/Tutz: Multivariate Statistical Modelling Based on Generalized Linear Models.
Farebrother: Fitting Linear Relationships: A History of the Calculus of Observations 1750 - 1900.
Farrell: Multivariate Calculation.
Federer: Statistical Design and Analysis for Intercropping Experiments, Volume I: Two Crops.
Federer: Statistical Design and Analysis for Intercropping Experiments, Volume II: Three or More Crops.
Fienberg/Hoaglin/Kruskal/Tanur (Eds.): A Statistical Model: Frederick Mosteller's Contributions to Statistics, Science and Public Policy.
Fisher/Sen: The Collected Works of Wassily Hoeffding.
Good: Permutation Tests: A Practical Guide to Resampling Methods for Testing Hypotheses.
Goodman/Kruskal: Measures of Association for Cross Classifications.
Gouriéroux: ARCH Models and Financial Applications.
Grandell: Aspects of Risk Theory.
Haberman: Advanced Statistics, Volume I: Description of Populations.
Hall: The Bootstrap and Edgeworth Expansion.
Härdle: Smoothing Techniques: With Implementation in S.
Hart: Nonparametric Smoothing and Lack-of-Fit Tests.
Hartigan: Bayes Theory.
Hedayat/Sloane/Stufken: Orthogonal Arrays: Theory and Applications.
Heyde: Quasi-Likelihood and its Application: A General Approach to Optimal Parameter Estimation.
Heyer: Theory of Statistical Experiments.
Huet/Bouvier/Gruet/Jolivet: Statistical Tools for Nonlinear Regression: A Practical Guide with S-PLUS Examples.
Jolliffe: Principal Component Analysis.
Kolen/Brennan: Test Equating: Methods and Practices.
Kotz/Johnson (Eds.): Breakthroughs in Statistics Volume I.

(continued after index)

Nozer D. Singpurwalla
Simon P. Wilson

Statistical Methods in Software Engineering

Reliability and Risk

With 55 Illustrations

 Springer

Nozer D. Singpurwalla
Department of Operations Research
The George Washington University
Washington, DC 20052
USA
nozer@research.circ.gwu.edu

Simon P. Wilson
Department of Statistics
Trinity College
Dublin 2
Ireland
swilson@stats.tcd.ie

Library of Congress Cataloging-in-Publication Data
Singpurwalla, Nozer D.
   Statistical methods in software engineering: reliability and risk
/Nozer D. Singpurwalla, Simon P. Wilson.
    p. cm. — (Springer series in statistics)
   Includes bibliographical references.
   ISBN 0-387-98823-8 (alk. paper)
   1. Software engineering. 2. Statistical methods. I. Wilson, Simon P.
II. Title. III. Series.
QA76.758.S535   1999
005.1—dc21   99-14737

Printed on acid-free paper.

© 1999 Springer-Verlag New York, Inc.
All rights reserved. This work may not be translated or copied in whole or in part without the written permission of the publisher (Springer-Verlag New York, Inc., 175 Fifth Avenue, New York, NY 10010, USA), except for brief excerpts in connection with reviews or scholarly analysis. Use in connection with any form of information storage and retrieval, electronic adaptation, computer software, or by similar or dissimilar methodology now known or hereafter developed is forbidden. The use of general descriptive names, trade names, trademarks, etc., in this publication, even if the former are not especially identified, is not to be taken as a sign that such names, as understood by the Trade Marks and Merchandise Marks Act, may accordingly be used freely by anyone.

Production managed by Frank McGuckin; manufacturing supervised by Nancy Wu.
Camera-ready copy provided by the author using EXP.
Printed and bound by Maple-Vail Book Manufacturing Group, York, PA.
Printed in the United States of America.

9 8 7 6 5 4 3 2 1

ISBN 0-387-98823-8 Springer-Verlag New York Berlin Heidelberg   SPIN 10034742

Preface

This preface pertains to three issues that we would like to bring to the attention of the readers: our objectives, our intended audience, and the nature of the material.

We have in mind several objectives. The first is to establish a framework for dealing with uncertainties in software engineering, and for using quantitative measures for decision making in this context. The second is to bring into perspective the large body of work having statistical content that is relevant to software engineering, which may not have appeared in the traditional outlets devoted to it. Connected with this second objective is a desire to streamline and organize our own thinking and work in this area. Our third objective is to provide a platform that facilitates an interface between computer scientists and statisticians to address a class of problems in computer science. It appears that such an interface is necessary to provide the needed synergism for solving some difficult problems that the subject poses. Our final objective is to serve as an agent for stimulating more cross-disciplinary research in computer science and statistics. To what extent the material here will meet our objectives can only be assessed with the passage of time.

Our intended audience is computer scientists, software engineers, and reliability analysts, who have some exposure to probability and statistics. Applied statisticians interested in reliability problems are also a segment of our intended audience. The content is pitched at a level that is appropriate for research workers in software reliability, and for graduate-level courses in applied statistics, computer science, operations research, and software engineering. Industrial scientists looking for a better understanding of the ideas behind the statistical tools that they use for addressing problems of software quality may also find the material of value. We have deliberately steered away from presenting techniques that are purely data analytic, since there are ample sources that do this.

Recognizing the diverse nature of our audience, and in keeping with our stated objectives, we have adopted an expository style and have striven to give as many illustrative examples as is possible; furthermore, we have endeavored to cast the examples in a context that may appeal to software engineers. Additional examples have been delegated to exercises. Readers who are formally trained in the statistical sciences will be familiar with the material of Chapter 2; they may find little that is new here, and may therefore be tempted to skip it. The same may also be true of the initial parts of Chapter 4. However, we urge such readers not to do so because of two reasons. The first is that the illustrative examples give a flavor of the nature of problems that we are trying to address. The second reason is that our interpretation of probability is personal (subjective); it therefore makes us look at the standard material in probability and statistics at a different angle. Of course, not all are willing to subscribe to this perspective. Computer scientists, operations research analysts, and software engineers should find the material of Chapter 2 and the initial parts of Chapter 4 as a useful review, but with a focus towards a specific application. The material in the other chapters is self-evident and so does not deserve special comment.

By way of a final admission, we are anticipating the criticism that any endeavor that attempts to fill a gap in the literature which is at the interface of computer science and statistics is necessarily incomplete. If this be so, then our hope is that the material here will stimulate the next generation of writers to expand the frontiers of the interface and to eliminate the pockets of incompleteness that we undoubtedly have created.

May 1999   Nozer D. Singpurwalla
                                                                        Simon P. Wilson

Acknowledgments

Nozer D. Singpurwalla acknowledges the support of The George Washington University, and the sponsorship of The Army Research Office, The Air Force Office of Scientific Research, and The Office of Naval Research for initiating and sustaining his research on the subject of this book. Drs. Jagdish Chandra, Robert Launer, Charles Holland, Jon Sjogren, Seymour Selig, Julia Abrahams, and Professors Edward Wegman, Harold Liebowitz, William Marlow, Donald Gross, and Richard Soland are singled out with heartful thanks. He also acknowledges the contributions to his learning of his many coauthors; Professors Richard Barlow, Frank Proschan, Dennis Lindley, and Jayaram Sethuraman deserve special mention. For ensuring quality during the final stages of this work, the keen eye and the skill of Karen Brady, Yuling Cui, Selda Kapan, Seung Byeon, Andrew Swift, and Chung-Wai Kong are acknowledged, with Chung-Wai and Yuling deserving a very singular recognition. Mrs. Teresita Abacan's infinite efforts to convert handwritten notes into a masterful manuscript, often at supersonic speeds, is gratefully admired. Professors Ingram Olkin and Stephen Feinberg cast the initial bait, a long time ago, to suggest the writing of this book, and Dr. John Kimmel of Springer Verlag ensured that the fish did not slip away. Finally, he acknowledges the support of his family members, wife, children, mother, mother-in-law, and sister, each of whom in their own subtle way contributed to his work by not imposing any demands on him. Like the hidden parameters of probability models, they are the invisible coauthors.

Simon P. Wilson acknowledges his parents, for all their support and encouragement.

CONTENTS

Preface	v
Acknowledgments	vii

1 Introduction and Overview — 1

1.1 What is Software Engineering? — 1
1.2 Uncertainty in Software Production — 2
    1.2.1 The Software Development Process — 2
    1.2.2 Sources of Uncertainty in the Development Process — 3
1.3 The Quantification of Uncertainty — 4
    1.3.1 Probability as an Approach for Quantifying Uncertainty — 4
    1.3.2 Interpretations of Probability — 6
    1.3.3 Interpreting Probabilities in Software Engineering — 9
1.4 The Role of Statistical Methods in Software Engineering — 9
1.5 Chapter Summary — 11

2 Foundational Issues: Probability and Reliability — 13

2.0 Preamble — 13
2.1 The Calculus of Probability — 14
    2.1.1 Notation and Preliminaries — 14
    2.1.2 Conditional Probabilities and Conditional Independence — 16
    2.1.3 The Calculus of Probability — 17
    2.1.4 The Law of Total Probability, Bayes' Law, and the Likelihood Function — 20
    2.1.5 The Notion of Exchangeability — 25
2.2 Probability Models and Their Parameters — 28
    2.2.1 What is a Software Reliability Model? — 28
    2.2.2 Some Commonly Used Probability Models — 29
    2.2.3 Moments of Probability Distributions and Expectation of Random Variables — 39
    2.2.4 Moments of Probability Models: The Mean Time to Failure — 41

x   Contents

    2.3 Point Processes and Counting Process Models   41
        2.3.1 The Nonhomogeneous Poisson Process Model   43
        2.3.2 The Homogeneous Poisson Process Model   45
        2.3.3 Generalizations of the Point Process Model   46
    2.4 Fundamentals of Reliability   52
        2.4.1 The Notion of a Failure Rate Function   53
        2.4.2 Some Commonly Used Model Failure Rates   54
        2.4.3 Covariates in the Failure Rate Function   57
        2.4.4 The Concatenated Failure Rate Function   58
    2.5 Chapter Summary   59

    Exercises for Chapter 2   61

3 Models for Measuring Software Reliability   67

    3.1 Background: The Failure of Software   67
        3.1.1 The Software Failure Process and Its Associated Randomness   68
        3.1.2 Classification of Software Reliability Models   70
    3.2 Models Based on the Concatenated Failure Rate Function   72
        3.2.1 The Failure Rate of Software   72
        3.2.2 The Model of Jelinski and Moranda (1972)   72
        3.2.3 Extensions and Generalizations of the Model by Jelinski and Moranda   75
        3.2.4 Hierarchical Bayesian Reliability Growth Models   76
    3.3 Models Based on Failure Counts   77
        3.3.1 Time Dependent Error Detection Models   77
    3.4 Models Based on Times Between Failures   80
        3.4.1 The Random Coefficient Autoregressive Process Model   80
        3.4.2 A Non-Gaussian Kalman Filter Model   81
    3.5 Unification of Software Reliability Models   82
        3.5.1 Unification via the Bayesian Paradigm   83
        3.5.2 Unification via Self-Exciting Point Process Models   84
        3.5.3 Other Approaches to Unification   86
    3.6 An Adaptive Concatenated Failure Rate Model   91
        3.6.1 The Model and Its Motivation   92

		Contents	xi

| | 3.6.2 | Properties of the Model and Interpretation of Model Parameters | 94 |
| 3.7 | Chapter Summary | | 95 |

Exercises for Chapter 3   97

4 Statistical Analysis of Software Failure Data   101

4.1 Background: The Role of Failure Data   101

4.2 Bayesian Inference, Predictive Distributions, and Maximization of Likelihood   103
- 4.2.1 Bayesian Inference and Prediction   104
- 4.2.2 The Method of Maximum Likelihood   105
- 4.2.3 Application: Inference and Prediction Using Jelinski and Moranda's Model   106
- 4.2.4 Application: Inference and Prediction Under an Error Detection Model   110

4.3 Specification of Prior Distributions   113
- 4.3.1 Standard of Reference—Noninformative Priors   114
- 4.3.2 Subjective Priors Based on Elicitation of Specialist Knowledge   115
- 4.3.3 Extensions of the Elicitation Model   117
- 4.3.4 Example: Eliciting Priors for the Logarithmic-Poisson Model   118
- 4.3.5 Application: Failure Prediction Using Logarithmic-Poisson Model   120

4.4 Inference and Prediction Using a Hierarchical Model   124
- 4.4.1 Application to NTDS Data: Assessing Reliability Growth   126

4.5 Inference and Predictions Using Dynamic Models   129
- 4.5.1 Inference for the Random Coefficient Exchangeable Model   131
- 4.5.2 Inference for the Adaptive Kalman Filter Model   141
- 4.5.3 Inference for the Non-Gaussian Kalman Filter Model   143

4.6 Prequential Prediction, Bayes Factors, and Model Comparison   145

	4.6.1	Prequential Likelihoods and Prequential Prediction	146
	4.6.2	Bayes' Factors and Model Averaging	148
	4.6.3	Model Complexity—Occam's Razor	150
	4.6.4	Application: Comparing the Exchangeable, Adaptive, and Non-Gaussian Models	151
	4.6.5	An Example of Reversals in the Prequential Likelihood Ratio	153
4.7	Inference for the Concatenated Failure Rate Model		154
	4.7.1	Specification of the Prior Distribution	155
	4.7.2	Calculating Posteriors by Markov Chain Monte Carlo	157
	4.7.3	Testing Hypotheses About Reliability Growth or Decay	159
	4.7.4	Application to System 40 Data	160
4.8	Chapter Summary		164

Exercises for Chapter 4 — 166

5 Software Productivity and Process Management — 169

5.1 Background: Producing Quality Software — 169

5.2 A Growth-Curve Model for Estimating Software Productivity — 170
    5.2.1 The Statistical Model — 171
    5.2.2 Inference and Prediction Under the Growth-Curve Model — 174
    5.2.3 Application: Estimating Individual Software Productivity — 176

5.3 The Capability Maturity Model for Process Management — 180
    5.3.1 The Conceptual Framework — 181
    5.3.2 The Probabilistic Approach for Hierarchical Classification — 183
    5.3.3 Application: Classifying a Software Developer — 186

5.4 Chapter Summary — 188

Exercises for Chapter 5 — 190

Contents   xiii

6 The Optimal Testing and Release of Software   191

6.1 Background: Decision Making and the
     Calculus of Probability   191
6.2 Decision Making Under Uncertainty   192
6.3 Utility and Choosing the Optimal Decision   194
    6.3.1 Maximization of Expected Utility   194
    6.3.2 The Utility of Money   195
6.4 Decision Trees   196
    6.4.1 Solving Decision Trees   197
6.5 Software Testing Plans   198
6.6 Examples of Optimal Testing Plans   202
    6.6.1 One-Stage Testing Using the Jelinski–Moranda
          Model   202
    6.6.2 One-and Two-Stage Testing Using the Model
          by Goel and Okumoto   206
    6.6.3 One-Stage Lookahead Testing Using the Model
          by Goel and Okumoto   211
    6.6.4 Fixed-Time Lookahead Testing for the
          Goel–Okumoto Model   212
    6.6.5 One-Bug Lookahead Testing Plans   214
    6.6.6 Optimality of One-Stage Look Ahead Plans   215
6.7 Application: Testing the NTDS Data   216
6.8 Chapter Summary   217

Exercises for Chapter 6   219

7 Other Developments: Open Problems   221

7.0 Preamble   221
7.1 Dynamic Modeling and the Operational Profile   222
    7.1.1 Martingales, Predictable Processes, and
          Compensators: An Overview   222
    7.1.2 The Doob–Meyer Decomposition of Counting
          Processes   224
    7.1.3 Incorporating the Operational Profile   227
7.2 Statistical Aspects of Software Testing: Experimental Designs   228
    7.2.1 Inferential Issues in Random and Partition Testing   229

xiv   Contents

      7.2.2   Comparison of Random and Partition Testing   231
      7.2.3   Design of Experiments in Software Testing   232
      7.2.4   Design of Experiments in Multiversion Programming   236
      7.2.5   Concluding Remarks   237
  7.3 The Integration of Module and System Performance   238
      7.3.1   The Protocols of Control Flow and Data Flow   239
      7.3.2   The Structure Function of Modularized Software   242

Appendices   247

Appendix A   Statistical Computations Using the Gibbs Sampler   249

  A.1 An Overview of the Gibbs Sampler   250
  A.2 Generating Random Variates—The Rejection Method   253
  A.3 Examples: Using the Gibbs Sampler   254
      A.3.1   Gibbs Sampling the Jelinski–Moranda Model   254
      A.3.2   Gibbs Sampling the Hierarchical Model   255
      A.3.3   Gibbs Sampling the Adaptive Kalman Filter Model   256
      A.3.4   Gibbs Sampling the Non-Gaussian Kalman Filter Model   258

Appendix B   The Maturity Questionnaire and Responses   261

  B.1 The Maturity Questionnaire   261
  B.2 Binary (Yes, No) Responses to the Maturity Questionnaire   265
  B.3 Prior Probabilities and Likelihoods   266
      B.3.1   The Maturity Levels $\mathcal{P}(M_i \mid M_{i-1})$ 266
      B.3.2   The Key Process Areas $\mathcal{P}(K_{ij})$ and $\mathcal{P}(K_{ij} \mid M_i)$ 266
      B.3.3   The Likelihoods $\mathcal{L}(K_{ij}; \underline{R}^{ij})$ 268

References   269

Author Index   283

Subject Index   287

1
INTRODUCTION AND OVERVIEW

1.1 What is Software Engineering?

Since the dawn of the computer age, in the 1940s, we have witnessed a prodigious increase in the performance and use of computers. Accompanying this evolution has been a steady shift in emphasis of computer systems development, from hardware—the physical components of the computer—to software—the process of instructing a computer to perform its tasks. Consequently, today only about 10% of the cost of a large computer system lies in the hardware, compared with over 80% in the 1950s. The reasons behind this trend are both the cause and the justification for the emergence of the field of software engineering. In essence, as is true of all mechanical technologies, the cost of hardware gets constantly driven down as new technologies of production come into play, whereas the cost of producing software, which involves harnessing the collective skills of several personnel, gets driven up. Further contributing to these costs are the nuances of delays and budget overruns [Charette (1989), Chapter 1].

The term *software engineering* was not coined until the late 1960s. At that time concerns about the "software crisis," with software being expensive, bug-ridden, and impossible to maintain, led to the notion that a move towards greater discipline in software development could resolve the problem. Hence "software engineering" was born. The IEEE glossary on the subject defines software engineering as the systematic approach to the development, operation, maintenance, and retirement of computer software.

Thus, contrary to common belief, software engineering is not limited to the efficient production of computer code. Indeed, according to Jalote (1991), only about one fifth of the cost of producing software can be attributed to coding. Coding is but one activity in a process that involves problem specification, requirements analysis, installation, and maintenance. Software engineering attempts to bring a systematic methodology to this entire four-phase process.

The definition of software engineering presumes an appreciation as to "what software is." Here again, the IEEE glossary provides an interpretation.

Software is the collection of computer programs, procedures, rules, and their associated documentation, and the last but not least, data.

Once again, contrary to common belief, software is not just computer code—it encompasses all the information necessary to instruct and to manage a computer system.

To summarize, software engineering can be viewed as the efficient management, of a cycle of activities involving the development, operation, maintenance, and retirement of software. By maintenance it is meant an upgrading of the system to respond to changing needs, and the elimination of any residual bugs. By retirement it is meant the designing of new software to replace the existing version.

1.2   Uncertainty in Software Production

As a general rule, uncertainty arises in any activity involving unknown factors. With software, uncertainty is inevitable in all four stages of the software engineering cycle. Despite this fact, attention to uncertainty has predominantly been focused towards the development phase. For this reason, we find it useful to start with a brief description of the development process; more details can be found in Jalote (1991).

1.2.1   *The Software Development Process*

A broadly agreed upon sequence of stages that constitute what is referred to as the software development process are: analysis and specification of requirements, design of the software, and finally, coding, testing, and debugging. There could be included a further stage, namely, installation; this stage involves implementing the software in a client's environment, training the client's staff, and changing the code to rectify bugs or other problems of implementation.

In the analysis and specification phase of the development process, the aim is to precisely define, in close partnership with the user, what the software is to accomplish. Mention of how this is to be done occurs at the next stage. From the point of view of the user, this phase may also include the selection of an organization to undertake the project. For small systems, the analysis and specifications phase may be relatively straightforward, but for large projects this phase will be difficult and prone to error. Techniques such as *data flow diagrams* have been developed to systematize the specification of requirements and to reduce the number of mistakes made. The aim is to produce an unambiguous specification of what the software is required to do; details are in Davis (1990).

In the design phase of the development process, a strategy is formulated to solve the problem that has been specified in the previous phase. The design phase progresses by splitting the original problem into subproblems that can be separately worked upon, and finally integrated. The design phase concludes with a *design document* that specifies how the problem is to be solved, what data structures and formats are to be used, the nature of the modules, and for each module its internal logic and the algorithms to be employed.

The coding and testing phase completes the development process. In coding, emphasis is placed on producing an easily understandable code that will aid greatly in reducing the costs of later testing and maintenance. Individual modules may be tested during coding, but the whole program is not. In the testing part of this phase, the modules are integrated to form the entire system which is then tested to see if it meets specifications. With testing, proper interaction between the modules is ensured. The purpose of testing is to detect the presence of *software faults* or *bugs* in the software code. A software fault is an error in the program source-text, which when the program is executed under certain conditions can cause a software failure.

By *software failure* we mean the deviation of the program output from what it should be according to our requirements. A software fault is generated the moment a programmer or system analyst makes a mistake. Once testing is completed, the system is demonstrated to its client. The nature of the tests given to the software is important. This is because the set of all possible inputs to the software is generally enormous and so is the sequence in which the inputs are received by the software; that is, the *operational profile* of the software is not unique. Therefore, exhaustive testing of the software is not possible and, consequently, the selection of appropriate tests is a crucial matter. Also critical is the manner in which information about the credibility of the software is assessed from the limited tests. By many accounts, the testing phase of the development process is viewed as being the most expensive. To date, statistical methods have played a key role with regard to only the testing phase of the development process.

1.2.2 Sources of Uncertainty in the Development Process

Conceptually, there are many sources of uncertainty in the analysis and specification phase of the process. However, one source that has received much attention pertains to the selection of the organization to be used to develop, install, and maintain the software. Here to make sensible decisions several factors, such as the abilities of the organizations to successfully undertake each phase of the development process, the technical and managerial qualifications of its staff, its track record, its ability to control quality, its responsiveness to changes, and the like must be considered.

Uncertainties in the design phase are those associated with the times required to complete the coding and the testing phases, those associated with changes in requirements, and those associated with the environment under which the software operates. Uncertainties associated with the testing phase pertain to the number of bugs observed, the time required to eliminate the bugs, the test bed, the testing strategy to use, and so on. Uncertainties do not disappear after the testing phase. Once the testing terminates and the software is released, uncertainties about the credibility of the software continue to persist as does the uncertainty about the time at which the software will be replaced.

Clearly, like other production processes, the software development process is besieged with uncertainties, uncertainties which interact (and propagate) with each other. All these impinge on the final cost of the project. For example, uncertainties about the selection of an organization for software development propagates to uncertainties about the quality of the code, which then affects the time for testing, and this has an influence on the reliability of the software. Within the code, the modules form a network of interacting programs, and the reliabilities of the modules combine to form the reliability of the system. The manner in which the uncertainties interact and propagate is generally complicated.

1.3 The Quantification of Uncertainty

Uncertainty is a common phenomenon that arises in almost all aspects of our lives. Here, we concern ourselves with ways of quantifying uncertainty and means by which we can cope with it, especially as it pertains to the specific field of software engineering. Two branches of mathematics play a role in approaches for quantifying and coping with uncertainty: *probability theory* for quantifying and combining uncertainties, and *statistical inference* for revising the uncertainties in the light of data. In what follows, and also in Chapter 2, we review key aspects of the former; later on, in Chapter 4, we expand the discussion to encompass aspects of the latter.

1.3.1 *Probability as an Approach for Quantifying Uncertainty*

The literature in mathematics and in philosophy discusses several approaches for quantifying uncertainty. All of these approaches, save possibility theory and fuzzy logic, have roots in the theory of probability. However, not all of them fully subscribe to the calculus of probability as the sole basis for treating uncertainty. This compromise in philosophy has occurred despite arguments which show that probability is a very defensible way for quantifying uncertainty. It is not our intention here to debate the various approaches for describing uncertainty. Rather, we start by stating that for our purposes, probability and its calculus are used as the sole means for quantifying the uncertainties in software engineering.

To start our discussion on probability, let us focus attention on some reference time, say τ, which for purposes of convenience is often taken to be zero. At time τ we have at our disposal two types of quantities, those known to us and those which are not. For example, with software, the known quantities would be the number of lines of code, the composition of the programming team, the amount of testing that the software has been subjected to, the cost of producing it, and so on. The unknown quantities are conceptually many, but the ones of greatest interest to us could be the number of bugs remaining in the software, the running time (measured in central processing unit time increments) of the software until failure, the ability of the software to perform a particular operation, and so on. The collection of known quantities is denoted $\mathcal{H}$, for history, and the unknowns, referred to as *random quantities,* are denoted by capital letters, such as T or X. The particular values that T and X can take, known as their *realizations*, are denoted by their corresponding small letters, t and x, respectively. If the realizations of a random quantity are numerical, that is, if t and x are numbers, then the random quantities are known as *random variables.* Of particular interest are some special random quantities, called *random events*. These are often denoted by E, and their distinguishing feature is that any E can take only two values, say e_1 and e_2. Random events are generally propositions, and these are either true or false. In the context of software, events could be propositions like, "this program contains no bugs," "this program will experience a failure when it is next used," "T will be greater than t, for some $t \geq 0$," and so on. Since a proposition is either true or false, $E = e_1$ could denote its truth, and $E = e_2$, otherwise. Often, the e_is are assigned numerical values, like 1 and 0, and in such cases the random events are known as *binary random variables*. Random variables are classified as being either *discrete* or *continuous*. Discrete random variables are those whose realizations are countable whereas continuous random variables are those whose realizations are not. For example, if the random variable N denotes the number of bugs that are remaining in the software, then N is discrete, whereas if T denotes the time to failure of the software, then T is continuous.

Probability theory deals with the quantification of uncertainty, at the reference time τ, our uncertain quantities being denoted by capital letters such as T, X, E, and the like. We need to quantify uncertainty, because to quantify is to measure, and measurement is necessary to bring to bear the full force of the logical argument. Thus, at time τ, we need to express (i.e., to *assess*) our uncertainty about a random quantity, or an event E, in the light of $\mathcal{H}$, the available history at time τ. But measurement means assigning numerical values, and following convention we denote this number by $\mathcal{P}^\tau(E \mid \mathcal{H})$, the superscript τ representing the time of assessment and the symbol $\mathcal{H}$ representing the fact that the assessment is made in the light of the history at time τ. The number $\mathcal{P}^\tau(E \mid \mathcal{H})$ is known as the *probability* of the event E (as assessed at τ in the light of $\mathcal{H}$). In the interest of brevity, it has become a practice to suppress both τ and $\mathcal{H}$ and to denote probability by simply $\mathcal{P}(E)$. However, it is very important, especially when describing the credibility of software, to bear in mind that at some future time $\tau + \gamma$, the history may change (because new information surfaces) and so $\mathcal{P}^{\tau+\gamma}(E \mid \mathcal{H})$ will not in general be the same as $\mathcal{P}^\tau(E \mid \mathcal{H})$. Having laid out the preceding framework, we next address several questions about the properties of $\mathcal{P}^\tau(E \mid \mathcal{H})$ that naturally arise.

- What does probability mean (that is, how should we interpret it)?
- How should we assign probabilities (that is, how should we make it operational)?
- What rules should govern probabilities (that is, what is the *calculus of probability*)?
- Who is supposed to be assessing these probabilities (that is, whose history is being taken into account)?

These questions are at the core of the several ongoing debates about the nature of probability. Following the attitude of Chebyshev (1821–1894), Markov (1856–1922), and Lyapunov (1857–1918), most mathematicians concentrate only on the calculus of probability, about which there is agreement, albeit not complete. Generally, the mathematicians have refrained from interpreting the remaining issues, and following the suggestion of Bernstein (1880–1968) [which culminated in Kolmogorov's (1933) famous work; Kolmogorov (1950)], view even the calculus of probability as being axiomatic. However, those interested in applications must come to terms with all the preceding issues. In response to this need, we next discuss the several interpretations of probability. The calculus of probability, to include the fundamentals of reliability and an overview of probability models, is reviewed later, in Chapter 2.

1.3.2 Interpretations of Probability

What does the number $\mathcal{P}^\tau(E \mid \mathcal{H})$ mean? For example, what does it mean to say that the probability of a coin landing heads on the next toss is 0.5, or that the probability is 0.999 that this piece of software is bug free? It turns out that the answer to this question is not unique, and that it depends on one's philosophical orientation. For example, the pioneers of probability theory, Bernoulli, DeMoivre, Bayes, Laplace, and Poisson, who like Newton were determinists, viewed probability as a measure of partial knowledge, or a degree of certainty, and used the "principle of indifference" (or insufficient reason) to invoke an argument of symmetry of outcomes to arrive at a number such as 0.5 for the probability of heads. However, for problems involving loaded coins symmetry could not be used, and the pioneers did not hesitate to use relative frequencies. Indeed, Bernoulli's law of large numbers describes conditions under which relative frequencies become stable, and the DeMoivre–Laplace central limit theorem describes the pattern of fluctuations of the relative frequencies from a central value. A relative frequency interpretation of probability may date back to Aristotle, but its beginnings can be traced to Quetlet, John Stewart Mill, and John Venn in 1866; its most prominent spokesperson was von Mises (1957). Difficulties with this interpretation of probability surfaced as early as 1860 with Maxwell's probabilistic description of the velocity of gas molecules, but the positivist sentiment of the early 20th century did not deter its growing importance. Consequently, much of statistical practice today is based on a relative frequency interpretation of probability. We show later, in Section 1.3.3, that this interpretation of probability poses difficulties in attaching meaning to a statement like "the probability that this software contains no bugs is 0.999." The most vigorous opponents of the frequency school have been the 20th century subjectivists such as Ramsey, de Finetti, and Savage who have sought a foundation for probability based on *personal* betting rates and *personal* degrees of belief. This is in slight contrast to the pioneers who sought a foundation for probability based on *fair* betting rates and warranted degrees of belief. The subjectivist or personal interpretation of probability forms a foundation for much of what is now practiced as *subjectivist Bayesian inference*. In what follows, we summarize the key features of the frequentist and the subjective interpretations of probability. In Section 1.3.3, we point out which of these two interpretations of probability is to be preferred for describing the credibility of software, and indicate the reasons behind our preference.

Before we close this section, it is useful to mention that there is another interpretation of probability which is due to Keynes (1883–1946) and also to Carnap (1891–1970). This is known as the "*a priori*" interpretation, and here probability describes a logical relationship between statements; consequently, every assigned probability is true, correct, and fixed. However, the assigned probabilities are relative to the evidence at hand and so an a priori probability is both objective and subjective. Harold Jeffreys was attracted to the a priori interpretation of probability but appears to have veered away from the notion that every assigned probability must be true, correct, and fixed.

Relative Frequency Theory of Probability

In the relative frequency theory of probability, also known as a *frequentist theory*, probability is defined as the limit of a relative frequency, expressed as an infinite series. Probability is metaphysically viewed, as something physical, and as an objective (i.e., consistently verifiable) property of the real world, such as weight or volume. Consequently, probabilities can only be assessed a posteriori (that is, upon observation). This is in contrast to some other theories which view probability as an index of human attitudes. The most important feature of the frequentist theory is that it can only be applied to scenarios wherein one can conceptualize indefinitely repetitive trials conducted under "almost identical conditions." That is, probability is a property of a *collective* or an *ensemble*. Individual and infrequent events are excluded from consideration, because they do not possess this repetitive character. Games of chance and social mass phenomena, such as insurance and demography, or production mass phenomena, such as those encountered in industrial quality control, are suitable collectives and well within the realm of application of the theory. Also suitable as a collective are the molecules of a gas undergoing Brownian motion; that is, the molecules collide with each other and with the walls of the container.

To summarize, in order to invoke the relative frequency theory of probability, we first need to establish the existence of a collective. Second, when we speak of the probability of a certain attribute, say heads, we mean the probability of *encountering* the attribute within the collective. Third, since probability is defined as the limit of a relative frequency expressed as an infinite series, such limits can only be proved to exist in a series that is mathematical. In applications there can be no assurance that a limit will exist, and if it does exist, its actual value can neither be verified nor disputed. The main virtues of this theory are psychological (on grounds of objectivity) and practical (it works in cases such as biased dice and loaded coins). It is appealing to physical scientists, to whom probability, like mass and volume, is a construct that cannot be directly observed but which serves a useful purpose.

Subjective or Personal Probability

The subjective or personal probability of an event, say E, is the degree of belief that a person (or a committee) has about the occurrence of E. Personal probabilities should therefore depend on $\mathcal{H}$, the background information that the person has about E. The probability need not be unique to all persons, and furthermore, can be different for the same person at different points in time. Clearly, subjective probability cannot be construed as being objective.

For example, suppose that event E denotes a coin landing heads on the next toss. Then by the probability of E, we mean a quantification of our belief about E. This belief could be guided by all our knowledge of the coin, such as its country of origin, its metallic composition and the like, our experience with tossing coins in general, and ultimately our judgment about the fairness of the coin. Suppose that based on all of the preceding considerations we declare $\mathcal{P}(E \mid \mathcal{H})$ to be 0.5. If our $\mathcal{H}$ were to change, perhaps because we flipped the coin several times and noted a preponderance of heads over tails, then we would be allowed to revise $\mathcal{P}(E|\mathcal{H})$ from 0.5 to a number larger than 0.5. Similarly, if E denotes the event that our software has no bugs, then $\mathcal{P}(E \mid \mathcal{H})$ denotes our personal belief about E based on $\mathcal{H}$, all our knowledge about the software, to include any testing we may have done on it. It is important to note that in order to declare $\mathcal{P}(E \mid \mathcal{H})$ we do not have to conceptualize an ensemble, nor do we have to think in terms of indefinite trials under almost identical conditions.

Subjective probability was made operational by de Finetti (1937) who thought of $P(E \mid \mathcal{H})$ as a *betting coefficient,* that is, the amount that the person declaring it is willing to stake in exchange of one monetary unit if E turns out to be true. If E turns out to be false, then the person is prepared to lose $P(E \mid \mathcal{H})$. Coherence (see Chapter 2) demands that a person willing to stake $P(E \mid \mathcal{H})$ for the occurrence of E, should also be prepared to stake an amount $1 - P(E \mid \mathcal{H})$ for the nonoccurrence of E. In avoiding the requirement of ensembles and the existence of unverifiable limits, subjective probability has a more universal scope of applicability than frequentist probability. Its main disadvantage stems from the thought that in actuality betting coefficients may not represent a person's true beliefs. (An indicator of the difference between the two is that betting coefficients are countably additive whereas subjective probabilities need only be finitely additive.)

1.3.3   Interpreting Probabilities in Software Engineering

Because the relative frequency theory of probability requires the conceptualization of a repeatable sequence of trials (or experiments) under almost identical conditions, it is not a suitable paradigm for quantifying uncertainty about software performance. There are several reasons for making this claim. The first is that software is a one-of-a-kind entity for which the notion of an infinite size ensemble is difficult to justify. Second, it is hard to foresee the repeated testing of a single piece of software under almost identical conditions; with computer applications, the notion of "almost identical conditions" is not precise. Finally, and perhaps more fundamentally, the objective nature of frequentist probability is anathema to the spirit of intuition and inspiration that is necessary for addressing software engineering problems. In all aspects of software development, the personal experience of the engineer or the manager is a vital source of information. The frequentist objectivist interpretation of probability forces us to ignore this knowledge. In contrast, the subjective interpretation allows us to discuss the uncertainty attached to a unique object, such as software, and also allows us to incorporate personal information and knowledge of the software development process by conditioning on $\mathcal{H}$.

The literature on statistical aspects of software engineering does not formally recognize the difference between objective and subjective probabilities. Consequently, the techniques used are a hybrid of those dictated by either school. In what follows, we strive to adhere to the subjective view.

1.4   The Role of Statistical Methods in Software Engineering

By statistical methods in software engineering we mean a unified framework for quantifying uncertainty, for updating it in the light of data, and for making decisions in its presence. Such methods have been developed and used for a wide variety of problems in software engineering. We close this chapter with an overview of the material that is described in the subsequent text.

By and large, the most widely appreciated use of statistical methods in software engineering is that pertaining to software credibility (or reliability). Here, the problem is to describe the quality of the software, usually in terms of the probability of not encountering any bugs over a specified period of time. A large number of probability models have been proposed to address this topic, and Chapter 3 describes some of the more popular ones. Many of these models have similar modeling strategies and assumptions, and in Section 3.5 we look at ways in which we may view these models as special cases of a more general type of models.

Related to the issue of software reliability are the analyses of software failure data. Here, observations on the detection of bugs are used to update the uncertainties about the software's credibility, and to make projections about future failures. The analyses of failure data are performed using the techniques of statistical inference, and an overview of one such technique is given in Chapter 4, where we also discuss the application of these techniques to some of the models of Chapter 3. An important purpose served by the models of Chapter 3 and their associated statistical inference is the problem of optimally testing software. Here, one needs to make a decision as to how much testing a piece of software must undergo before it is released for use. Such decisions are based on both the reliability of the software and a tradeoff between the costs of testing versus the costs of in-service failures. Optimum testing is discussed in Chapter 6. An essential aspect of optimal testing is the design of an effective test plan, that is, the design of the software testing experiment. Since the number of possible inputs to a piece of software is necessarily limited, the choice of inputs that maximize the information which can be gleamed from them is a central issue.

The material described pertains to the role played by statistical methods at the end of the software development process, when the software has been created. Statistical methods can also play a role at the beginning of the development process. Often, after establishing specifications, one of the first decisions to be made is the selection of a software house, or a programming team, to develop the code. A deterministic scheme for classifying software development houses into one of five classes has been developed by the Software Engineering Institute of Carnegie Mellon University. In Chapter 5 we describe a probabilistic version of this scheme wherein the classifications made have associated with them a measure of uncertainty. That is, instead of classifying a software house into exactly one of the five categories, as is done by the Software Engineering Institute's procedure, we assign a weight to each category, with the weights reflecting our strength of belief regarding a software house's membership in each category. Chapter 5 also discusses techniques to assess the productivity of programming teams. Such assessments are useful for project planning, wherein it is necessary to have good estimates of the time and effort required to complete programming and coding tasks.

The final chapter pertains to some recent developments on the use of statistical methods in software engineering. We anticipate that the impact of such methods will continue to be felt, and our aim is to give the reader a feel for the direction in which the subject is heading. Naturally, our choice of material is highly subjective and is limited to what we are aware of at the time of this writing.

1.5   Chapter Summary

In this chapter we have attempted to define what is software, and what is software engineering. We have described the software engineering cycle as being composed of the four stages of development, operation, maintenance, and retirement, and have pointed out the nature of uncertainty that arises at each of these stages. We have said that uncertainty arises when we have to select an organization to develop the software, when we have to assess the times required to code and test the software, when we have to assess the quality of software via the number of bugs it contains, and when we have to decide on a testing strategy.

By far, the most important message of this chapter is the thesis that, for the purposes of this book, probability and its calculus are used as the sole basis of quantifying uncertainties in software engineering. This is followed by a brief discussion of the different types of probability, and the position that the subjective interpretation of probability is the one that is most suitable for addressing the problems that are posed here. The chapter ends with a discussion of the key role played by statistical methods in software engineering, and an overview of the remaining chapters.

2
FOUNDATIONAL ISSUES: PROBABILITY AND RELIABILITY

2.0 Preamble

In Chapter 1 we have drawn attention to some scenarios in software engineering where uncertainty is encountered, and have discussed the need for its quantification. We mentioned that there are many approaches for quantifying uncertainty, but that in our view, probability is the most comprehensive one. We have also discussed the notions of random quantities, random variables, random events, and the importance of the background information $\mathcal{H}$. The role of a reference time τ at which probabilities were assessed was mentioned, and finally, it was argued that for any random quantity $\mathcal{E}$, a subjective interpretation of its probability $\mathcal{P}^\tau(\mathcal{E} \mid \mathcal{H})$ was an appropriate paradigm for dealing with the kind of problems that we are involved with here.

We start this chapter with details about the properties of $\mathcal{P}^\tau(\mathcal{E} \mid \mathcal{H})$, that is, about the *calculus of probability*, and give some reasons that justify it. We have mentioned before that whereas the interpretation of probability is subject to debate, its calculus is by and large universal. Possibility theory [see Zadeh (1981)] has often been proposed as an alternate way of quantifying uncertainty; its calculus is very different from the calculus of probability, and we have yet to see arguments that justify it. We have singled out for mention here possibility theory, because many engineers seem to be attracted to it and also to its precursor, fuzzy logic.

Our discussion of the calculus of probability is followed by its consequences, such as the *law of total probability* and *Bayes' Law*; these play a

14   2. Foundational issues: Probability and Reliability

central role in developing probability models and incorporating the effect of new information in our appreciation of uncertainty. The first section includes a discussion about the notions of *independence, likelihood,* and *exchangeability*; these are important ideas that play a key role in developing probability models and in updating them. The next section introduces the idea of *probability models* and *parameters*; it ends with examples of some commonly used models in probability and statistics, models that are also of relevance to us here. Section 2.3 deals with what are known as *counting process models*; such models are useful in software engineering because they are a natural vehicle for describing events, such as failures, that occur over time. Indeed, some of the most commonly used models for assessing the reliability of software are counting process models, also known as point process models. The chapter ends with an introduction to the key concepts of component reliability theory and their role in assessing software reliability. There is a large body of literature on system reliability theory that is discussed later in Chapter 7. This postponement is due to the fact that the ideas of system reliability theory have not as yet permeated the current mainstream work on software reliability. Nonetheless, we feel that its impact is yet to come, especially in dealing with modularized software, and thus have chosen to include it for later discussion. Readers specializing in probability and statistics may choose to skip to Section 2.4. Others who could benefit from a review may prefer to continue. Wherever feasible, the preliminaries introduced here have been reinforced by describing scenarios from software engineering.

2.1   The Calculus of Probability

2.1.1   Notation and Preliminaries

In what follows, we assume τ to be zero and suppress it. For a discrete random variable X taking values x, let $\mathcal{E}$ denote the event that $X = x$, so that $\mathcal{P}(\mathcal{E} \mid \mathcal{H})$ is $\mathcal{P}(X = x \mid \mathcal{H})$; it is abbreviated as $\mathcal{P}_X(x \mid \mathcal{H})$. If at any value of X, say x, $\mathcal{P}_X(x \mid \mathcal{H}) > 0$, then X is said to have a *point mass* at x. If $\mathcal{E}$ is the event that $X \leq x$, then $\mathcal{P}(X \leq x \mid \mathcal{H})$ is known as the *distribution function* of X, and is denoted as $F_X(x \mid \mathcal{H})$. If X is continuous and takes all values in an interval, say $[0,\infty)$, and if $F_X(x \mid \mathcal{H})$ is differentiable with respect to x, for (almost) all x in $[0,\infty)$, then $F_X(x \mid \mathcal{H})$ is said to be *absolutely continuous*, and its derivative at x, denoted by $f_X(x \mid \mathcal{H})$, is called the *probability density function* of X at x. Irrespective of whether X is discrete or continuous, $F_X(x \mid \mathcal{H})$ is nondecreasing in x, and ranges from 0 to 1. If X is continuous, $F_X(x \mid \mathcal{H})$ increases in x smoothly, whereas if X is discrete, it increases as a step function taking jumps at those values of x at which X has a point mass.

Whereas the interpretation of $\mathcal{P}_X(x \mid \mathcal{H})$ is clear, namely, that it is the probability that X takes the value x, the interpretation of $f_X(x \mid \mathcal{H})$ needs explanation. Specifically, $f_X(x \mid \mathcal{H})dx$ is *approximately* the probability that X takes a value between x and $x + dx$. Since $f_X(x \mid \mathcal{H})dx \downarrow 0$, as $dx \downarrow 0$, the

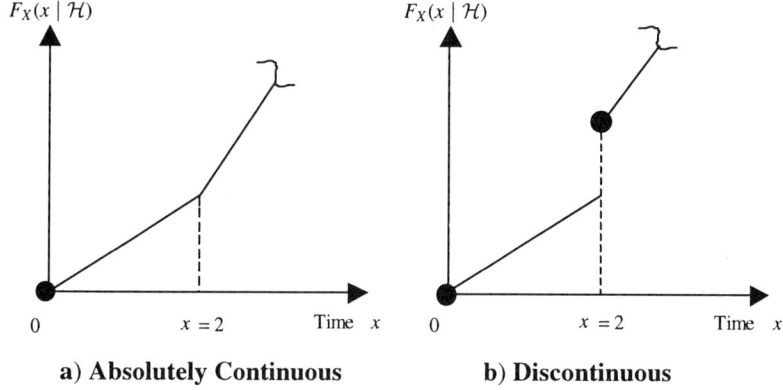

FIGURE 2.1. Illustration of Absolutely Continuous and Discontinuous Distribution Functions.

probability that a continuous random variable takes any particular value is zero. Finally, suppose that X is a *mixed random variable*; that is, it is both discrete and continuous, with a point mass, at say x^*. Then its $F_X(x \mid \mathcal{H})$ increases smoothly for all values of X at which it is continuous, and takes a jump of size $\mathcal{P}_X(x^* \mid \mathcal{H})$ at x^*. Mixed random variables are of interest in reliability, particularly software reliability, wherein there is a nonzero probability of failure at specified time points. Such time points are dictated by the *operational profile* of the software.

To illustrate the preceding notions we consider the following idealized scenario. Suppose that a piece of software has an operating cycle of three hours, the first two of which are under a normal user environment and the last one under a more demanding one. That is, the software experiences a change in the operational profile two hours after its inception. Let X be the time, measured in CPU units, at which the software experiences a failure, either because of the presence of a bug or from other causes. If we assume that the transition from the normal to the more demanding environment does not pose any instantaneous shocks to the software, then $F_X(x \mid \mathcal{H})$ could be of the form shown in Figure 2.1a). The main aspect of this figure is the change of shape at $x = 2$. Observe that $F_X(x \mid \mathcal{H})$ is continuous in x but not differentiable at $x = 2$; it is therefore absolutely continuous. By contrast, suppose that the transition in the operational profile imposes a shock to the software so that there is a nonzero probability, say p, that the software will fail at $x = 2$. In this case $F_X(x \mid \mathcal{H})$ takes an upward jump at $x = 2$; see Figure 2.1b). Now $F_X(x \mid \mathcal{H})$ is not absolutely continuous, and X is a mixed random variable.

The conventions mentioned before generalize when we are interested in two (or more) random variables, say X_1 and X_2; now, $F_X(x \mid \mathcal{H})$ is replaced by $F_{X_1, X_2}(x_1, x_2 \mid \mathcal{H})$, and $f_X(x \mid \mathcal{H})$ by $f_{X_1, X_2}(x_1, x_2 \mid \mathcal{H})$. Note that $F_{X_1, X_2}(x_1, x_2 \mid \mathcal{H})$

abbreviates $P(X_1 \leq x_1$ and $X_2 \leq x_2 \mid \mathcal{H})$, and $f_{X_1, X_2}(x_1, x_2 \mid \mathcal{H})dx_1dx_2$ approximates $P(x_1 \leq X_1 \leq x_1 + dx_1$ and $x_2 \leq X_2 \leq x_2 + dx_2 \mid \mathcal{H})$. When there is no cause for ambiguity, the subscripts associated with F and f are often omitted.

2.1.2 Conditional Probabilities and Conditional Independence

Perhaps one of the more subtle notions in probability theory is that of conditional probability. For two random variables X_1 and X_2, and background $\mathcal{H}$, the probability that X_1 takes the value x_1, *were it to be so* that X_2 takes the value x_2, is called the *conditional probability of X_1 given X_2*; it is denoted $P(X_1 = x_1 \mid X_2 = x_2, \mathcal{H})$ or $P_{X_1 \mid X_2}(x_1 \mid x_2, \mathcal{H})$, the vertical bar representing a separation between the event of interest $X_1 = x_1$, and the conditioning event $X_2 = x_2$. If the event of interest is $X_1 \leq x_1$, and the conditioning event $X_2 = x_2$, then $P(X_1 \leq x_1 \mid X_2 = x_2, \mathcal{H})$ is abbreviated $F_{X_1 \mid X_2}(x_1 \mid x_2, \mathcal{H})$; it is known as the *conditional distribution function of X_1 given X_2*. If X_1 is continuous, and $F_{X_1 \mid X_2}(x_1 \mid x_2, \mathcal{H})$ differentiable for all values x_2, then the derivative of the latter is called the *conditional probability density of X_1 given X_2*; it is denoted by $f_{X_1 \mid X_2}(x_1 \mid x_2, \mathcal{H})$.

It is important to bear in mind that all probability statements, including those of conditional probability, are made at the reference time τ, when both X_1 and X_2 are unknown. Thus conditional probability statements are in the "subjunctive." In other words, when we are making a conditional probability assessment, we are *assuming* (or pretending) that $X_2 = x_2$; in actuality we do not know as yet if $X_2 = x_2$. Indeed, had X_2 been observed as taking the value x_2, then it would become a part of the background history $\mathcal{H}$ and the notion of a conditional probability would be moot. Conditional probabilities reflect the importance of the reference time in making probability assessments.

From a subjective point of view, how should we interpret conditional probabilities and how do we make its numerical value operational? From a subjective point of view, a conditional probability represents our strength of belief about X_1, at time τ, had the background history been expanded (but in actuality was not) from $\mathcal{H}$ to ($\mathcal{H}$ and X_2). Since numerical values of probabilities reflect our disposition to betting in the face of uncertainty, a conditional probability represents the amount that we are willing to stake on X_1, but now under the provision that all bets are off if the conditioning event turns out, in the future, to be untrue, that is, if $X_2 \neq x_2$. A conditional probability is a useful device for assessing probabilities, because it incorporates the notion of "what if" in the process of interrogating personal beliefs about uncertain events.

The notion of conditional probabilities leads us to another idea in probability, the *judgment* of *conditional independence*. Consider two discrete random variables X_1 and X_2, and suppose that

$$P(X_1 = x_1 \mid X_2 = x_2, \mathcal{H}) = P(X_1 = x_1 \mid \mathcal{H});$$

then X_1 and X_2 are said to be mutually independent, conditional on the background history $\mathcal{H}$.

We emphasize that like probability, independence is always conditional because had $\mathcal{H}$ been different, say $\mathcal{H}^*$, then the preceding equality may not hold. From a subjective point of view, the equality displayed here says that our assessment of uncertainty about $X_1 = x_1$ will not be affected by any added (future) knowledge about X_2. X_1 and X_2 are *dependent* if they are not independent. Like probability, independence and dependence are judgments, and may or may not be supported by the physics of the situation. The idea of independence generalizes for a collection of uncertain quantities; it is often assumed because it simplifies the process of probability assessments by removing the need to think about relationships between the various random variables. It is a strong assumption, very idealistic in software reliability assessment.

To illustrate the ideas of conditional independence suppose that software to perform a certain function is developed by two separate teams, A and B. Let X_A be the time, measured in CPU units, at which the software developed by team A experiences a failure, similarly X_B. An analyst studies the two codes and assesses the reliabilities of the two codes as $\mathcal{P}(X_A \geq \tau \mid \mathcal{H})$ as p_A, and $\mathcal{P}(X_B \geq \tau \mid \mathcal{H})$ as p_B. We say that the analyst judges X_A and X_B independent, if the analyst is unwilling to change p_B were he or she to be informed that the software developed by team A experiences a failure at, say some time τ^*. That is, to this analyst, $\mathcal{P}(X_B \geq \tau \mid X_A = \tau^*, \mathcal{H})$ continues to be the previously assessed p_B.

Clearly, the judgment of independence assumed here is not realistic. Even though the software has been developed by two separate teams, they have presumably worked from a common specification; the two codes therefore are likely to have some commonalities. Consequently, the knowledge (admittedly conjectural) that $X_A = \tau^*$ should cause the analyst to revise his assessment from p_A to a value smaller (larger) than p_A, if $\tau^* < (>) \tau$. Indeed experiments on software development by several teams conducted by Knight and Levenson (1986) verify the lack of independence mentioned previously.

The literature in hardware reliability mentions several models for describing dependent lifelengths of two-component systems; particularly discussed are the models of Freund (1961), Marshall and Olkin (1967), and Lindley and Singpurwalla (1986b), to name a few. Their appropriateness for describing the failure of software codes remains to be explored.

2.1.3 The Calculus of Probability

The calculus of probability is a set of rules that tells us how uncertainties about different events combine. For keeping the discussion general, consider two events $\mathcal{E}_1$ and $\mathcal{E}_2$, and background $\mathcal{H}$. Then, the following three rules can be viewed as being basic to the calculus.

Convexity: For any event $\mathcal{E}$,

$$0 \leq \mathcal{P}(\mathcal{E} \mid \mathcal{H}) \leq 1;$$

Additivity: If both $\mathcal{E}_1$ and $\mathcal{E}_2$ cannot occur simultaneously (i.e., if they are *mutually exclusive*), then

$$\mathcal{P}(\mathcal{E}_1 \text{ or } \mathcal{E}_2 \mid \mathcal{H}) = \mathcal{P}(\mathcal{E}_1 \mid \mathcal{H}) + \mathcal{P}(\mathcal{E}_2 \mid \mathcal{H});$$

Multiplicativity:

$$\mathcal{P}(\mathcal{E}_1 \text{ and } \mathcal{E}_2 \mid \mathcal{H}) = \mathcal{P}(\mathcal{E}_1 \mid \mathcal{E}_2, \mathcal{H}) \, \mathcal{P}(\mathcal{E}_2 \mid \mathcal{H}).$$

The first rule says that the probability of an event can take any value in the convex set [0, 1]. Since probabilities are assessed only for those events whose outcome is unknown to us, the value 1 can be meaningfully assigned only to events (propositions) that can be *logically* proven to be true; such events are called *certain events*. Similarly, the value 0 should be assigned only to events that are logically false; such events are called *impossible events*. It does not make sense to talk about probabilities of events whose outcomes are already known. If $\mathcal{E}$ pertains to the disposition of a continuous random variable, say X, then the convexity rule says that the probability density function of X, say $f_X(x \mid \mathcal{H})$, must be nonnegative. However, the function itself $f_X(x \mid \mathcal{H})$ may take values greater than 1; recall that $f_X(x \mid \mathcal{H})$ has a probabilistic interpretation only when it is multiplied by dx.

By a repeated application of the preceding rules, both the additivity and the multiplicativity laws can be generalized. For n events $\mathcal{E}_i$, $i = 1, 2, \ldots, n$, the additivity law takes the form

$$\mathcal{P}(\mathcal{E}_1 \text{ or } \mathcal{E}_2 \text{ or}, \ldots, \text{ or } \mathcal{E}_n \mid \mathcal{H}) = \sum_{i=1}^{n} \mathcal{P}(\mathcal{E}_i \mid \mathcal{H}),$$

provided that the $\mathcal{E}_i$s are mutually exclusive; the multiplicative law takes the form

$$\mathcal{P}(\mathcal{E}_1 \text{ and } \mathcal{E}_2 \text{ and}, \ldots, \text{ and } \mathcal{E}_n \mid \mathcal{H}) = \mathcal{P}(\mathcal{E}_1 \mid \mathcal{E}_2, \ldots, \mathcal{E}_n, \mathcal{H}) \times$$
$$\mathcal{P}(\mathcal{E}_2 \mid \mathcal{E}_3, \ldots, \mathcal{E}_n, \mathcal{H}) \times \ldots \times \mathcal{P}(\mathcal{E}_n \mid \mathcal{H}).$$

When n is finite, the additivity law is said to obey the property of *finite additivity*; when n is infinite it is said to obey *countable additivity*. Subjectivists like de Finetti claim that all that is needed is finite additivity; mathematicians demand countable additivity for rigor and generality.

The convexity and the additivity laws ensure that the probabilities of all mutually exclusive events should sum to 1, and if the $\mathcal{E}_i$s pertain to the dispositions of a continuous random variable, then these laws ensure that the integral of the probability density function over all possible values of the random variable must be 1.

When $\mathcal{E}_1$ and $\mathcal{E}_2$ are not mutually exclusive, then it can be shown (see Exercise 1) that

$$\mathcal{P}(\mathcal{E}_1 \text{ or } \mathcal{E}_2 \mid \mathcal{H}) = \mathcal{P}(\mathcal{E}_1 \mid \mathcal{H}) + \mathcal{P}(\mathcal{E}_2 \mid \mathcal{H}) - \mathcal{P}(\mathcal{E}_1 \text{ and } \mathcal{E}_2 \mid \mathcal{H});$$

furthermore, if $\mathcal{E}_1$ and $\mathcal{E}_2$ are judged independent, then the preceding becomes

$$\mathcal{P}(\mathcal{E}_1 \text{ or } \mathcal{E}_2 \mid \mathcal{H}) = \mathcal{P}(\mathcal{E}_1 \mid \mathcal{H}) + \mathcal{P}(\mathcal{E}_2 \mid \mathcal{H}) - \mathcal{P}(\mathcal{E}_1 \mid \mathcal{H})\mathcal{P}(\mathcal{E}_2 \mid \mathcal{H}).$$

As an illustration as to how these rules play a useful role, consider the hardware and software components of a typical computer system. Let $\mathcal{E}_H$ denote the event that the hardware experiences a failure during the next hours of operation, and $\mathcal{E}_S$ the event that the software experiences a failure in the same time interval. The computer system is therefore a *series system*, whose *unreliability* for a mission of eight hours duration is given by the addition rule as

$$\mathcal{P}(\mathcal{E}_H \text{ or } \mathcal{E}_S \mid \mathcal{H}) = \mathcal{P}(\mathcal{E}_H \mid \mathcal{H}) + \mathcal{P}(\mathcal{E}_S \mid \mathcal{H}) - \mathcal{P}(\mathcal{E}_H \text{ and } \mathcal{E}_S \mid \mathcal{H}),$$

where $\mathcal{P}(\mathcal{E}_i \mid \mathcal{H})$ is the probability that event $\mathcal{E}_i$, $i = H, S$, occurs, and if $\mathcal{E}_H$ and $\mathcal{E}_S$ are judged independent (this judgment being realistic for the scenario considered) the unreliability of the computer system becomes

$$\mathcal{P}(\mathcal{E}_H \text{ or } \mathcal{E}_S \mid \mathcal{H}) = \mathcal{P}(\mathcal{E}_H \mid \mathcal{H}) + \mathcal{P}(\mathcal{E}_S \mid \mathcal{H}) - \mathcal{P}(\mathcal{E}_H \mid \mathcal{H})\,\mathcal{P}(\mathcal{E}_S \mid \mathcal{H}),$$

Suppose now that the hardware component is supported by a backup system that operates in parallel (that is, simultaneously) with main system. If $\mathcal{E}_B$ denotes the event that the backup system fails in the time interval of interest, then the unreliability of the hardware system is

$$\mathcal{P}(\mathcal{E}_H \text{ and } \mathcal{E}_B \mid \mathcal{H}) = \mathcal{P}(\mathcal{E}_H \mid \mathcal{E}_B, \mathcal{H})\,\mathcal{P}(\mathcal{E}_B \mid \mathcal{H})\,.$$

The preceding expression is not further simplified because it is generally unrealistic to assume that a hardware system and its backup have independent lifelengths.

Continuing with this theme, the unreliability of the computer system becomes (upon suppressing the $\mathcal{H}$)

2. Foundational issues: Probability and Reliability

$$\mathcal{P}[(\mathcal{E}_H \text{ and } \mathcal{E}_B) \text{ or } \mathcal{E}_S] = \mathcal{P}(\mathcal{E}_H \text{ and } \mathcal{E}_B) + \mathcal{P}(\mathcal{E}_S) - \mathcal{P}(\mathcal{E}_S \text{ and } \mathcal{E}_B \text{ and } \mathcal{E}_H)$$

$$= \mathcal{P}(\mathcal{E}_B \mid \mathcal{E}_H)\,\mathcal{P}(\mathcal{E}_H) + \mathcal{P}(\mathcal{E}_S) -$$

$$\mathcal{P}(\mathcal{E}_S \mid \mathcal{E}_B \text{ and } \mathcal{E}_H)\,\mathcal{P}(\mathcal{E}_B \mid \mathcal{E}_H)\,\mathcal{P}(\mathcal{E}_H),$$

upon an application of the multiplication rule. Since the software and the hardware systems are assumed to have independent lifelengths, the preceding simplifies to

$$= \mathcal{P}(\mathcal{E}_B \mid \mathcal{E}_H)\,\mathcal{P}(\mathcal{E}_H) + \mathcal{P}(\mathcal{E}_S) - \mathcal{P}(\mathcal{E}_S)\,\mathcal{P}(\mathcal{E}_B \mid \mathcal{E}_H)\,\mathcal{P}(\mathcal{E}_H)$$

$$= \mathcal{P}(\mathcal{E}_B \mid \mathcal{E}_H)\,\mathcal{P}(\mathcal{E}_H)\,[1 - \mathcal{P}(\mathcal{E}_S)] + \mathcal{P}(\mathcal{E}_S).$$

Why should we subscribe to a calculus for uncertainty that is based on the preceding rules? A simple answer to this question is that the laws were enunciated and proved to be useful since the times of Cardano, and that they were adhered to by the founders like Bernoulli, de Moivre, Bayes, and Laplace. Indeed one of Bayes' major contributions was his discourse on conditional probability and the multiplication rule. A more formal answer is that the mathematical theory of probability takes these laws as axioms, although Kolmogorov (1950) argues for them based on relative frequency considerations. A more convincing answer would be that subjectivists, like Ramsey and Savage, deduce the laws from primitive considerations, such as a person's ability to compare any two events based on their likelihoods of occurrence [cf. DeGroot (1970), p. 70], and that de Finetti (1974) uses the idea of *scoring rules* to claim the inevitability of these laws [see Lindley (1982a)]. Further support for these rules also comes from the argument that if *betting coefficients* do not obey the calculus of probability, then one can be trapped into the situation of a *Dutch book* [cf. Howson and Urbach (1989), p.56]. A Dutch book is a gamble in which you lose irrespective of the outcome; a person who engages in a Dutch book is declared to be *incoherent*. Because of the preceding arguments the claim is made that using a calculus different from the calculus of probability, such as that of possibility theory, leads to incoherence.

2.1.4 The Law of Total Probability, Bayes' Law, and the Likelihood Function

A simple application of the three laws of probability yields two other important laws. The first is the *law of total probability*, and the second is *Bayes' Law*. The law of total probability, also known as the *law of the extension of conversation*, is a useful device for developing probability models; see Section 2.2. Bayes' Law provides a vehicle for coherently revising probabilities in the light of new information; it becomes a tool for incorporating the effect of data in our assessment of uncertainty.

The Law of Total Probability

Suppose that X_1 and X_2 are two discrete random variables for which we have assessed their *joint probability* $P(X_1 = x_1$ and $X_2 = x_2 \mid \mathcal{H})$, for all possible values x_1 and x_2 that X_1 and X_2 can respectively take. Then, by the additivity rule, our uncertainty about X_1 alone (known as the *marginal of* X_1) is given by

$$P(X_1 = x_1 \mid \mathcal{H}) = \sum_{x_2} P(X_1 = x_1, X_2 = x_2 \mid \mathcal{H}); \qquad (2.1)$$

the summation is over all possible values that X_2 can take. Were X_1 and X_2 to be continuous, then the summation would be replaced by an integral and the probabilities by their corresponding densities; consequently, the *marginal density of* X_1 is:

$$f(x_1 \mid \mathcal{H}) = \int_{x_2} f(x_1, x_2 \mid \mathcal{H}) dx_2. \qquad (2.2)$$

The law of total probability now follows from the multiplicative rule; in the discrete case

$$P(X_1 = x_1 \mid \mathcal{H}) = \sum_{x_2} P(X_1 = x_1 \mid X_2 = x_2, \mathcal{H}) P(X_2 = x_2 \mid \mathcal{H}), \quad (2.3)$$

and mutatis mutandis, for the continuous case.

The law of total probability shows how one can coherently assess the uncertainty about X_1 via its appropriate conditional assessments in the light of X_2. It illustrates the role of conditional probability as a facilitator of uncertainty assessment. A use of this law presumes that conditional probabilities are easier to assess than the unconditional ones, which in most cases is generally true.

Bayes' Law

Bayes' Law, also known as the *law of inverse probability*, has been attributed to the Reverend Thomas Bayes (1702–1761). However, it is often claimed that it was Laplace who was responsible for discovering its current form, independent of Bayes, and for popularizing its use. Both Bayes and Laplace were interested in assessing the probabilities of the causes of an event, the causes having occurred at a time prior to the occurrence of the event; thus the term inverse probability. For the case of discrete random variables X_1 and X_2, the multiplicative rule and the marginalization rule give

2. Foundational issues: Probability and Reliability

$$\mathcal{P}(X_1 = x_1 \mid X_2 = x_2, \mathcal{H}) = \frac{\mathcal{P}(X_1=x_1, X_2=x_2 \mid \mathcal{H})}{\sum_{x_1} \mathcal{P}(X_1=x_1, X_2=x_2 \mid \mathcal{H})}. \tag{2.4}$$

An application of the multiplicative rule to the numerator and to the denominator now gives us Bayes' Law as

$$\mathcal{P}(X_1 = x_1 \mid X_2 = x_2, \mathcal{H}) = \frac{\mathcal{P}(X_2=x_2 \mid X_1=x_1, \mathcal{H})\, \mathcal{P}(X_1=x_1 \mid \mathcal{H})}{\sum_{x_1} \mathcal{P}(X_2=x_2 \mid X_1=x_1, \mathcal{H})\, \mathcal{P}(X_1=x_1 \mid \mathcal{H})}. \tag{2.5}$$

When the random variables are continuous, the replacement of probabilities by densities and the sum by the integral occurs, so that (2.5) becomes

$$f(x_1 \mid x_2, \mathcal{H}) = \frac{f(x_2 \mid x_1, \mathcal{H}) f(x_1 \mid \mathcal{H})}{\int_{x_1} f(x_2 \mid x_1, \mathcal{H}) f(x_1 \mid \mathcal{H}) dx_1}.$$

As an illustration of how Bayes' Law can be used to address problems of interest to us here, consider a computer system comprised of a hardware and a software component. Let $\mathcal{E} = 1(0)$ denote the event that the computer system experiences a failure (survival) for a specified interval of time. The failure of the computer system can be attributed to either a hardware or a software failure, or both. Let $\mathcal{E}_S = 1(0)$ denote the event that the software experiences a failure (survival) during the time interval specified previously. Similarly, let $\mathcal{E}_H = 1(0)$ denote the failure (survival) of the hardware. Note that the events $\mathcal{E}_S = 1$ and $\mathcal{E}_S = 0$ are mutually exclusive so that $\mathcal{P}(\mathcal{E}_S = 1 \mid \mathcal{H}) = 1 - \mathcal{P}(\mathcal{E}_S = 0 \mid \mathcal{H})$, by the convexity rule.

Bayes' Law is useful for addressing questions pertaining to the cause of failure of the computer system. For example, we may be interested in knowing the probability that software failure was the cause of failure of the computer system, if the system experiences failure. That is, we may want to know $\mathcal{P}(\mathcal{E}_S = 1 \mid \mathcal{E} = 1)$, which by Bayes' Law takes the form (upon suppressing $\mathcal{H}$)

$$\mathcal{P}(\mathcal{E}_S = 1 \mid \mathcal{E} = 1) = \frac{\mathcal{P}(\mathcal{E}=1 \mid \mathcal{E}_S=1)\, \mathcal{P}(\mathcal{E}_S=1)}{\mathcal{P}(\mathcal{E}=1 \mid \mathcal{E}_S=1)\, \mathcal{P}(\mathcal{E}_S=1) + \mathcal{P}(\mathcal{E}=1 \mid \mathcal{E}_S=0)\, \mathcal{P}(\mathcal{E}_S=0)}.$$

But $\mathcal{P}(\mathcal{E} = 1 \mid \mathcal{E}_S = 1) = 1$, since the computer system is a series system, and $\mathcal{P}(\mathcal{E} = 1 \mid \mathcal{E}_S = 0) = \mathcal{P}(\mathcal{E}_H = 1)$, since the computer system can only fail if there is either a hardware or a software failure (or both). Thus

$$\mathcal{P}(\mathcal{E}_S = 1 \mid \mathcal{E} = 1) = \frac{\mathcal{P}(\mathcal{E}_S=1)}{\mathcal{P}(\mathcal{E}_S=1) + \mathcal{P}(\mathcal{E}_H=1)\, \mathcal{P}(\mathcal{E}_S=0)}.$$

Similarly, we can show that

$$P(\mathcal{E}_H = 1 \mid \mathcal{E} = 1) = \frac{P(\mathcal{E}_H=1)}{P(\mathcal{E}_H=1) + P(\mathcal{E}_S=1)\,P(\mathcal{E}_H=0)}$$

is the probability that hardware was the cause of the system failure.

Since the events $(\mathcal{E}_S = 1 \mid \mathcal{E} = 1)$ and $(\mathcal{E}_S = 0 \mid \mathcal{E} = 1)$ are mutually exclusive, $P(\mathcal{E}_S = 1 \mid \mathcal{E} = 1) = 1 - P(\mathcal{E}_S = 0 \mid \mathcal{E} = 1)$, a result that can also be verified by a direct application of Bayes' Law to $P(\mathcal{E}_S = 1 \mid \mathcal{E} = 1)$. The same is also true of $P(\mathcal{E}_H = 1 \mid \mathcal{E} = 1)$.

Clearly, for this example, all that we need to know for answering the questions posed is to assess $P(\mathcal{E}_S = 1)$ and $P(\mathcal{E}_H = 1)$; the conditional probabilities are either 1, or one of the preceding two. In many other applications of Bayes' Law, the conditional probabilities are not that simple. For example, suppose that the event $\mathcal{E}_S$ is redefined, so that now $\mathcal{E}_S^* = 1$ denotes the fact that the software has at least one bug in its code. Then, $P(\mathcal{E} = 1 \mid \mathcal{E}_S^* = 1)$ need not necessarily be 1, since the bugs could reside in a region of the code that is not always visited during an application. Thus now $P(\mathcal{E}_S^* = 1 \mid \mathcal{E} = 1)$ represents the probability that the bugs in the software were the cause of the computer system's failure, and to evaluate it we must assess $P(\mathcal{E} = 1 \mid \mathcal{E}_S^* = 1)$ in addition to evaluating $P(\mathcal{E}_S^* = 1)$ and $P(\mathcal{E}_H = 1)$. Recall that

$$\begin{aligned}P(\mathcal{E}_S^* = 1 \mid \mathcal{E} = 1) &= \frac{P(\mathcal{E}=1\mid\mathcal{E}_S^*=1)P(\mathcal{E}_S^*=1)}{P(\mathcal{E}=1\mid\mathcal{E}_S^*=1)P(\mathcal{E}_S^*=1) + P(\mathcal{E}=1\mid\mathcal{E}_S^*=0)P(\mathcal{E}_S^*=0)} \\ &= \frac{P(\mathcal{E}=1\mid\mathcal{E}_S^*=1)P(\mathcal{E}_S^*=1)}{P(\mathcal{E}=1\mid\mathcal{E}_S^*=1)P(\mathcal{E}_S^*=1) + P(\mathcal{E}_H=1)P(\mathcal{E}_S^*=0)}.\end{aligned}$$

The Likelihood Function

An examination of (2.5) reveals some interesting features. First, note that the left-hand side is a function of the realizations of X_1 alone, because X_2 is assumed fixed at x_2, and $\mathcal{H}$ is a known entity. This function, being a conditional probability, satisfies the calculus of probability. The same is also true of the second term of the numerator of the right-hand side of (2.5). The denominator of the right-hand side is a constant because all the values x_1 have been summed out. Thus we may write (2.5) as

$$P(X_1 = x_1 \mid X_2 = x_2, \mathcal{H}) \propto P(X_2 = x_2 \mid X_1 = x_1, \mathcal{H})\, P(X_1 = x_1 \mid \mathcal{H}). \quad (2.6)$$

The middle term of (2.6), namely, $P(X_2 = x_2 \mid X_1 = x_1, \mathcal{H})$, remains to be interpreted. Why have we singled out this term? By all accounts, since it has arisen via an application of the multiplicativity rule to (2.4), should it therefore not be anything more than a conditional probability? This is indeed so, as long as both X_1 and X_2 are uncertain quantities; recall that all conditional probability statements are in the subjunctive, so that $\mathcal{P}(X_1 = x_1 \mid X_2 = x_2, \mathcal{H})$ refers to our uncertainty about X_1, if $X_2 = x_2$. However, if X_2 were *actually observed* as being x_2, then $\mathcal{P}(X_2 = x_2 \mid X_1 = x_1, \mathcal{H})$ cannot be interpreted as a probability; probabilities make sense only for those events about which we are uncertain. How then should we interpret $\mathcal{P}(X_2 = x_2 \mid X_1 = x_1, \mathcal{H})$?

When X_2 is known to equal x_2, $\mathcal{P}(X_2 = x_2 \mid X_1 = x_1, \mathcal{H})$ is referred to as the *likelihood of X_1* for X_2 observed and fixed at x_2, and $\mathcal{P}(X_2 = x_2 \mid X_1 = x_1, \mathcal{H})$ as a function of x_1, is known as the *likelihood function of X_1* for X_2 fixed at x_2. The likelihood function not being a probability need not obey the laws of probability; that is, the function when summed (or integrated) over all values x_1 need not equal one. In fact there is a well-known example in the analysis of software failure data [cf. Forman and Singpurwalla (1977)] wherein the likelihood function integrates to infinity. Because of the preceding, the likelihood function has been interpreted as one that provides a relative degree of support given by the data (i.e., for the fixed value x_2) to the various values x_1 that X_1 can possibly take. When X_1 and X_2 are continuous, (2.6) will then take the form $f(x_1 \mid x_2, \mathcal{H}) \propto f(x_2 \mid x_1, \mathcal{H}) f(x_1 \mid \mathcal{H})$, with $f(x_2 \mid x_1, \mathcal{H})$ the likelihood function, and the other terms the probability densities.

For the situation in which X_2 is known to equal x_2, the term $\mathcal{P}(X_1 = x_1 \mid \mathcal{H})$ of (2.6) quantifies our uncertainty about X_1 based on $\mathcal{H}$ alone, whereas the term $\mathcal{P}(X_1 = x_1 \mid X_2 = x_2, \mathcal{H})$ quantifies our uncertainty about X_1 based on both $\mathcal{H}$ and $X_2 = x_2$. Because of this, the left-hand side of (2.6) is referred to as the *posterior probability* of X_1, posterior to observing x_2, and the second term on the right-hand side of (2.6), the *prior probability* of X_1. Bayes' Law shows us how the likelihood connects the prior and the posterior probabilities. Alternatively viewed, Bayes' Law facilitates the incorporation of new information in our assessments of uncertainty, and thus becomes a tool of experimental science.

To better appreciate the essential import of the notion of a likelihood, let us revisit our example illustrating Bayes' Law and focus on the last expression preceding (2.6), namely,

$$\mathcal{P}(\mathcal{E}_S^* = 1 \mid \mathcal{E} = 1) = \frac{\mathcal{P}(\mathcal{E}=1 \mid \mathcal{E}_S^*=1)\,\mathcal{P}(\mathcal{E}_S^*=1)}{\mathcal{P}(\mathcal{E}=1 \mid \mathcal{E}_S^*=1)\,\mathcal{P}(\mathcal{E}_S^*=1) + \mathcal{P}(\mathcal{E}_H=1)\,\mathcal{P}(\mathcal{E}_S^*=0)}.$$

Since conditional probabilities are in the subjunctive, the left-hand side of the preceding expression is to be interpreted as the probability that the presence of bugs in the software is the cause of system failure *were* it be true that the system has failed. When this probability is assessed, it is not known if the system has indeed failed; that is, the true disposition of the system is unknown to the probability assessor. For definitiveness, suppose that $\mathcal{P}(\mathcal{E}_S^* = 1) = 0.01$, $\mathcal{P}(\mathcal{E}_H = 1) = 0.05$, and that $\mathcal{P}(\mathcal{E} = 1 \mid \mathcal{E}_S^* = 1) = 0.7$. This implies that the software is relatively free of bugs, that the hardware component is very reliable, but that the computer system has a high probability of failure should the software contain one or more bugs. When such is the case, the probability that the software will be the cause of failure, *should* the system experience a failure is $((0.7)(0.01))/((0.7)(0.01) + (0.05)(0.99)) = 0.12$.

Now suppose that it is known for a fact—that is, it is actually observed—that the computer system has failed, but it is not known whether the software or the hardware triggered the failure. What now is our probability that the software is the cause of the system failure? Must it still be 0.12? To answer this question, we formally proceed as before (according to Bayes' Law) because this is what we said we would do should $\mathcal{E} = 1$, but now $\mathcal{P}(\mathcal{E} = 1 \mid \mathcal{E}_S^* = 1)$ cannot be interpreted as a probability. Recall, probability is meaningful for only those events that have yet to occur (or are unknown to us), and ($\mathcal{E} = 1$) has indeed occurred. $\mathcal{P}(\mathcal{E} = 1 \mid \mathcal{E}_S^* = 1)$ is therefore a likelihood, more clearly written as $\mathcal{L}(\mathcal{E}_S^* = 1; \mathcal{E} = 1)$, and the likelihood being the degree of support provided by the observed data $\mathcal{E} = 1$, to the unknown event $\mathcal{E}_S^* = 1$, may or may not be assigned the value 0.7. What really matters now are the relative values assigned to $\mathcal{L}(\mathcal{E}_S^* = 1; \mathcal{E} = 1)$ and $\mathcal{L}(\mathcal{E}_S^* = 0; \mathcal{E} = 1)$, although all that we need to know for computing $\mathcal{P}(\mathcal{E}_S^* = 1 \mid \mathcal{E} = 1)$ is the former. Since the likelihood is not a probability, it is perfectly all right to have $\mathcal{L}(\mathcal{E}_S^* = 1; \mathcal{E} = 1) + \mathcal{L}(\mathcal{E}_S^* = 0; \mathcal{E} = 1) \neq 1$.

Commentary

We have seen that Bayes' Law is just a theorem in probability. However, because of its having given birth to the notion of a likelihood, it has become associated with a set of techniques called *Bayesian statistics*. What does one mean by the term Bayesian statistics? For one, Bayesian statistics is not merely a use of Bayes' Law for making statistical inferences. To some, it also encompasses a subjective interpretation of probability, but to all it requires a strict adherence to what is known as the *likelihood principle* [cf. Berger and Wolpert (1984)]. Loosely speaking, the likelihood principle says that the contribution made by the data (new information) is solely embodied in the likelihood function, and nothing more. This dictum makes many of the well-known statistical procedures such as those based on confidence limits, significance levels, goodness of fit testing, and hypotheses tests with Type I and Type II errors, and the method of maximum likelihood, not acceptable. These procedures subscribe to the frequentist view of probability, and in so doing are unable to express uncertainty solely via the calculus of probability.

2.1.5   The Notion of Exchangeability

Like independence, exchangeability helps us simplify the assessment of probabilities. As before, consider two discrete random variables X_1 and X_2, taking values x_1 and x_2, respectively. Then, X_1 and X_2 are said to be *exchangeable*, if for all values of x_1 and x_2, and background $\mathcal{H}$,

$$\mathcal{P}(X_1 = x_1 \text{ and } X_2 = x_2 \mid \mathcal{H}) = \mathcal{P}(X_1 = x_2 \text{ and } X_2 = x_1 \mid \mathcal{H}); \quad (2.7)$$

that is, the assessed probabilities are unchanged (invariant) by switching (permuting) the indices. Because permuting the indices does not affect the assessed probabilities, we may think of exchangeable quantities as being similar to each other. One can also view the judgment of exchangeability as a judgment of *indifference* between the random quantities; we do not care what values each random variable takes. All that we care about is the set of values that the two random variables can take.

Like independence, exchangeability is a judgment about two (or more) uncertain quantities, based on $\mathcal{H}$. It is weaker than independence, because, in general, exchangeable random variables are dependent. Independent random variables having identical probability distributions are exchangeable (but not vice versa). To see why, observe that if X_1 and X_2 are independent and identically distributed, then suppressing $\mathcal{H}$,

$$\mathcal{P}(X_1 = x_1 \text{ and } X_2 = x_2) = \mathcal{P}(X_1 = x_1)\mathcal{P}(X_2 = x_2)$$

$$= \mathcal{P}(X_1 = x_2)\mathcal{P}(X_2 = x_1)$$

$$= \mathcal{P}(X_1 = x_2 \text{ and } X_2 = x_1),$$

implying that they are exchangeable. Finally, a collection of random variables $X_1, X_2, \ldots, X_n$ is said to be exchangeable, if every subset of $X_1, \ldots, X_n$ is an exchangeable collection. Exchangeability was introduced by de Finetti (1937), (1972), on grounds that it is more meaningful in practice than independence. The assumption of independence implies, de facto, an absence of learning.

To illustrate the nature of the roles played by the assumptions of independence and exchangeability, suppose that software code to perform a certain operation is developed by four different teams, all working from a common set of specifications. Let $X_i = 1(0)$ denote the event that team i's code results in a correct (erroneous) output, $i = 1, \ldots, 4$. The four codes are to be used in a *fault-tolerant* system, and we are required to assess the credibility (reliability) of the system. A fault-tolerant system will produce a response if three or more of its outputs agree with each other, and the response is a correct response if $\sum_{i=1}^{4} X_i \geq 3$. Thus, we are required to assess $\mathcal{P}(\sum_{i=1}^{4} X_i \geq 3 \mid \mathcal{H})$. For purpose of illustration, suppose that we judge $\mathcal{P}(X_i = 1 \mid \mathcal{H}) = 0.5, i = 1, \ldots, 4$. Then, under the judgment of independence (of the X_is), and suppressing the $\mathcal{H}$s,

$$\mathcal{P}(\sum_{1}^{4} X_i = 4) = \mathcal{P}(X_1 = 1, X_2 = 1, X_3 = 1, X_4 = 1),$$

which by the multiplication rule,

$$= \mathcal{P}(X_1 = 1 \mid X_2 = X_3 = X_4 = 1) \times \mathcal{P}(X_2 = 1 \mid X_3 = X_4 = 1) \times$$

$$\mathcal{P}(X_3 = 1 \mid X_4 = 1) \times \mathcal{P}(X_4 = 1)$$

$$= \mathcal{P}(X_1 = 1) \times \mathcal{P}(X_2 = 1) \times \mathcal{P}(X_3 = 1) \times \mathcal{P}(X_4 = 1)$$

$$= (0.5)^4 .$$

The practical importance and significance of the statement $\mathcal{P}(X_1 = 1 \mid X_2 = X_3 = X_4 = 1) = \mathcal{P}(X_1 = 1)$, is that under independence, the added knowledge that were $X_2 = X_3 = X_4 = 1$, our assessment of the probability that $X_1 = 1$ remains unchanged from its previous value of 0.5. Surely, one would expect that the event $X_2 = X_3 = X_4 = 1$ would cause an upward revision of $\mathcal{P}(X_1 = 1)$ from the value 0.5. Similarly, it can be easily seen that under independence

$$\mathcal{P}(\sum_1^4 X_i = 3) = 4(0.5)^3 (0.5) = 4(0.5)^4 ,$$

so that the credibility of the fault tolerant system is given by

$$\mathcal{P}(\sum_1^4 X_i = 3) + \mathcal{P}(\sum_1^4 X_i = 4) = 4(0.5)^4 + (0.5)^4 = 5(0.5)^4 = 0.3125;$$

the events $\sum_{i=1}^4 X_i = 3$ and $\sum_{i=1}^4 X_i = 4$ are mutually exclusive.

Analogous calculations would show that the probability of the fault-tolerant system producing an erroneous response is 0.3125. Thus the probability that the fault-tolerant system produces a response (correct or erroneous) is $2(0.3125) = 0.6250$, and that it produces no response (that is, the four codes do not arrive at a consensus) is $(1 - 0.6250) = 0.3750$.

How do these answers compare with those obtained through the assumption that the X_is, $i = 1, \ldots, 4$, are exchangeable? The main matter to note here is that under exchangeability, all that we need to assume is permutation invariance. Thus, for example, to assess $\mathcal{P}(\sum_{i=1}^4 X_i = 3)$ we must require that:

$$\mathcal{P}(X_1 = 1, X_2 = 1, X_3 = 1, X_4 = 0)$$

$$= \mathcal{P}(X_1 = 1, X_2 = 1, X_3 = 0, X_4 = 1)$$

$$= \mathcal{P}(X_1 = 1, X_2 = 0, X_3 = 1, X_4 = 1)$$

$$= \mathcal{P}(X_1 = 0, X_2 = 1, X_3 = 1, X_4 = 1),$$

and it does not matter how each individual probability is assessed, similarly, for the event $\sum_{i=1}^{4} X_i = 1$. For the event $\sum_{i=1}^{4} X_i = 2$, we must have:

$$\mathcal{P}(X_1 = X_2 = 0, X_3 = X_4 = 1) = \mathcal{P}(X_1 = X_2 = 1, X_3 = X_4 = 0)$$

$$= \mathcal{P}(X_1 = 0, X_2 = 1, X_3 = 0, X_4 = 1)$$

$$= \mathcal{P}(X_1 = 0, X_2 = X_3 = 1, X_4 = 0)$$

$$= \mathcal{P}(X_1 = 1, X_2 = X_3 = 0, X_4 = 1)$$

$$= \mathcal{P}(X_1 = 1, X_2 = 0, X_3 = 1, X_4 = 0).$$

The events $\sum_{i=1}^{4} X_i = 4$ and $\sum_{i=1}^{4} X_i = 0$ being unique, permutation invariance is not an issue.

In order to make our probability assessments here compatible with our previous assumption that $\mathcal{P}(X_i = 1) = 0.5$, $i = 1, \ldots, 4$, we need to have the assumptions that $\mathcal{P}(X_1 = X_2 = X_3 = X_4 = 1) = \mathcal{P}(X_1 = X_2 = X_3 = X_4 = 0) = 0.2$, $\mathcal{P}(X_1 = X_2 = X_3 = 1, X_4 = 0) = 0.05$. Furthermore, we must also have $\mathcal{P}(X_1 = X_2 = 1, X_3 = X_4 = 0) = 0.0333$. With this assignment of probabilities, it follows that under exchangeability the credibility of the fault-tolerant system is 0.4, and the probability that the fault tolerant system produces a response (correct or incorrect) is 0.8. These numbers being greater than their counterparts obtained via independence, we may conjecture that for fault-tolerant systems, the assumption of independence tends to exaggerate the assessed probability of non-response.

2.2 Probability Models and Their Parameters

2.2.1 What is a Software Reliability Model?

We have seen that for any random quantity $\mathcal{E}$, our uncertainty based on background $\mathcal{H}$ is expressed by $\mathcal{P}(\mathcal{E} \mid \mathcal{H})$. In actuality $\mathcal{H}$, being everything that we know, is large, very complex, and of high dimension. Furthermore, much of $\mathcal{H}$ may be irrelevant to $\mathcal{E}$. What is therefore suitable is a way to abridge $\mathcal{H}$ so that it is manageable.

Suppose that there is a new random quantity, say Θ, scalar or vector. Then, assuming Θ to be discrete taking values θ, we can use the law of total probability to write

$$\mathcal{P}(\mathcal{E} \mid \mathcal{H}) = \sum_{\theta} \mathcal{P}(\mathcal{E} \mid \theta, \mathcal{H}) \, \mathcal{P}(\theta \mid \mathcal{H}). \tag{2.8}$$

For Θ continuous, an integral replaces the sum, and probability density functions replace the $\mathcal{P}$s.

Now suppose that were we to know Θ, we would judge $\mathcal{E}$ independent of $\mathcal{H}$, so that for all θ, $\mathcal{P}(\mathcal{E} \mid \theta, \mathcal{H}) = \mathcal{P}(\mathcal{E} \mid \theta)$. Then (2.8) would become

$$\mathcal{P}(\mathcal{E} \mid \mathcal{H}) = \sum_{\theta} \mathcal{P}(\mathcal{E} \mid \theta) \, \mathcal{P}(\theta \mid \mathcal{H}), \tag{2.9}$$

suggesting that our uncertainty about $\mathcal{E}$ can be expressed via two probability distributions, $\mathcal{P}(\mathcal{E} \mid \theta)$ and $\mathcal{P}(\theta \mid \mathcal{H})$. The distribution $\mathcal{P}(\mathcal{E} \mid \theta)$ is called a *probability model* for $\mathcal{E}$, and $\mathcal{P}(\theta \mid \mathcal{H})$ the *prior distribution* of Θ. If $\mathcal{E}$ denotes a lifelength, then $\mathcal{P}(\mathcal{E} \mid \theta)$ is called a *failure model* [cf. Singpurwalla (1988a)], and if $\mathcal{E}$ denotes the time to failure of a piece of software, then $\mathcal{P}(\mathcal{E} \mid \theta)$ is called a *software reliability model*. In making the judgment of independence between $\mathcal{E}$ and $\mathcal{H}$ given Θ, we are interpreting Θ as a device for summarizing the background information $\mathcal{H}$. Θ is known as the *parameter* of the probability model. The manner in which we have introduced Θ suggests that it is an unobservable quantity that simplifies the assessment process; to de Finetti, it (often) is just a Greek symbol! Its role is to impart independence between $\mathcal{E}$ and $\mathcal{H}$. Because Θ is unknown, its uncertainty must also be expressed by probability; thus the appearance of a prior distribution is inevitable, whenever probability models are introduced.

A consequence of (2.9) is the appearance of probabilities that are easier to assess than $\mathcal{P}(\mathcal{E} \mid \mathcal{H})$. The choice of a probability model and the prior distribution is a subjective one, although there is often a natural probability model to choose; some examples are given in the following section. The choice of $\mathcal{P}(\theta \mid \mathcal{H})$ is a contentious issue. Various approaches have been proposed: the use of "objective" priors is one [Berger (1985), Chapter 3]; another is using "expert opinion" [Lindley and Singpurwalla (1986a)]. For a unified perspective on statistical modeling, see Singpurwalla (1988a).

2.2.2 *Some Commonly Used Probability Models*

In this section we briefly present some natural probability models (or distributions) that can be used for addressing generic problems in many applications, including those in software engineering. The list is not complete, and some models that appear later in the book are not described here. For a more comprehensive list, see Bernardo and Smith (1994), or Johnson and Kotz (1970). We attempt to motivate many of the models using software testing as the application scenario. The others presented here are for completeness and their usefulness in the subsequent text.

The Bernoulli Distribution

Suppose that a piece of software is subjected to a certain input. Our uncertainty here pertains to the event $\mathcal{E}$, where $\mathcal{E}$ is the proposition that the software provides a correct output. Define a binary random variable X that takes the value 1 if $\mathcal{E}$ is true, and zero otherwise. Such a random variable is called a *Bernoulli* random variable, after James Bernoulli who gave us the famous (weak) law of large numbers. Let $\mathcal{H}$ be the background information we have about the software. Then, the *input specific reliability* of the software is $\mathcal{P}(X = 1 \mid \mathcal{H})$, and our aim is to assess this quantity. To do this, suppose we introduce (*extend the conversation to*) a parameter P that takes values p, with $0 \leq p \leq 1$, and invoke the law of total probability; then

$$\mathcal{P}(X = 1 \mid \mathcal{H}) = \int_p \mathcal{P}(X = 1 \mid p, \mathcal{H}) f(p \mid \mathcal{H}) dp.$$

Now suppose that given P, we judge X to be independent of $\mathcal{H}$. Then, the preceding simplifies as

$$\mathcal{P}(X = 1 \mid \mathcal{H}) = \int_p \mathcal{P}(X = 1 \mid p) f(p \mid \mathcal{H}) dp,$$

where $\mathcal{P}(X = 1 \mid p)$ is the probability model and $f(p \mid \mathcal{H})$ the prior density function of P. In what follows, we focus attention on only the probability model. Bernoulli's proposal was to let $\mathcal{P}(X = 1 \mid p) = p$; then the calculus of probability requires that $\mathcal{P}(X = 0 \mid p) = 1 - p$. Such a probability model is called the *Bernoulli distribution*, and as stated before, X is a Bernoulli random variable. The experiment (or act) of subjecting the software to an input and observing its success or failure is known as a *Bernoulli trial*. A compact way to express a Bernoulli distribution is

$$\mathcal{P}(X = x_i \mid p) = p^{x_i}(1-p)^{(1-x_i)}, \text{ for } x_i = 0, 1. \tag{2.10}$$

Thus, when the probability model is a Bernoulli, the input-specific reliability of the software is

$$\mathcal{P}(X = 1 \mid \mathcal{H}) = \int_p p f(p \mid \mathcal{H}) dp.$$

2.2 Probability Models and Their Parameters

If in our judgment all the values p that P can take are equally likely, that is, we have no basis for preferring one value of p over another, then P is said to have a *uniform distribution* over the interval $(0, 1)$ and $f(p \mid \mathcal{H}) = 1, 0 < p < 1$. When such is the case, it is easy to verify that $\mathcal{P}(X = 1 \mid \mathcal{H}) = \frac{1}{2}$.

Binomial Distribution

Suppose now that the software is subjected to N distinct inputs, and our uncertainty is about X, the number of inputs for which the software produces a correct output. The proportion of correct outputs is a measure of the reliability of the software. Clearly, X can take values $x = 0, 1, 2, \ldots, N$, and we need to know $\mathcal{P}(X = x \mid \mathcal{H})$. There are many ways in which one can address this problem. The simplest is to assume that each input is a Bernoulli trial leading to a Bernoulli random variable X_i, $i = 1, 2, \ldots, N$, with $X_i = 1$, if the ith input results in a correct output, and $X_i = 0$, otherwise.

Since $X = \sum X_i$, there are $\binom{N}{x}$ mutually exclusive ways in which $X = x$; one possibility is that the first x trials result in a correct output and the remaining do not. To assess the probability of such an event, namely,

$$\mathcal{P}(X_1 = \ldots = X_x = 1, \text{ and } X_{x+1} = \ldots = X_N = 0 \mid \mathcal{H}),$$

we extend the conversation to a parameter P taking values p, with $0 < p < 1$, invoke the multiplicative law, assume that given p the X_is are independent of each other and also of $\mathcal{H}$, and assume a Bernoulli model for each X_i. Then

$$\mathcal{P}(X_1 = \ldots = X_x = 1, \text{ and } X_{x+1} = \ldots = X_N = 0 \mid p, \mathcal{H})$$
$$= \int_p p^x (1-p)^{N-x} f(p \mid \mathcal{H}) dp,$$

where $f(p \mid \mathcal{H})$ is the density function of p. Since the $\binom{N}{x}$ possibilities are mutually exclusive and each has probability $p^x(1-p)^{N-x}$, we invoke the additivity law of probability to obtain

$$\mathcal{P}(X = x \mid \mathcal{H})$$
$$= \int_p \mathcal{P}(X = x \mid p) f(p \mid \mathcal{H}) dp$$
$$= \int_p \binom{N}{x} p^x (1-p)^{N-x} f(p \mid \mathcal{H}) dp. \tag{2.11}$$

The probability model

$$P(X = x \mid p) = \binom{N}{x} p^x (1-p)^{N-x}, \quad x = 0, \ldots, N$$

is called the *binomial distribution;* the notation $\binom{N}{x}$ denotes the quantity $N!/(x!(N-x)!)$, with $x! \stackrel{\text{def}}{=} x \cdot (x-1) \cdot (x-2) \cdots 2 \cdot 1$.

Poisson's Approximation to the Binomial Distribution

In many applications involving Bernoulli trials, it can happen that N is large and $(1-p)$ is small, but their product $N \times (1-p)$ is moderate. In the case of software testing, this situation arises when software that is almost bug free is subjected to a large number of inputs, so that $(1-p)$ is small and N very large so that $N \times (1-p)$ is moderate. When such is the case it is convenient to use an approximation to the binomial distribution, which is due to Poisson. Specifically, if we let $\lambda = N \times (1-p)$, then using a Taylor series expansion and the inductive hypothesis, it can be shown (see Exercise 4) that

$$\binom{N}{x} p^x (1-p)^{N-x} \approx e^{-\lambda} \frac{\lambda^x}{x!}. \tag{2.12}$$

The probability model

$$P(X = x \mid \lambda) = e^{-\lambda} \frac{\lambda^x}{x!}, \quad x = 0, 1, 2, \ldots,$$

is known as the *Poisson distribution.*

The Geometric Distribution

Now suppose that a piece of software is subjected to an indefinite sequence of *distinct* inputs, each resulting in a correct or an incorrect output. We are interested in X, the number of inputs at which the software experiences its first failure—this could be a meaningful measure of the software's reliability. We are uncertain about X, and so need to know $P(X = x \mid \mathcal{H})$, where $x = 1, 2, \ldots, \infty$. As before, we start by assuming that each input is a Bernoulli trial leading to a Bernoulli random variable X_i, $i = 1, 2, \ldots$, with $X_i = 1$, if the ith input results in a correct output, and $X_i = 0$, otherwise.

Clearly, $P(X = x \mid \mathcal{H}) = P(X_1 = X_2 = , \ldots, X_{x-1} = 1, X_x = 0 \mid \mathcal{H})$, and to assess this probability we introduce a parameter P, taking values $0 < p < 1$, invoke the multiplicative law, assume that given p the X_is are independent of each other and of $\mathcal{H}$, and assume a *common* Bernoulli model for each X_i. Both here, and also in our discussion of the binomial distribution, the assumption of a

2.2 Probability Models and Their Parameters

Bernoulli model with a common parameter P for each X_i is idealistic. It suggests that all the inputs have the same impact on the software. We should weaken this assumption, but for now keep it as such to motivate a geometric distribution. Under the preceding assumptions, it is easy to see that

$$\mathcal{P}(X_1 = X_2 = , \ldots, X_{x\text{-}1} = 1, X_x = 0 \mid p, \mathcal{H})$$

$$= \int_p p^{x\text{-}1}(1-p)f(p \mid \mathcal{H})dp,$$

where, as before, $f(p \mid \mathcal{H})$ is the probability density function of P. Thus to conclude:

$$\mathcal{P}(X = x \mid \mathcal{H}) = \int_p \mathcal{P}(X = x \mid p)f(p \mid \mathcal{H})dp$$

$$= \int_p p^{x\text{-}1}(1-p)f(p \mid \mathcal{H})dp; \quad (2.13)$$

The probability model

$$\mathcal{P}(X = x \mid p) = p^{x\text{-}1}(1-p), \quad x = 1, 2, \ldots,$$

is called a *geometric distribution*.

Discussion

The models described thus far pertain to a discrete random variable X, and arise in the context of evaluating $\mathcal{P}(X = x \mid \mathcal{H})$ for x taking values in some subset of $\{0, 1, \ldots, \}$. We have attempted to motivate each model by considering the scenario of assessing the reliability of software by testing it against several inputs. Our motivating arguments can be labeled idealistic, and this is perhaps true; however, they set the stage for subsequent more realistic developments. For example, we could expand on our setup by assuming that each Bernoulli random variable X_i has an associated parameter P_i, and that the sequence of P_is is exchangeable; see, for example, Chen and Singpurwalla (1996). In all cases we focused only on probability models and left open the question of specifying $f(p \mid \mathcal{H})$, the prior probability density function of P. This is a much debated issue which can trace its origins to the work of Bayes and Laplace; a recent reference is Geisser (1984). A natural choice is the *beta density function*

$$f(p \mid \alpha, \beta) = \frac{\Gamma(\alpha+\beta)}{\Gamma(\alpha)\Gamma(\beta)} p^{\alpha\text{-}1}(1-p)^{\beta\text{-}1}, \quad 0 < p < 1,$$

34   2. Foundational issues: Probability and Reliability

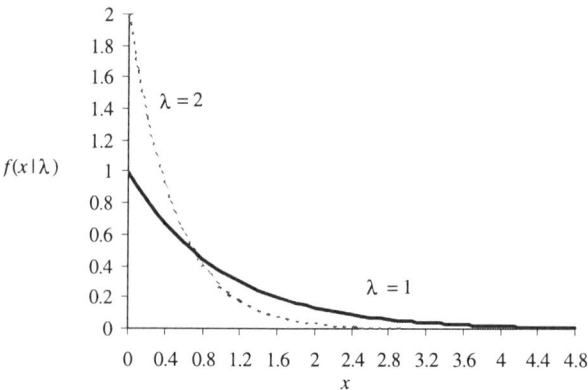

FIGURE 2.2. Exponential Density Function.

where $\Gamma(a)$ is the gamma function. A random variable X having a beta distribution with parameters α and β is denoted "$X \sim \mathcal{B}(\alpha, \beta)$." The uniform distribution is a special case of the beta distribution with $\alpha = \beta = 1$.

The Exponential Distribution

We have seen by now, that with software testing we may count the number of correct outputs in a series of N Bernoulli trials, as in the case of the binomial distribution, or we may count X the number of inputs at which we encounter the first incorrect output, as in the case of the geometric distribution. The exponential distribution, introduced here, can be viewed as the continuous analogue of the geometric distribution. Roughly speaking, suppose that the sequence of inputs to the software occurs continuously over time; that is, the software receives a distinct input at every instant of time. Alternatively viewed, suppose that a Bernoulli trial (with a common Bernoulli model) is performed at every instant of time. Then the X of our geometric distribution will be continuous, and is to be interpreted as the time to the first occurrence of an incorrect output; that is, the time to failure of the software. As before, we are uncertain about X, and are interested in a measure of the reliability of the software $\mathcal{P}(X \geq x \mid \mathcal{H})$, where $x \geq 0$. If we extend the conversation to a parameter Λ, with Λ taking values $0 < \lambda < \infty$, and invoke the assumption that X is independent of $\mathcal{H}$ were Λ known, then

$$\mathcal{P}(X \geq x \mid \mathcal{H}) = \int_\lambda \mathcal{P}(X \geq x \mid \lambda) f(\lambda \mid \mathcal{H}) d\lambda,$$

2.2 Probability Models and Their Parameters

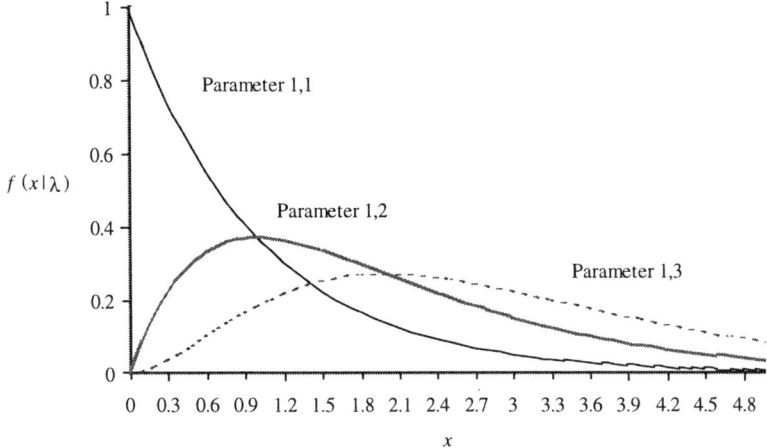

FIGURE 2.3. Gamma Density Function.

where $\mathcal{P}(X \geq x \mid \lambda)$ is the probability model, and $f(\lambda \mid \mathcal{H})$ the prior probability density function of Λ. How do these quantities relate to their analogues in (2.13)? After all, they have been motivated by similar considerations. What is the relationship between the p of (2.13) and the λ given previously?

Using limiting arguments, and supposing that Bernoulli trials are performed at times $1/n, 2/n, \ldots$, it can be shown (See Exercise 4) that as $n \to \infty$, $i/n \to x$, and with $p = 1 - \lambda/n$,

$$\mathcal{P}(X \geq x \mid \lambda) = e^{-\lambda x}, \text{ for both } x, \lambda > 0. \tag{2.14}$$

Since X is continuous, it has a density $f(x \mid \lambda) = \lambda e^{-\lambda x}$; see Figure 2.2. The probability model (2.14) is known as the *exponential distribution* with a scale parameter λ. It has found widespread applications in applied probability, notably reliability theory and queueing theory. A random variable X having an exponential distribution with scale parameter λ is denoted "$X \sim \mathcal{E}(\lambda)$."

The Gamma Distribution

The setup described previously shows how the time to first failure of the software can be described by an exponential distribution. In many applications, once a software failure is detected, its cause is identified and the software debugged. The debugged software is now viewed as a new product and the cycle of subjecting it to a sequence of distinct inputs repeats. However, there are scenarios in which a failed piece of software is not debugged until after several failures, say k, for $k = 1, 2, \ldots$ The failed software is simply reinitialized and

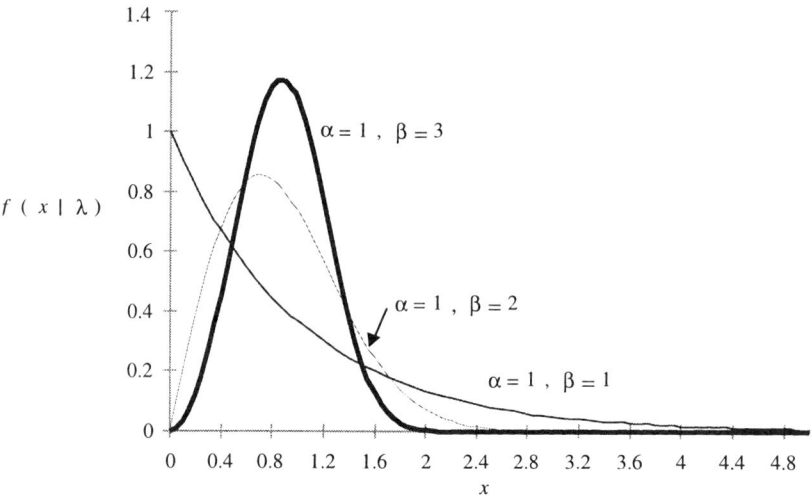

FIGURE 2.4. Weibull Density Function.

continued to being subjected to the sequence of distinct inputs. When such is the case, we may be interested in $X(k)$, the time to occurrence of the kth failure.

The gamma distribution is a generalization of the exponential, and can be motivated as the distribution of $X(k)$, the time to occurrence of the kth failure, in the software testing process. If we make the kind of assumptions that resulted in the exponential distribution for $X(1)$, then we can show (see Section 2.3.3) that for any specified k, $X(k)$ has a probability density function of the form

$$f_{X(k)}(x \mid \lambda, k) = \frac{e^{-\lambda x} \lambda^k x^{k-1}}{\Gamma(k)}, \text{ for } x \geq 0, \text{ and } \lambda > 0; \qquad (2.15)$$

see Figure 2.3. The function $\Gamma(u) = \int_0^\infty e^{-s} s^{(u-1)} ds$, is known as the *gamma function*; it generalizes the factorials, as for integer values of u, $\Gamma(u+1) = u!$

The model (2.15) is known as a *gamma distribution*, with *scale (shape) parameter* $\lambda(k)$. A random variable X having a gamma distribution with scale (shape) $\alpha(\beta)$ is denoted "$X \sim \mathcal{G}(\alpha, \beta)$." When $k = 1$, (2.15) becomes the density function of an exponential distribution. Even though our motivation here implies that k should be an integer, it need not in general be so.

The Weibull Distribution

Another generalization of the exponential is the Weibull distribution, famous for its wide range of applicability in many problems of hardware

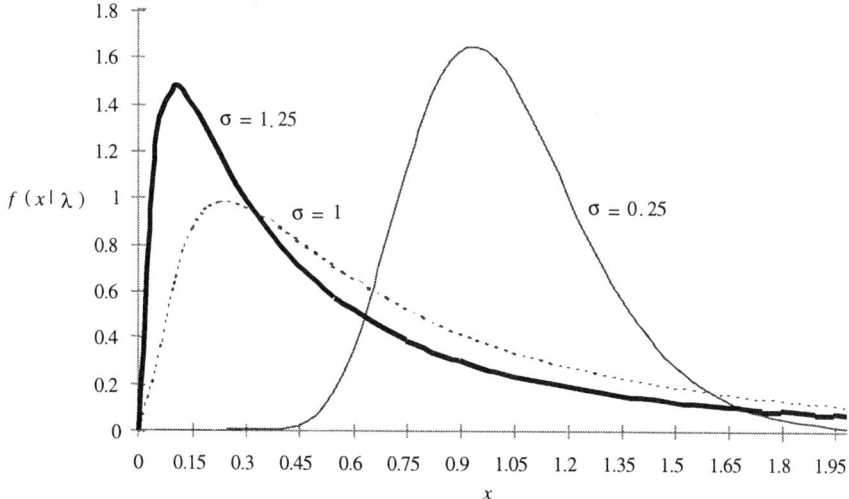

FIGURE 2.5. Lognormal Density Function.

reliability. A motivation for invoking this distribution in the context of assessing software reliability is given later, once we introduce the notion of the failure rate of a probability model; see Section 2.4.1. For now, we just introduce this distribution.

A continuous random variable X is said to have a *Weibull distribution*, with a scale parameter $\alpha > 0$, and a shape parameter $\beta > 0$, denoted "$X \sim \mathcal{W}(\alpha,\beta)$," if

$$\mathcal{P}(X \geq x \mid \alpha, \beta) = e^{-\alpha x^\beta}, \text{ for } x \geq 0. \tag{2.16}$$

The density function of X (Figure 2.4) is $f(x \mid \alpha, \beta) = \alpha\beta x^{\beta-1} e^{-\alpha x^\beta}$; for $\beta = 1$, it is an exponential.

The Lognormal Distribution

The Weibull distribution was introduced under the pretext that it was a generalization of the exponential distribution, the latter having been motivated as the time to first failure of software subjected to a series of instantaneous but distinct inputs. The gamma distribution was introduced as another generalization of the exponential, but it also had the motivation of being the time to the kth failure of software that is initialized upon failure. Both the gamma and the Weibull have another common feature. Their density functions are skewed to the right (i.e., they have long tails) suggesting that under their regimes large failure

38   2. Foundational issues: Probability and Reliability

times can occur, but rarely so, more (less) rarely under a Weibull with $\beta >(<) 1$, than under a gamma. Another probability distribution whose density is skewed to the right is the lognormal distribution, sometimes used to describe the times between software failure.

An introduction to the lognormal and the truncated normal distributions is greatly facilitated if we start with the normal, or the Gaussian, distribution. The Gaussian distribution, discovered by De Moivre, is one of the most frequently encountered distributions in applied and theoretical statistics. Its popularity stems from the fact that it has been used, since the time of Gauss, as the distribution of observational errors, which are both positive and negative, or in general the distribution of symmetric fluctuations about a central tendency. Consequently the Gaussian is useful for describing several random phenomena such as the deviations of heights and of IQs from their central values, the deviations of material strengths from their nominal values, the vibrations of a rotating shaft around its axis, and so on.

A continuous random variable X, taking values x, is said to have a *Gaussian distribution* with *mean* μ and *variance* σ^2, denoted "$X \sim \mathcal{N}(\mu, \sigma^2)$," if for parameters $-\infty < \mu < +\infty$, and $\sigma > 0$, the probability model for X has a probability density function of the form

$$f(x \mid \mu, \sigma) = \frac{1}{(2\pi\sigma^2)^{\frac{1}{2}}} \exp\left(-\frac{(x-\mu)^2}{2\sigma^2}\right), \text{ for } -\infty < x < +\infty. \quad (2.17)$$

When $\mu = 0$, and $\sigma^2 = 1$, the Gaussian distribution is known as the *Standard Normal distribution*. Failure times are often skewed and rarely symmetric around a nominal value. Thus, the Gaussian has not been used as a probability model for lifelengths. Why then our interest in the Gaussian?

For one, this distribution has properties that are attractive for modeling and inference. The Gaussian distribution is a consequence of many limit theorems in probability. A more pragmatic reason is that we are able to generate skewed distributions by suitable transformations of the Gaussian. For example, if X is a lifetime, and if it is reasonable to assume [cf. Singpurwalla and Soyer (1992)] that the deviations of $\log_e X$ from a central value, say μ, are symmetric, so that $\log_e X \sim \mathcal{N}(\mu, \sigma^2)$, then X has a skewed distribution, called the *lognormal distribution*, denoted "$X \sim \Lambda(\mu, \sigma)$" (see Figure 2.5). The probability density function of a lognormal distribution function is

$$f(x \mid \mu, \sigma) = \frac{1}{x\sqrt{(2\pi\sigma^2)}} \exp\left(-\frac{(\log x - \mu)^2}{2\sigma^2}\right), \text{ for } 0 < x < \infty. \quad (2.18)$$

The Truncated Normal Distribution

Another skewed distribution, which is derived from the Gaussian and which has applications in software quality assessment is the *truncated normal distribution*. This is a normal distribution whose range is restricted, so that x belongs to an interval $[a, b]$. Its density function is of the form

$$f(x \mid \mu, \sigma) = \frac{1}{K\sqrt{2\pi\sigma^2}} \exp\left(-\frac{(x-\mu)^2}{2\sigma^2}\right), \text{ for } a \leq x \leq b, \quad (2.19)$$

where the *normalizing constant* K is such that $\int_a^b f(x \mid \mu, \sigma)dx = 1$. The truncated normal distribution has been used by Campodónico and Singpurwalla (1994) for incorporating and modeling expert opinion in software reliability assessment.

2.2.3 *Moments of Probability Distributions and Expectation of Random Variables*

Moments and expected values are convenient ways of summarizing probability models. Indeed some of the most commonly used statistics in day-to-day operations have their genesis in the notion of moments. Examples are the mean, the variance, the correlation, the mean time to failure, and so on. Such statistics are often the mainstay of much of the data analyses done in software engineering. The aim of this subsection is to put the commonly used statistics in their proper perspective.

The notion of the first moment of a probability distribution takes its roots from kinetics where it is used to represent any object by a point. Similarly, the second moment of a distribution finds analogy with the moment of inertia that describes how the mass of the object is distributed about an axis of rotation. Thus, were we to conceptualize the probability distribution of a random variable as an object having a unit mass that is distributed along its realization, then its moments can be viewed as summary measures of uncertainty. Related to the idea of moments, but finding its origin in games of chance, is the notion of an expectation; it indicates the payoff expected in repeated plays of a game.

Following the notation of Section 2.1.1, consider a discrete random variable X taking value x. Let $\mathcal{P}_X(x \mid \mathcal{H}) = P(X = x \mid \mathcal{H})$; then the *k*th moment of $\mathcal{P}_X(x \mid \mathcal{H})$ about the origin 0 is defined as

$$E(X^k \mid \mathcal{H}) = \sum_{x=0}^{\infty}(x-0)^k \, \mathcal{P}_X(x \mid \mathcal{H}) < \infty;$$

when $k = 1$, the first moment $E(X \mid \mathcal{H})$, is also known as the *mean* of X, or the *expected value* of X with respect to $\mathcal{P}_X(x \mid \mathcal{H})$. The second moment of $\mathcal{P}_X(x \mid \mathcal{H})$

about its mean $E(X \mid \mathcal{H})$ is known as the *variance* of X, and is denoted $V(X \mid \mathcal{H})$; its square root is called the *standard deviation* of X. Verify that

$$V(X \mid \mathcal{H}) = \sum_{x=0}^{\infty} \left(x - E(X \mid \mathcal{H})\right)^2 \mathcal{P}_X(x \mid \mathcal{H}) = E(X^2 \mid \mathcal{H}) - E^2(X \mid \mathcal{H}).$$

If X is absolutely continuous with a probability density function $f_X(x \mid \mathcal{H})$, then

$$E(X^k \mid \mathcal{H}) = \int_0^{\infty} (x-0)^k f_X(x \mid \mathcal{H}) dx < \infty;$$

similarly, $V(X \mid \mathcal{H})$.

With two random variables X_1 and X_2, taking values x_1 and x_2, respectively, the *product moment* of $\mathcal{P}(X_1 = x_1, X_2 = x_2 \mid \mathcal{H})$, or the *joint expectation* of $X_1 X_2$ is defined as

$$E(X_1 X_2 \mid \mathcal{H}) = \sum_{x_1=0}^{\infty} \sum_{x_2=0}^{\infty} x_1 x_2 \, \mathcal{P}(X_1 = x_1, X_2 = x_2 \mid \mathcal{H}) < \infty.$$

The *covariance* of X_1 and X_2, denoted $\text{Cov}(X_1, X_2 \mid \mathcal{H})$, is defined as $E(X_1 X_2 \mid \mathcal{H}) - E(X_1 \mid \mathcal{H}) E(X_2 \mid \mathcal{H})$. And finally, $\rho(X_1, X_2 \mid \mathcal{H})$, the *correlation* between X_1 and X_2, is defined as $(\text{Cov}(X_1, X_2 \mid \mathcal{H})) / (S(X_1) S(X_2))$, where $S(X)$ is $\sqrt{V(X \mid \mathcal{H})}$, the standard deviation of X. The correlation $\rho(X_1, X_2 \mid \mathcal{H})$ is a measure of the extent of the *linear* relationship between the X_is; it is zero when they are independent. However, $\rho(X_1, X_2 \mid \mathcal{H}) = 0$ does not necessarily imply the independence; indeed when $X_1^2 + X_2^2 = R^2$, a constant, $\rho(X_1, X_2 \mid \mathcal{H}) = 0$.

The kth moment (about 0) of $\mathcal{P}(X_1 = x_1 \mid X_2 = x_2, \mathcal{H})$, the conditional distribution of X_1, were $X_2 = x_2$, is defined as

$$E(X_1^k \mid X_2 = x_2, \mathcal{H}) = \sum_{x_1=0}^{\infty} (x_1 - 0)^k \mathcal{P}(X_1 = x_1 \mid X_2 = x_2, \mathcal{H}) < \infty,$$

when $k = 1$, $E(X_1 \mid X_2 = x_2, \mathcal{H})$, is known as the *conditional expectation*, or *conditional mean* of X_1, with respect to $\mathcal{P}(X_1 = x_1 \mid X_2 = x_2, \mathcal{H})$. Similarly, the *conditional variance* $V(X_1 \mid X_2 = x_2, \mathcal{H})$ is seen to be

$$V(X_1 \mid X_2 = x_2, \mathcal{H}) = E(X_1^2 \mid X_2 = x_2, \mathcal{H}) - E^2(X_1 \mid X_2 = x_2, \mathcal{H}).$$

2.2.4 *Moments of Probability Models: The Mean Time to Failure*

The notion of conditional means and conditional variances enables us to discuss the moments of probability models. Recall that probability models are conditional probability statements, conditioned on unknown parameters, that are usually denoted by Greek symbols. Thus, for example, it is easy to verify that were we to suppose that P is p, then the first moment of the Bernoulli distribution (2.10) is simply p, and the variance of a random variable having this distribution is $p(1-p)$. Similarly, the first moment of the binomial distribution is np and the variance of a binomial random variable is $np(1-p)$. The mean and the variance of a random variable having the Poisson distribution (2.12) are both λ. A verification of these is left as an exercise for the reader.

The first moment of a probability model that is a failure model (see Section 2.2.1) is of particular interest. It is known as the *mean time to failure*, abbreviated MTTF, and is one of the most frequently encountered terms in reliability; in fact to many it is a measure of an item's reliability. For example, if the failure model is the exponential (2.14), then the mean time to failure is $1/\lambda$, and the variance is $1/\lambda^2$. Similarly, if the failure model is a gamma (2.15), then were we to know both k and λ, then the mean time to failure is k/λ, and the variance is $k/(\lambda^2)$. Note that in all these cases, we are supposing that the unknown parameters are known, and thus when we talk of the mean time to failure, we are really talking about the conditional means and variances. We later show (see Section 2.4) that the MTBF (mean time between failures) can be used as a proxy for the reliability of an item only when its failure model is the exponential. When an item's failure model has two or more parameters, the MTBF alone does not describe the item's reliability. This elementary but important fact is often overlooked by those in practice.

2.3 Point Processes and Counting Process Models

Counting process models have played a key role in the analysis of software failure data, and it appears that this role will continue to expand. By way of some motivation, suppose that we are interested in observing the occurrences of a repeatable event over a period of time. The simplest example is the arrival of customers at a service station, such as a bank. Another example is the occurrence of earthquakes of a specified magnitude at a particular location. An example that is of interest to us here is the points in time at which a piece of software fails. In all such cases, the event of interest does not occur with any regularity and is therefore unpredictable. Consequently, we are not sure about the times at which the event will occur, and also about the number of events that will occur in any time interval. Such a phenomenon is called a *point process*, because, as its name suggests, it can be depicted by points on a horizontal line, the line representing time, and the points the occurrences of events over time. It is not essential that the horizontal line denote time; it could, for example, represent the length of a

42   2. Foundational issues: Probability and Reliability

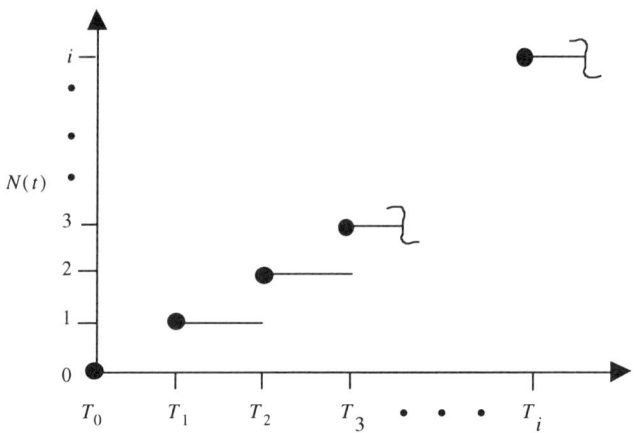

FIGURE 2.6. The Sample Path of a Counting Process.

strand of wire, and the points, the position of defects along its length. Many other examples are possible.

A *counting process*, as the name implies, is simply a count of the number of events that have occurred in any specified interval of time. Since the horizontal line has been designated to represent time, the vertical line is used to indicate the number of counts over time. Specifically, if we let $T_1 \leq T_2 \leq \cdots \leq T_i \leq \cdots$ denote the points in time at which an event of interest occurs, and $N(t)$ the number of events that occur by time t, then a plot of the T_is versus $N(t)$ (see Figure 2.6) traces the evolution of the counts over time; it is known as the *sample path* of the point process. It is a step function starting at zero, and taking jumps of size one at each T_i. Since we are uncertain about both the T_is and $N(t)$, the sample path of the point process is not known and should therefore be viewed as an unknown step function. Once the process is observed, the sample path becomes known and the probabilistic aspects of the problem are not relevant. In practical applications, we may observe both the T_is and the $N(t)$, or simply $N(t)$.

Since $N(t)$ is unknown for any value of t, $t \geq 0$, we are faced with the problem of describing our uncertainty about an infinite collection of random variables, one for each value of t. Any indexed collection of random variables is called a *stochastic process*, and when interest is focused on counts, the process is called a *stochastic counting process*; it is denoted by $\{N(t); t \geq 0\}$. In our case the index has been time t, but in other applications it could be length, or simply the set of integers. For example, the collection of random variables

$T_1 \leq T_2 \leq \cdots \leq T_i \leq \cdots$, or the collection of interarrival times $X_i \stackrel{\text{def}}{=} T_i - T_{i-1}$, $i = 1, 2, \ldots$, are also stochastic processes, but they are indexed on the set of integers; such processes are commonly referred to as *time series processes*.

The purpose of this section is expository; its aim is to introduce probability models for the counting process $\{N(t); t \geq 0\}$, especially those models that have proved to be of value for describing software failures. In the sequel we are also able to deduce probability models for the sequences $\{T_i\}$ and $\{X_i\}$ previously defined. Indeed, there is one commonly used model for counting software failures. It is the nonhomogeneous Poisson process model discussed next. However, the potential for using other models exists, and this needs to be explored.

2.3.1 The Nonhomogeneous Poisson Process Model

The Poisson process model for describing our uncertainty about the process $\{N(t); t \geq 0\}$ is one of the simplest and perhaps the best known of all counting process models. Experience has shown it to be a satisfactory description of commonly occurring phenomena in an assortment of applications. However, there are assumptions underlying this model, and these may not be realistic in any particular application. To introduce Poisson process models, we start with the problem of assessing $\mathcal{P}(N(t) = k \mid \mathcal{H}, \mathcal{H}_t)$, for any fixed t, $t \geq 0$; $\mathcal{H}$ denotes any background information that we may have about the physical scenario that generates the process, and $\mathcal{H}_t$ denotes observations on the process itself up to, but not including, time t. That is, $\mathcal{H}_t = \{N(u); 0 \leq u \leq t^-\}$. As before, we extend the conversation to a parameter $\Lambda^*(t)$, which can take the value $\Lambda(t) \geq 0$, with $\Lambda^*(0) = 0$; we next invoke an assumption of independence with respect to $\mathcal{H}$ to write

$$\mathcal{P}(N(t) = k \mid \mathcal{H}, \mathcal{H}_t)$$

$$= \int_{\Lambda(t)} \mathcal{P}(N(t) = k \mid \mathcal{H}_t, \Lambda(t)) f(\Lambda(t) \mid \mathcal{H}, \mathcal{H}_t) d\Lambda(t), \tag{2.20}$$

where $\mathcal{P}(N(t) = k \mid \mathcal{H}_t, \Lambda(t))$ is a probability model for $N(t)$, and $f(\Lambda(t) \mid \mathcal{H}, \mathcal{H}_t)$ is the probability density of $\Lambda^*(t)$ at $\Lambda(t)$. In writing the preceding, we have not followed our convention of denoting unknown quantities by capital letters and their realized values by the corresponding small letters. The reason for this departure is that the derivative of $\Lambda^*(t)$, assuming that it exists, is a quantity of interest, and it is common to denote it by $\lambda^*(t)$. The parameters $\Lambda^*(t)$ and $\lambda^*(t)$ are functions of time; the former is known as the *mean value function* of the process $\{N(t); t \geq 0\}$, and $\lambda^*(t)$ is known as the *intensity function* (or the *rate*) of the process. It can be shown that given $\Lambda^*(t)$, if $N(t)$ is independent of $\mathcal{H}_t$, then

$E(N(t))$, the expected number of events by time t, is indeed $\Lambda^*(t)$. The proof is left as an exercise for the reader; see Exercise 8.

The specification of a probability structure for the mean value function is an active area of research. One approach is to assume a functional form for $\Lambda^*(t)$, and then to endow its parameters with a joint distribution; see, for example, Campodónico and Singpurwalla (1995), who use this strategy for analyzing some software failure data. Another approach is to assume that for $t \geq 0$, $\lambda^*(t)$ is itself a stochastic process called an *intensity process*; this is discussed in Sections 2.3.3 and 7.1.2.

Suppose now that $\Lambda^*(\bullet)$ is a finite valued, nonnegative, and nondecreasing function of t. Then a probability model for $N(t)$ is said to be a *nonhomogeneous Poisson process model*, if for all $t \geq 0$, the following "postulates" are invoked.

(i) $\mathcal{P}(N(t) = k \mid \mathcal{H}_t, \Lambda^*(\bullet)) = \mathcal{P}(N(t) = k \mid \Lambda^*(\bullet))$;

(ii) $\mathcal{P}(N(0) = 0 \mid \Lambda^*(\bullet)) = 1$; and

(iii)   for any $0 \leq s < t$, the number of events that occur in $[s,t]$ has a Poisson distribution with a parameter $(\Lambda^*(t) - \Lambda^*(s))$; that is, for $k = 0, 1, 2, \ldots$,

$$\mathcal{P}((N(t) - N(s) = k \mid \Lambda^*(\bullet)) = \frac{(\Lambda^*(t) - \Lambda^*(s))^k}{k!} \exp(-(\Lambda^*(t) - \Lambda^*(s))).$$

The essential import of postulate (i) is that, were we to know the mean value function $\Lambda^*(\bullet)$, then a knowledge of the past behavior of the process is irrelevant for our assessment of uncertainty about future occurrences. As a consequence, given $\Lambda^*(\bullet)$, the number of events occurring in disjoint time intervals are independent random variables. This property is known as the *independent increments* property, and is a defining characteristic of all Poisson process models. An advantage of having such a property is the ease with which statistical inference for Poisson process models can be done; a specification of the likelihood function is straightforward. However, assumption (i) is very strong and often unrealistic. Despite this, Poisson process models have been used to describe software failures [cf. Musa and Okumoto (1984)]. Finally, since $\Lambda^*(0) = 0$, postulate (iii) says that $N(t)$ has a Poisson distribution with paramter $\Lambda^*(t)$.

It is useful to note that $\Lambda^*(t)$ need not be continuous, and even if it is continuous, it need not be differentiable. Jump discontinuities in $\Lambda^*(t)$ correspond to points at which events in a Poisson process occur at predetermined times, and the number of events that occur at such points has a Poisson distribution with a parameter equal to the size of the jump of the intensity function.

The Distribution of Interarrival and Waiting Times

When point process models are used to describe uncertainties associated with the number of software failures, two other quantities are also of interest. The first is the *interarrival times* X_i, that is, the times between consecutive failures, and the second is T_i, the *waiting time* to the ith failure, for $i = 1, 2, \ldots$. In general, it is possible to describe uncertainties about both these quantities conditional on their previous values. Specifically, in the case of the nonhomogeneous Poisson process, were we to know $\Lambda(t)$, its derivative $\lambda(t)$, the realizations x_i of X_i, and t_i of T_i, $i = 1, 2, \ldots, i-1$, then the density of X_i at x_i is [from postulate (iii)] of the form

$$f(x_i \mid x_1, \ldots, x_{i-1}, \Lambda(\bullet))$$

$$= \lambda(x_i + \sum_{j=1}^{i-1} x_j) \times \exp\left(\Lambda(\sum_{j=1}^{i-1} x_j) - \Lambda(x_i + \sum_{j=1}^{i-1} x_j)\right), \text{ for } x_i \geq 0, \quad (2.21)$$

and the probability density function of T_i at t_i, for $t_i \geq t_{i-1}$, is of the form

$$f(t_i \mid t_1, \ldots, t_{i-1}, \Lambda(\bullet)) = \lambda(t_i) \, e^{-(\Lambda(t_i) - \Lambda(t_{i-1}))}. \quad (2.22)$$

The preceding results are intriguing, especially in the light of postulate (i) which says that future occurrences of $N(t)$ are independent of its past. Specifically, (2.22) says that the distribution of T_i, the next time to failure depends on T_{i-1}, the last time to failure. Such a property, namely, dependence on only the last event, is known as the *Markov property*. More interestingly, (2.21) says that the distribution of X_i, the ith interarrival time, depends on the entire previous history of the process.

2.3.2 The Homogeneous Poisson Process Model

A special case of the nonhomogeneous Poisson process model is when $\lambda^*(t)$ [the derivative of $\Lambda^*(t)$] is a constant, say λ^*, so that if λ^* takes a value λ, then $\Lambda(t)$ takes the value λt, and

$$\mathcal{P}(N(t) = k \mid \lambda) = e^{-\lambda t} \frac{(\lambda t)^k}{k!}, \text{ for } k = 0, 1, 2, \ldots, \quad (2.23)$$

this is called the *homogeneous Poisson process model*. It is perhaps the most commonly used point process model.

The interarrival and the waiting times of a homogeneous Poisson process take a simple and attractive form. Verify that with $\Lambda(t) = \lambda t$, (2.21) becomes

$$f(x_i \mid x_1, \ldots, x_{i-1}, \lambda) = \lambda e^{-\lambda x_i}, \text{ for } x_i \geq 0, \quad (2.24)$$

which is the density function of an exponential distribution with scale λ. Since the distribution of X_i is independent of $X_1, \ldots, X_{i-1}$, we say that the interarrival times in a homogeneous Poisson process are independently and identically exponentially distributed. Similarly, with $\Lambda(t) = \lambda t$, (2.22) becomes

$$f(t_i \mid t_1, \ldots, t_{i-1}, \lambda) = \lambda e^{-\lambda(t_i - t_{i-1})}, \text{ for } t_i \geq t_{i-1}. \quad (2.25)$$

Equation (2.25) can be used to obtain the probability density function of T_i, were we to know only λ, that is, the *unconditional* density function of T_i, given λ, for $i = 1, 2, \ldots$. To do this, we first note that if $t_0 \overset{\text{def}}{=} 0$, then $f(t_1 \mid \lambda) = \lambda \exp(-\lambda t_1)$, and using the law of the extension of conversation, it is seen that

$$f(t_2 \mid \lambda) = \int_0^{t_2} f(t_2 \mid t_1, \lambda) f(t_1 \mid \lambda) dt_1 = \int_0^{t_2} \lambda e^{-\lambda(t_2 - t_1)} \lambda e^{-\lambda t_1} dt_1,$$

or that

$$f(t_2 \mid \lambda) = \lambda^2 e^{-\lambda t_2} t_2, \text{ for } t_2 > 0,$$

which is a gamma density with a scale parameter λ and a shape parameter 2. Continuing in this manner, we can deduce that in general

$$f(t_i \mid \lambda) = \frac{\lambda^i (e^{-\lambda t_i}) t_i^{i-1}}{(i-1)!}, \text{ for } t_i \geq 0, \quad (2.26)$$

which is a gamma density with a scale parameter $\lambda > 0$, and a shape parameter i, $i = 1, 2, \ldots$. The simplicity of these results makes the homogeneous Poisson process model attractive for use when all that one wishes to do is an expedient data analysis.

2.3.3 *Generalizations of the Point Process Model*

There are several generalizations of the preceding point process model, each of which could be a suitable candidate for describing the points generated by software failures. With some of these generalizations it is not possible to retain the defining characteristic of the Poisson process models, namely, that of independent increments.

The Compound Poisson Process

The simplest generalization is to allow the point process to take jumps of random size; recall that the sample path shown in Figure 2.5 pertains to jumps that are only of unit size. A point process that retains all the characteristics of a Poisson process (homogeneous or nonhomogeneous), save the one which restricts jumps to be of a unit size, is a compound Poisson process. In the context of software failures, a compound Poisson process may be an appropriate model if, upon the occurrence of failure, a random number of bugs are detected and corrected; see, for example, Sahinoglu (1992). With such processes, we have two sources of uncertainty: the times at which the software fails, and the number of bugs that are identified and corrected upon each failure. As before, if we let $T_1, T_2, \ldots, T_i, \ldots$, denote the times at which an event of interest (say software failure) occurs, and if associated with each T_i there is a random variable Z_i denoting some attribute of interest (say the number of bugs that are detected, or the time to debug the software and put it back in operation), then the process $\{\mathcal{N}(t)\,;\, t \geq 0\}$, where

$$\mathcal{N}(t) \stackrel{\text{def}}{=} \sum_{i=1}^{N(t)} Z_i,$$

and $N(t)$ is the number of events that occur in time $[0, t]$, is called a *compound Poisson process*.

To describe our uncertainty about $\mathcal{N}(t)$ we need to know, in addition to $\Lambda^*(t)$, the probability distributions of the Z_is. When such is the case, it is easy to see, using the law of the extension of conversation by conditioning on k events, that if $\Lambda^*(t) = \Lambda(t)$, and if

$$F^k(z) \stackrel{\text{def}}{=} \mathcal{P}(\sum_1^k Z_i \leq z),$$

then

$$\mathcal{P}(\mathcal{N}(t) \leq \nu \mid \Lambda(t), F^\bullet(\nu)) = \sum_{k=0}^{\infty} \exp(-\Lambda(t)) \frac{(\Lambda(t))^k}{k!} F^k(\nu),$$

$$\text{for } 0 \leq \nu < \infty. \tag{2.27}$$

The distribution function $F^k(z)$ is known as the *k-fold convolution* of the Z_is. The derivation of (2.27) is left as an exercise for the reader.

Simplifications occur if we assume that the Z_is are independent and identically distributed. For example, if Z_i represents the debugging time subsequent to the ith failure, then we may assume that the Z_is are independent and identically exponentially distributed with scale δ. In this case $F^k(z)$ is a gamma distribution with scale δ and shape k. With the preceding interpretation, our uncertainty about $\mathcal{N}(t)$, the *total debugging time*, or the *software's downtime* in the interval $[0, t]$, is described by a compound Poisson process, and the model (2.27) is helpful for assessing the "availability" of software.

Commentary

There are other aspects of compound Poisson processes that need to be discussed. The first is that of independent increments, and the second that of computability. It is easy to see that the compound Poisson process model (2.27) retains the independent increments property only if the Z_is are independent; otherwise, $\mathcal{N}(t)$ inherits the dependence between the Z_is. The assumption of independence may be unrealistic, because it implies two things: the absence of an increase in debugging efficiency with time, and a failure to account for the fact that typically, later failures are harder to detect and rectify than the earlier ones. But why are independent increments important? Can we not do without them? The answer to these questions has to do with the likelihood function which is needed for making statistical inferences. Independence simplifies a specification of the likelihood function; we can do without it, but only at the price of computational difficulties. Indeed, the popularity of the Poisson process model is largely driven by its property of independent increments. Finally, on the matter of computability, even under the assumption of independent increments, (2.27) is difficult to compute; it involves an infinite sum over k of the distribution function $F^k(z)$. One strategy would be to approximate using limiting arguments involving $t \to \infty$; another would be a Monte Carlo simulation.

The Doubly Stochastic Poisson Process

The notion of a *doubly stochastic Poisson process* or a *Poisson process with a random environment* was introduced to describe those situations wherein there is a *physical* motivation for supposing that the mean value function $\Lambda^*(t)$ of a Poisson process model is itself a stochastic process. Furthermore, it is assumed that a knowledge of the history of the process does not change the probabilistic structure of $\Lambda^*(t)$; that is, the probability structure of $\Lambda^*(t)$ is assumed to be *preassigned* [cf. Cox and Isham (1980), p. 10]. This is a very strong assumption. Its consequence is that in (2.20), the $f(\Lambda(t) \mid \mathcal{H}, \mathcal{H}_t)$ simplifies to $f(\Lambda(t) \mid \mathcal{H})$, so that under a doubly stochastic Poisson process model for $\mathcal{P}(N(t) = k \mid \mathcal{H}, \mathcal{H}_t)$, with a preassigned probability structure for $\Lambda^*(t)$,

$$\mathcal{P}(N(t) = k \mid \mathcal{H}, \mathcal{H}_t) = \int_{\Lambda(t)} \mathcal{P}\Big(N(t) = k \mid \Lambda(t)\Big) f(\Lambda(t) \mid \mathcal{H}) d\Lambda(t).$$

Of course, from a Bayesian point of view, all parameters are unknown, and hence all Poisson process models should really be regarded as being doubly stochastic. The main point of departure is the assumption of a preassigned probability structure for $\Lambda^*(t)$. It is more realistic to suppose that a knowledge of the past occurrence of the process influences our assessment of uncertainty about

2.3 Point Processes and Counting Process Models

$\Lambda(t)$, so that under the assumption of a Poisson process model, a proper Bayesian approach would result in (2.20) taking the form

$$P(N(t) = k \mid \mathcal{H}, \mathcal{H}_t) = \int_{\Lambda(t)} \frac{e^{-\Lambda(t)}(\Lambda(t))^k}{k!} f(\Lambda(t) \mid \mathcal{H}, \mathcal{H}_t) d\Lambda(t). \quad (2.28)$$

Observe that under (2.28) the process $\{N(t); t \geq 0\}$ will lose its independent increments property.

To further appreciate the arguments that motivate a consideration of the doubly stochastic feature of Poisson processes, we introduce an alternate, but equivalent, specification of the postulates of a Poisson process model for $N(t)$ given $\Lambda^*(t)$.

Suppose that in (2.20), $\Lambda^*(t)$ takes the value $\Lambda(t)$, and that $\lambda(t)$, the derivative of $\Lambda(t)$, exists. Then, given $\lambda(t)$, a probability model for $N(t)$ is a *nonhomogeneous Poisson process model* if:

(i) for any time t, and a small interval of time Δt,

$$P(N(t + \Delta t) - N(t) = 1 \mid \lambda(t), \mathcal{H}_t)$$

$$= P(N(t + \Delta t) - N(t) = 1 \mid \lambda(t)) = \lambda(t)\Delta t + o(\Delta t),$$

and

$$P(N(t + \Delta t) - N(t) > 1 \mid \lambda(t), \mathcal{H}_t)$$

$$= P(N(t + \Delta t) - N(t) > 1 \mid \lambda(t)) = o(\Delta t),$$

so that

$$P(N(t + \Delta t) - N(t) = 0) = 1 - \lambda(t)(\Delta t) + o(\Delta t);$$

(ii) $P(N(0) = 0 \mid \lambda(\bullet)) = 1.$

The quantity $o(h)$ is a correction term; it denotes a function of h such that

$$\lim_{h \to 0} \frac{o(h)}{h} = 0.$$

Its role is to ensure that $P(N(t + h) - N(t) = 1 \mid \lambda(t))$ does not exceed 1 when h is large. The independent increments property of the Poisson process model is a consequence of (i).

With the preceding specification, the intensity function $\lambda(t)$ of the process can be given a physical interpretation. Specifically, since $\lambda(t)(\Delta t)$ approximately equals the probability that an event will occur in a small interval of time in the vicinity of t, we are able to relate $\lambda(t)$ to the actual process that generates the events of interest, for example, the underlying stress in the case of structural failures, or geological factors in the case of earthquakes. In the context of software failure $\lambda(t)$ would be determined by the underlying operational profile of the software. Recall that a software's *operational profile* is a description of the environment under which it operates. Consequently, $\lambda(t)$ tends to be large (small) when the software executes complex (simple) operations. When the workload on the software is uncertain, or changes randomly over time, so that the operational profile is itself a stochastic process, then so will $\lambda^*(t)$, and a doubly stochastic Poisson process model for $\mathcal{P}(N(t) = k \mid \mathcal{H}, \mathcal{H}_t)$ will arise naturally. Observe that given $\lambda(t)$, the probability model for $N(t)$ will retain the independent increments property, and if the probability structure of $\lambda^*(t)$ is pre-assigned, then the process $\{N(t); t \geq 0\}$ itself will also possess the independent increments property. When such is the case, the stochastic process $\{N(t); t \geq 0\}$ is called a *doubly stochastic Poisson process*.

The Self-Exciting Point Process

A prime motivation for introducing self-exciting point processes is the need to relax the independent increments feature of Poisson process models for $\{N(t); t \geq 0\}$. In the context of software failures, since a software code can conceptually consist of only a finite number of bugs, the independent increments property is not tenable. A knowledge of the past occurrences of the process must influence our uncertainty about future occurrences. There are several strategies for introducing dependence among the increments, one of which is via the prescription (2.28), which de facto is a Bayesian model for a doubly stochastic Poisson process. A closely related approach is via the mechanism of a self-exciting point process model for $\{N(t); t \geq 0\}$; this is described in the following.

Suppose that $\mathcal{H}_t$ comprises $N(t^-)$, and the waiting times $T_1, T_2, \ldots, T_{N(t^-)}$; that is, $\mathcal{H}_t$ is the progress (or history) of the process up to but not including t. We start by recalling (2.20); suppose that $\lambda^*(t)$ the derivative of $\Lambda^*(t)$ exists and, that given $\lambda(t)$ a model for $N(t)$, is of the form

$$\mathcal{P}\Big(N(t + \Delta t) - N(t) = 1 \mid \mathcal{H}_t, \lambda(t)\Big) = \mathcal{P}(N(t + \Delta t) - N(t) = 1 \mid \lambda(t))$$

$$= \lambda(t)(\Delta t) + o(\Delta t).$$

2.3 Point Processes and Counting Process Models

Then, it follows [see (2.20)] that

$$P(N(t + \Delta t) - N(t) = 1 \mid \mathcal{H}, \mathcal{H}_t) = E(\lambda^*(t) \mid \mathcal{H},\mathcal{H}_t) + o(\Delta t),$$

where $E(\lambda^*(t) \mid \mathcal{H},\mathcal{H}_t)$ is the conditional expectation of $\lambda^*(t)$.

Motivated by the preceding line of reasoning, we say that $\{N(t); t \geq 0\}$ is a *self-exciting point process* (SEPP) if the following postulates can be invoked.

(i) For any time t, and a small interval of time Δt,

$$P(N(t + \Delta t) - N(t) = 1 \mid \mathcal{H}, \mathcal{H}_t) = E(\lambda^*(t) \mid \mathcal{H},\mathcal{H}_t) + o(\Delta t).$$

(ii) For any subset Q_t of $\mathcal{H}_t$, and a function $g(x)$ with $\lim_{x \to 0} g(x) = 0$,

$$P(N(t + \Delta t) - N(t) \geq 2 \mid \mathcal{H}, Q_t)$$
$$= P(N(t + \Delta t) - N(t) = 1 \mid \mathcal{H}, Q_t) \, g(\Delta t);$$

and

(iii) $P(N(0) = 0 \mid \mathcal{H}_t) = 1$.

Note that the conditioning on $\mathcal{H}_t$ ensures that $\{N(t); t \geq 0\}$ does not have the independent increments property.

The second of the preceding properties is known as *conditional orderliness*. In essence, it guarantees that the probability of the process increasing by more than one, in a short interval of time, is small; thus $N(t)$ is well behaved and does not suddenly explode to infinity. If $Q(t) = \phi$, the empty set, then the second property reduces to what is known as *unconditional orderliness,* and now

$$P(N(t + \Delta t) - N(t) \geq 2) = P(N(t + \Delta t) - N(t) = 1) \, g(\Delta t);$$

if $Q(t) = \mathcal{H}(t)$, then the second and third of the preceding properties lead to the result that

$$P(N(t + \Delta t) - N(t) \geq 2 \mid \mathcal{H}_t)$$
$$= P(N(t + \Delta t) - N(t) = 1 \mid \mathcal{H}_t) \, g(\Delta t)$$
$$= o(\Delta t).$$

There are different degrees to which $\lambda^*(t)$ depends on $\mathcal{H}_t$, and this idea is formalized by the notion of the *memory of the self-exciting Poisson process*. Specifically, a SEPP is of *memory m*, if

for $m = 0$: $\quad \lambda^*(t)$ depends only on $N(t)$; that is,
$\qquad\qquad\quad$ $E(\lambda^*(t) \mid \mathcal{H}_t) = E(\lambda^*(t) \mid N(t))$;

for $m = 1$: $\quad \lambda^*(t)$ depends only on $N(t)$ and $T_{N(t)}$; that is,
$\qquad\qquad\quad$ $E(\lambda^*(t) \mid \mathcal{H}_t) = E(\lambda^*(t) \mid N(t), T_{N(t)})$;

for $m \geq 2$: $\quad \lambda^*(t)$ depends on $N(t)$, $T_{N(t)}$, and at most the last $(m-1)$ inter-arrival times; and

$m = -\infty$: $\quad \lambda^*(t)$ is independent of the entire progress of the process.

Note that the case $m = -\infty$ corresponds to the doubly stochastic Poisson process (DSPP). Also, since the special case of the DSPP, when $\lambda^*(t)$ takes the value $\lambda(t)$ with probability 1, is the nonhomogeneous Poisson process (NHPP), we have the following, as a chain of implications for the point process models we have discussed,

$$\text{HPP} \subset \text{NHPP} \subset \text{DSPP} \subset \text{SEPP},$$

where HPP abbreviates the homogeneous Poisson process.

In Chapter 3 we point out that almost all of the proposed models for software reliability are special cases of the SEPP. Indeed the current research in analyzing software failure data focuses heavily on point process models with intensities described as stochastic processes [cf. Gokhale, Lyu, and Trivedi (1998)].

2.4 Fundamentals of Reliability

Much of the literature on statistical aspects of software engineering has been devoted to the topic of software credibility, or reliability. By credibility, we mean the risk of an in-process software failure. Even though a lot has been written about the differences between hardware and software reliability, it is useful to bear in mind that the general principles by which reliability problems are addressed are common to both applications. What distinguishes reliability problems from the others in which probability and statistics are used is that here the event of interest is failure, and the uncertain quantity the time to failure T.

Since T is continuous and takes values in $[0, \infty)$, our aim is to assess $\mathcal{P}(T \geq t \mid \mathcal{H})$ for some $t \geq 0$. When viewed as a function of t, the quantity $\mathcal{P}(T \geq t \mid \mathcal{H})$ is called the *reliability function*, or the *survival function* of T; it is denoted by $\bar{F}_T(t \mid \mathcal{H})$. Note that $\bar{F}_T(t \mid \mathcal{H})$ decreases in t, from 1 at $t = 0$, to 0 at $t = \infty$. The argument t of $\bar{F}_T(t \mid \mathcal{H})$ is called the *mission time*. Also, if $F_T(t \mid \mathcal{H}) = \mathcal{P}(T \leq t \mid \mathcal{H})$, then from the laws of probability $\bar{F}_T(t \mid \mathcal{H}) = 1 - F_T(t \mid \mathcal{H})$.

2.4.1 The Notion of a Failure Rate Function

One of the key notions in reliability theory is that of the failure rate function of T (or of the distribution function of T). Suppose that $F_T(t \mid \mathcal{H})$ is absolutely continuous so that $f_T(t \mid \mathcal{H})$, the density function of T at t, exists. Then the *predictive failure rate function* of T, at $t \geq 0$, is defined as

$$r_T(t \mid \mathcal{H}) = \frac{f_T(t \mid \mathcal{H})}{\bar{F}_T(t \mid \mathcal{H})} \; .$$

The failure rate function derives its importance because of its interpretation as a conditional probability. Specifically, $r_T(t \mid \mathcal{H})dt$ approximates the conditional probability that an item fails in the time interval $[t, t + dt]$ were we to suppose that it is surviving at t; that is,

$$r_T(t \mid \mathcal{H})dt \approx \mathcal{P}(t \leq T \leq t + dt \mid T \geq t, \mathcal{H}).$$

Whereas a direct specification of $\bar{F}_T(t \mid \mathcal{H})$ is often difficult, specifying conditional probabilities, like $r_T(t \mid \mathcal{H})dt$, is generally easier. There may be physical and/or subjective features that help guide this choice. Since the failure rate at t is the instantaneous probability of failure of an item that is assumed to survive until t, the failure rate of items that age, such as machinery and humans, will increase with t. Similarly, the failure rate of software, were it not to experience failure, will decrease with time, since the absence of failure enhances our opinion of the software's quality. Recall that, subjectively, a conditional probability is the informed opinion of a particular individual at a particular time.

Analogous to the notion of a predictive failure rate is the notion of a model failure rate. Specifically, suppose that to assess $\mathcal{P}(T \geq t \mid \mathcal{H})$, a parameter θ is introduced, the law of total probability with its paraphernalia of conditional independence is invoked, and a probability model for T, $\mathcal{P}(T \geq t \mid \theta)$, is obtained. Then, assuming that $f_T(t \mid \theta)$, the probability density of $(t \mid \theta)$, exists for all t, the *model failure rate* of T, at $t \geq 0$, is defined as

$$r_T(t \mid \theta) = \frac{f_T(t \mid \theta)}{\bar{F}_T(t \mid \theta)} \; , \tag{2.29}$$

where $\bar{F}_T(t \mid \theta)$ is $\mathcal{P}(T \geq t \mid \theta)$. As before, $r_T(t \mid \theta)$ is interpreted as

$$r_T(t \mid \theta)dt \approx \mathcal{P}(t \leq T \leq t + dt \mid T \geq t, \theta).$$

Since both the predictive and the model failure rates are probabilities, and since probabilities are personal, we may state that failure rates do not exist outside our minds [cf. Singpurwalla (1988a)]. Also, it is helpful to recall that since all probabilities are assessed at some reference time τ, and since the failure rate is a conditional probability, it too is assessed at τ. The conditioning argument $T \geq t$, in $\mathcal{P}(t \leq T \leq t + dt \mid T \geq t, \mathcal{H})$, is to be interpreted in the subjunctive; that is, it is the probability of failure in $[t, t + dt]$, were the item to survive to t. If the item is observed to actually survive to t, then this information becomes a part of the history $\mathcal{H}$ and our uncertainty assessment process now commences at the reference time $\tau + t$.

To see how a specification of the failure rate, predictive or model, facilitates the assessment of reliability, we concentrate on (2.29) and start with the observation that

$$r_T(t \mid \theta) = \frac{f_T(t\mid\theta)}{\overline{F}_T(t\mid\theta)} = -\frac{d}{dt} \log(\overline{F}_T(t \mid \theta));$$

integrating and exponentiating both sides of the preceding gives us the *exponentiation formula* of reliability

$$\overline{F}_T(t \mid \theta) = \exp\left(-\int_0^t r_T(u \mid \theta)du\right). \tag{2.30}$$

It is because of the preceding formula that the failure rate function is often used as a proxy for the reliability of an item.

The development so far assumes that $\overline{F}_T(t \mid \theta)$ is absolutely continuous so that $f_T(t \mid \theta)$ exists (almost) everywhere. When such is not the case because $f_T(t \mid \theta)$ has a jump at, say t^*, then $r_T(t \mid \theta)$ is given by (2.29) for all $t \neq t^*$, and is

$$r_T(t^* \mid \theta) = \frac{f_T(t^{*})-f_T(t^{*-})}{\overline{F}_T(t\mid\theta)}, \text{ at } t = t^*.$$

2.4.2 Some Commonly Used Model Failure Rates

The simplest failure model is the exponential, with $\overline{F}_T = e^{-\lambda t}$, $t \geq 0$; see (2.14). From (2.29) it is easy to verify that the failure rate of $(t \mid \lambda)$ is a constant, λ; furthermore, from (2.30), it is easily seen that if the failure rate of $(t \mid \lambda)$ is λ, then $\overline{F}_T(t \mid \lambda) = e^{-\lambda t}$, thus the claim that the exponential failure model is the only one for which the *model* failure rate is a constant, and vice versa. Also, recall (see Section 2.2.4) that for the exponential failure model the MTBF is $1/\lambda$. Thus a knowledge of the MTBF is sufficient for a specification of both the failure rate function and the reliability function. The constant model failure rate

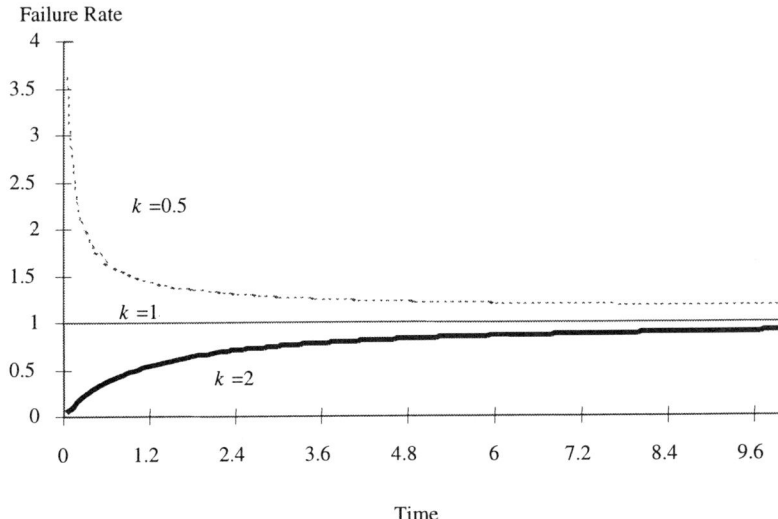

FIGURE 2.7. The Failure Rate of a Gamma Distribution ($\lambda = 1$).

assumption implies that were the parameter Λ to be known to us as λ, then the lifelength of the item does not reflect the property of aging. That is, our assessment of failure in the vicinity of t is not influenced by the knowledge of survival at t. This assumption, also known as *lack of memory*, is a strong one to make because it implies the absence of learning.

The next failure model to consider is the gamma distribution whose probability density is given in (2.15). Its distribution function is not in closed form and so a closed form expression for the failure rate function is not available. However, it can be numerically shown that the failure rate function of $(t \mid \lambda, k)$ is a constant equal to (the scale parameter) λ, when the shape parameter $k = 1$; it is decreasing for $k < 1$, and is increasing for $k > 1$, asymptoting to λ when $t \to \infty$; see Figure 2.7. Similarly, when the failure model is the Weibull distribution function (2.16), the failure rate of $(t \mid \alpha, \beta)$ is the constant α when the shape parameter $\beta = 1$, and increases (decreases) when $\beta > (<)1$; see Figure 2.8. It is important to note that the exponential failure model is a special case of both the gamma and the Weibull models.

The failure rate of a lognormal distribution is also not available in closed form. But unlike the monotone failure rates of the gamma and the Weibull distributions, the failure rate of the lognormal distribution can be made to initially increase and then decrease to zero, depending on the choice of the parameters μ and σ; see Figure 2.9 wherein $e^{\mu} = 1000$, and $\sigma = 1$ and 3.

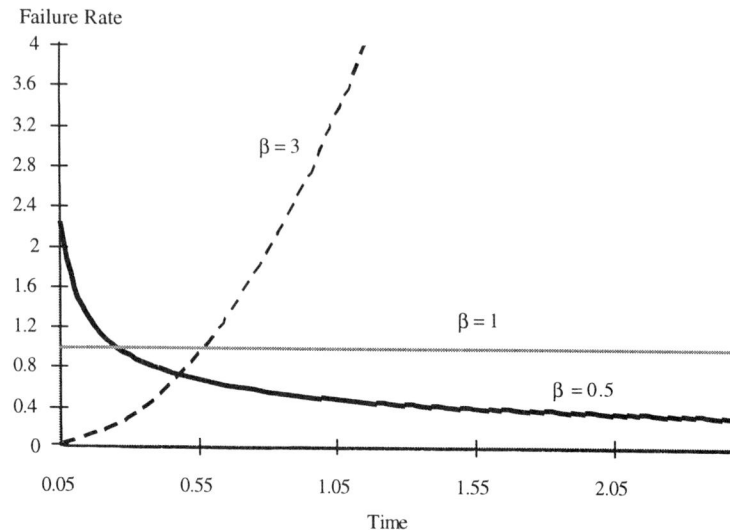

FIGURE 2.8. The Failure Rate of a Weibull Distribution ($\alpha = 1$).

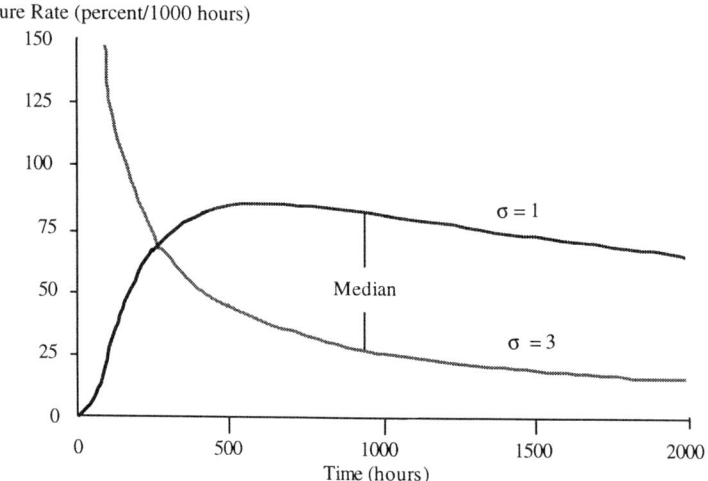

FIGURE 2.9. The Failure Rate of a Lognormal Distribution ($e^{\mu} = 1000$, and $\sigma = 1$ and 3).

2.4.3 Covariates in the Failure Rate Function

In any given scenario, there are many factors that influence our assessment of the lifelength of an item. For example, with software it could be the known factors such as the size of the code, the number of modules, the quality of the programming team, the operational profile, and the like. Such factors are referred to as *covariates*. Those covariates whose values are known at the time -of assessment τ become a part of the background history $\mathcal{H}$. But how should we treat covariates, like the operational profile, whose future values are unknown at τ? Can we adopt here a strategy that parallels the one we used in the context of doubly stochastic Poisson processes?

To address these questions, suppose that to incorporate the effect of an uncertain covariate we introduce a parameter $\mathcal{Z}$, taking values ζ, into the failure rate function. The parameter $\mathcal{Z}$ should bear some interpretive relationship to the covariate. For example, if the covariate pertains to an unknown presence or absence of a certain attribute, say *fault tolerance* in the case of software, then $\mathcal{Z}$ could take the value one; zero, otherwise. Having introduced $\mathcal{Z}$ we are also required to assess $\mathcal{P}(\mathcal{Z} \leq \zeta)$. Thus the $r_T(t \mid \theta)$ of (2.29) is replaced by $r_T(t \mid \theta, \zeta)$, and the exponentiation formula now takes the form

$$\bar{F}_T(t \mid \theta, \zeta) = \exp\left(-\int_0^t r_T(u \mid \theta, \zeta) du\right). \qquad (2.31)$$

The left-hand side of (2.31) now represents our assessment of the reliability of the item were we to know, besides θ, also ζ, the value taken by the parameter $\mathcal{Z}$ which is our proxy for the covariate.

When the covariate of interest changes over time, it is best described by a stochastic process $\{\mathcal{Z}(t); t \geq 0\}$, and if $\mathcal{Z}(u)$ takes the value $\zeta(u)$, then (2.31) takes the form

$$\bar{F}_T(t \mid \theta, \zeta(u); 0 \leq u \leq t) = \exp\left(-\int_0^t r_T(u \mid \theta, \zeta(u)) du\right). \qquad (2.32)$$

Since we are uncertain about the progression of the covariate over time, we must average the right-hand side of (2.32) over all the sample paths of $\mathcal{Z}(u)$, $0 \leq u \leq t$, to obtain its expectation

$$\bar{F}_T(t \mid \theta) = E\left(\exp\left(-\int_0^t r_T(u \mid \theta, \zeta(u)) du\right)\right); \qquad (2.33)$$

its evaluation can be a formidable task: (2.33) follows from (2.32) by the law of total probability.

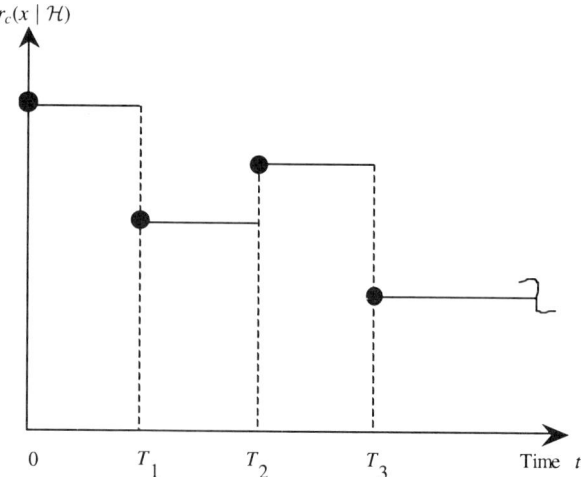

FIGURE 2.10. A Concatenated Failure Rate Function.

There are circumstances, perhaps not in software reliability, where the exponentiation formula is invalid. This occurs with *internal covariates*. Internal covariates are those whose disposition indicates the failure or the survival of the item; for example, the blood pressure if zero is an indicator of sure failure. When such is the case, (2.31) should not be used; see Singpurwalla and Wilson (1995).

2.4.4 The Concatenated Failure Rate Function

What has been discussed thus far pertains to the failure rate of the distribution of a single random variable T. With software, which presumably undergoes the constant process of testing and debugging, we are interested in the behavior of a collection of interfailure times $X_1, X_2, \ldots, X_i, \ldots$, each interfailure time representing the time to failure of a debugged version of the software. Associated with each X_i is $r_{X_i}(x \mid \mathcal{H})$, the failure rate function of its marginal distribution function. Generally, the interfailure times are not independent; thus the individual failure rates must bear some relationship to each other. Because the notion of the failure rate of a joint distribution function has not been sufficiently well articulated, especially its intuitive import [see, e.g., Basu (1971) or Marshall (1975)], the notion of the failure rate of software is an elusive one. However, by most accounts, when investigators refer to the term "the failure rate of software" what they mean is a *concatenation* (or the side-by-side placement) of the failure rates of the conditional distribution of X_i, given X_1, $\ldots, X_{i-1}$, for $i = 2, 3, \ldots$; see Singpurwalla (1995). This practice, first advocated by Jelinski and Moranda (1972), prompts us to introduce the notion of a *concatenated failure rate function* $r_c(x \mid \mathcal{H})$ as

$$r_c(x \mid \mathcal{H}) = r_{X_i \mid X_1, \ldots, X_{i-1}}(x - T_{i-1} \mid \mathcal{H}), \text{ for } T_{i-1} \leq x < T_i, \quad (2.34)$$

with $T_1 \leq T_2 \leq \ldots \leq T_i \leq \ldots$, denoting the times to failure, and $X_1, X_2, \ldots, X_i, \ldots$, the interfailure times; by convention, $T_0 \equiv 0$.

In most cases the ith segment of $r_c(x \mid \mathcal{H})$ depends only on T_{i-1}, so that it simplifies as $r_{X_i \mid T_{i-1}}(x - T_{i-1} \mid \mathcal{H})$. An illustration of this simplified version is shown in Figure 2.10; it consists of several constant segments, each segment corresponding to an interfailure time.

It is important to emphasize that a concatenated failure rate function is not the elusive failure rate function of the joint distribution function $F_{X_1, \ldots, X_n}(x_1, \ldots, x_n \mid \mathcal{H})$. In fact, since $r_c(x \mid \mathcal{H})$ is defined in terms of the successive failure times $T_1 \leq T_2 \leq \ldots \leq T_i \leq \ldots$, it is a random function. In $r_c(x \mid \mathcal{H})$, each segment starts from the time of the last failure rather than the usual 0 from which the failure rate of the marginal distribution of X_i would commence. Most important, neither the concatenated failure rate function, nor its segments like $r_{X_i \mid X_1, \ldots, X_{i-1}}(x - T_{i-1} \mid \mathcal{H})$, for $T_{i-1} \leq x < T_i$, can be used in the exponentiation formula (2.30) to obtain the conditional distribution of X_i given $X_1, \ldots, X_{i-1}$. This is because with the condition $T_{i-1} \leq x < T_i$, T_i becomes an internal covariate rendering (2.30) invalid. In order to use (2.30) we must not constrain x so that it is less than T_i. Thus a purpose served by $r_c(x \mid \mathcal{H})$ is a graphical display of the behavior of the successive failure rates of the conditional distributions of the interfailure times during the test–debug phase of the software. However, and more important, it has been shown by Chen and Singpurwalla (1997) that $r_c(x \mid \mathcal{H})$ is the intensity function of the self-exciting point process that generates the T_is; see Section 3.5.2.

By way of a final remark, we note that it is possible [cf. Al-Mutairi, Chen, and Singpurwalla (1998)] that the value taken by the ith segment of $r_c(x \mid \mathcal{H})$, could depend on X_{i-1}, the preceding interfailure time. Consequently, the ith segment is written as $r_{X_i \mid T_{i-1}, T_{i-2}}(x - T_{i-1} \mid \mathcal{H})$. In general, the ith segment could depend on any function of all the preceding interfailure times. When such is the case, the modeling effort tends to get very complicated.

2.5 Chapter Summary

We started this chapter with an overview of the calculus of probability, to include topics such as conditional probability, conditional independence, the law of total probability, and Bayes' Law. This was followed by an articulation of the likelihood function and the notion of exchangeability. All these topics constitute the foundational material for quantifying, combining, and updating uncertainties, and are presented from the point of view of an expository overview.

We then used the law of total probability to introduce the notions of a software reliability model and the parameters of such models. Examples of models that can be used to address several generic problems faced by software engineers were motivated and introduced. These include the Bernoulli, the binomial, the geometric, the Poisson, the exponential, the Weibull, the gamma, and the lognormal.

This was followed by a discussion of an important class of probability models, namely, the point process models and their role in software engineering. Such models include the popular homogeneous, the nonhomogeneous, and the compound Poisson process models. Independent increments, the defining characteristic of Poisson process models, was discussed and its limitations for modeling problems of software failure were pointed out.

In response to the preceding concern two new types of point process models were introduced, namely, the doubly stochastic Poisson process and the self-exciting point processes. The former arises in software testing wherein the operational profile is itself a stochastic process. The latter is natural in testing, since the software code consists of a finite number of bugs so that the assumption of independent increments is untenable. The hierarchical structure of point process models was noted; specifically, it was pointed out that

$$\text{HPP} \subset \text{NHPP} \subset \text{DSPP} \subset \text{SEPP}.$$

In this chapter we also introduced some fundamentals of reliability theory, namely, the survival function, and the predictive and the model failure rate functions. This was followed by a discussion of some commonly used model failure rates such as the exponential, the Weibull, the gamma, and the lognormal. The treatment of covariates by conditioning on the failure rate function was described, and the chapter ended with the introduction of the concatenated failure rate function as a way to model the interfailure times of software that is evolving over time.

Exercises for Chapter 2

1. Verify the additivity rule for nonmutually exclusive events $\mathcal{E}_1$ and $\mathcal{E}_2$.

2. Suppose that you have three coins. Coin A has a 50% probability of landing heads, coin B has a 25% probability of landing heads, and coin C is two-headed. A friend picks one of the coins at random and tosses it, telling you that it landed heads.

    (a) By conditioning on which coin is picked and applying the law of total probability, show that the probability of a head is 7/12.

    (b) Using Bayes' Law, calculate the probability that coin C was picked given a head was thrown. Repeat this calculation for coins A and B.

    (c) Now, suppose the coin that was picked was thrown again. By conditioning on which coin was picked and applying the law of total probability, show that the probability of obtaining a head on the second throw given that the first throw was a head is 3/4.

3. During testing, a piece of software is subjected to a sequence of N inputs, each judged to have the same probability of a successful output, which we denote p; thus the probability model for the number of successful outputs is binomial.
   A prior distribution on p is assessed to be a *beta distribution* with parameters α and β; that is,

   $$\pi(p \mid \alpha, \beta) = \frac{\Gamma(\alpha+\beta)}{\Gamma(\alpha)\Gamma(\beta)} \, p^{\alpha-1} (1-p)^{\beta-1}, \quad 0 < p < 1.$$

   Note that the mean of a beta distribution is $\alpha/(\alpha+\beta)$ (i.e. the ratio of the first parameter to the sum of both). We observe that x of the N inputs resulted in a successful output.

    (a) Apply Bayes' Law to show that the distribution of p in the light of x and N, also known as the *posterior distribution* of p, is of the form:

    $$\pi(p \mid x, N) = \frac{\Gamma(\alpha+\beta+N)}{\Gamma(\alpha+x)\,\Gamma(\beta+N-x)} \, p^{\alpha+x-1} (1-p)^{\beta+N-x-1},$$
    $$0 < p < 1,$$

that is, another beta distribution with parameters $\alpha + x$ and $\beta + N - x$.

(b) What is the mean of this new distribution of p?

(c) If little is known about p a priori, a possible prior is the uniform distribution on [0, 1], which is a special case of the beta distribution with $\alpha = \beta = 1$. Show that the posterior mean under this prior can be written as

$$E(p \mid x, N) = \frac{2}{2+N} \times \frac{1}{2} + \frac{N}{2+N} \times \frac{x}{N}.$$

(d) Show that, for the beta prior in general, the posterior mean can be written as a convex combination of the prior mean $\alpha/(\alpha+\beta)$ and the proportion of successes in the data x/N.

(e) What happens to the posterior mean as the number of tests N gets large?

(f) Another series of N inputs is to be tested on the software. Assuming that the performance of the software has not changed, we are interested in the number of successful outputs Y in this new set. By conditioning on p, and using the law of total probability, show that the distribution of Y given x,

$$\mathcal{P}(Y = y \mid x) = \binom{N}{y} \frac{\Gamma(\alpha+\beta+N)\Gamma(y+\alpha+x)\Gamma(2N-y-x+\beta)}{\Gamma(\alpha+x)\Gamma(\beta+N-x)\Gamma(\alpha+\beta+2N)},$$

for $y = 0, 1, \ldots, N$. This distribution is called the *beta-binomial distribution*.

4. Verify Equations (2.12) and (2.14).

5. Suppose we are testing software with a large number of inputs, each taking roughly the same short time to compute, and each judged to have the same high probability of success. We have argued in this chapter that the time until the first incorrect output can be modeled by a continuous random variable X that is approximately exponentially distributed; thus $\mathcal{P}(X \geq x \mid \lambda) = e^{-\lambda x}$, for $x \geq 0$ and a parameter $\lambda > 0$; the density function of X at x is

$$f_X(x \mid \lambda) = \lambda e^{-\lambda x}, \; x \geq 0.$$

Note that the mean of X is $1/\lambda$.
Suppose that an exponential prior distribution with parameter ℓ is assigned to λ.

(a) Show that the posterior distribution of λ, given an observed time to first incorrect output x, is a gamma distribution of the form:

$$\pi(\lambda \mid x, \ell) = (\ell+x)^2 \, \lambda e^{-(\ell+x)\lambda}.$$

(b) After observing x, and under the assumption that the software is still performing as before the failure, you are interested in the predictive distribution for the time to the next failure Y. By conditioning on λ and applying the law of total probability, show that the density of Y at y, given x (and ℓ) is

$$f(y \mid x, \ell) = \frac{2(\ell+x)^2}{(\ell+x+y)^3}, \quad y \geq 0.$$

6. New software is being tested at a telephone exchange for routing calls. At each call, the software succeeds in routing with a probability p, independently of other calls.

(a) Assuming p known, what is the distribution of the number of calls taken until one fails to be routed correctly?

(b) A uniform prior distribution on $[0, 1]$ is assessed on p. In a test, the software first failed to route the 20th call. Calculate the posterior distribution of p, and the mean of this distribution.

(c) If testing continues any further, the software development firm will incur a late delivery penalty of $50,000. However, the developer will also pay a penalty for faulty performance of the software. It will pay $40,000 for every one out of a hundred calls that is not routed successfully.

    i. What is the *expected* penalty the company will pay, based on the results of testing so far?

    ii. Should the company release the software now or test further? You may assume, somewhat ideally, that further testing results in nearly faultfree software.

2. Foundational issues: Probability and Reliability

7. The arrival, over time, of calls to a telephone exchange is, at least for short periods of time, well modeled by a homogeneous Poisson process. Suppose that such an exchange is known to receive calls in a certain part of the day according to a Poisson process with the rate of λ per minute. As in the previous question, at each call the software succeeds in routing the call with a probability p, independently of other calls.

   (a) The distribution of the number of calls taken until one fails to be routed correctly by the software is geometric. Use this information to show that the form of the distribution of the time to the first failure is exponential with parameter $\lambda(1-p)$.

   Hint: Recall the distribution of time to the nth event in a Poisson process.

   (b) Now suppose that you are given the information that the first failure occurred after T units of time, and that this was the kth call to arrive.

   (i) Write down the likelihood [of the parameter(s)] given these data.

   (ii) An exponential prior with parameter ℓ is assessed on λ, and a uniform prior on $[0, 1]$ is assessed on p. The priors are assumed independent. Show that the posterior distributions of p and λ are beta and gamma distributions respectively, and that p and λ are a posteriori independent.

   (c) In a sequence of tests, N failures are observed. The times between each failure are $t_1, t_2, \ldots, t_N$, and the number of calls between each failure is $k_1, k_2, \ldots, k_N$. As in the last part of the question, write down the posterior distribution of (p, λ).

8. Show that the expected number of events that occur by time τ, in a nonhomogeneous Poisson process with a mean value function $\Lambda^*(t)$, is $\Lambda^*(\tau)$.

9. In the next chapter, we look at models for software failure. One such model is that of Goel and Okumoto, which assumes that failures occur as a nonhomogeneous Poisson process with mean value function $\Lambda(t) = \alpha(1 - e^{-bt})$, for parameters $\alpha, b > 0$.

   (a) Write down the probability of observing N failures by a time t.

(b) Suppose that b is known, and you are interested in describing your uncertainty about α. An exponential prior with parameter ℓ is assessed on α. Write down the posterior distribution of α given that N failures are observed by time t.

(c) A software company has tested software for 10 weeks, during which time it failed 24 times. Now the company must decide whether to release the software. If it decides to test further, it will incur a late delivery penalty of \$100,000. If it releases now, it will incur a penalty of \$5,000 for every in-service failure. Assume that:

- at each failure, a new bug will be discovered and fixed perfectly;

- continued testing will lead to almost failure-free software;

- and that $\ell = 0.02$.

Also note that $\lim_{t \to \infty} \Lambda(t) = \alpha$, so that α can be interpreted as the expected number of bugs to be discovered over the entire life of the software.

(i) What is the expected number of bugs remaining in the software, given testing to time 10 revealed 24 bugs?

(ii) Should the company release now, or test further?

10. Two software engineers, S_o and S_P, are contemplating the failure rate of a piece of software. S_o, the optimist, is of the opinion that the software contains no bugs—but of course is not sure of this opinion—and conceptualizes the failure rate as an exponentially decaying function of time of the form $\omega e^{-\omega t}$, for some $\omega > 0$, and $t \geq 0$. That is, the longer the software survives, the stronger is S_o's conviction of no bugs. S_P, on the contrary, is a pessimist who feels that the software consists of residual bugs. Like S_o, S_P is also not sure of S_P's conviction. Consequently, S_P conceptualizes the failure rate as a linearly increasing function of time of the form $\alpha + \beta t$, for some $\alpha \geq 0$ and $\beta > 0$, and $t \geq 0$. S_P's view is that the longer you wait the larger is the possibility of encountering a bug.

(a) Assuming that ω, α, and β are known, how do S_o and S_P assess the reliability of the software for a mission of time τ, for some

$\tau > 0$? Verify that S_o's assessed reliability is greater than that of S_P's.

(b) Let $\bar{F}_o(\tau \mid w)$ be S_o's assessed reliability in (a). Show that S_o's *mean time to software failure* is given by $\int_0^\infty \bar{F}_o(\tau \mid w) d\tau$.

Hint: Use integration by parts; $\bar{F}_o(\tau \mid w)$ is absolutely continuous.

Using the preceding formula, find S_o's and S_P's mean times to failure.

(c) How would you proceed if in (a), w, α, and β were unknown? That is, find $\bar{F}_o(\tau \mid \bullet)$ and $\bar{F}_P(\tau \mid \bullet)$.

(d) Implement your proposal in (c) by making suitable choices and describe the circumstances under which S_P's assessed reliability is greater than S_o's.

Hint: An uncertain optimist can be more pessimistic than a better informed skeptic.

(e) Suppose that after time $t^* > 0$, the operational profile of the software changes, so that more demands are made on the software, and the possibilities of encountering hidden bugs (if any) greatly increase. Describe how S_o and S_P will account for this feature when addressing (a).

3
MODELS FOR MEASURING SOFTWARE RELIABILITY

3.1 Background: The Failure of Software

In Chapter 2 we introduced the general idea of a software reliability model and that of the failure rate of software. Over the last two decades, a considerable amount of effort has been devoted to developing software reliability models—by some counts, there appear to be over one hundred. The aim of this chapter is to give an overview of a few of the most commonly used models by software engineers, and to introduce the reader to some of the more recent developments in the overall enterprise of model development. The issue of how to use these models in applications involves the topic of statistical inference, and this has been delegated to Chapter 4.

Like hardware reliability, *software reliability* is defined as the probability of failure-free operation of a computer code for a specified mission time in a specified input environment (the *operational profile*). With this definition, there are two terms that need explanation. By *failure-free operation*, we mean that the code is producing output which agrees with specifications. Software failure is caused by faults or "bugs" that reside in the code; when an input to the software activates a module where a fault is present, a failure can occur. There may be faults in the code that are never activated and our definition says that, since these will never manifest themselves as failure, they can be ignored. In other words, all bugs do not necessarily cause failures, but all failures are caused by bugs. Since we can observe failures but cannot hope to directly observe bugs, software reliability models usually pertain to the former. Secondly, by *mission time* we mean computer time or CPU time, that is, time over which the software is operational and is ready to receive, is receiving, or is active on inputs.

The causes of software failure are different from those of hardware failure. A consequence is that it is possible to have software that is bug-free and so will never experience failure for any mission time, whereas hardware experiences deterioration with use and is thus prone to failure over time. Software fails because of bugs in the logic of the code; these bugs are introduced due to human error. Hardware fails because of material defects and/or wear, both of which initiate and propagate microscopic cracks that lead to failure. With hardware failures, the random element is, most often, the time for a dominant crack to propagate beyond a threshold. Thus meaningful probability models for the time to hardware failure take cognizance of the rates at which the cracks grow in different media and under different loadings. With the failure of software, the situation is different. To see this, we first need to obtain an appreciation of the random elements in the software failure process. For this, the following idealization, prompted by the initial work of Jelinski and Moranda (1972), has been proposed [see Singpurwalla and Soyer (1996)].

3.1.1 The Software Failure Process and its Associated Randomness

A program is viewed as a "black box," or a "logic engine," that consists of statements bearing a logical relationship to each other. The engine receives, over time, different *types* of inputs (i.e., inputs that travel on different paths through the code), some of which may not be compatible with its design. If each compatible input traverses its intended path, then all its outputs are the desired ones, and the program is said to be perfect; that is, it is 100% reliable. If there are any errors in the logic engine, clerical or conceptual, then it is possible that a certain (compatible) input will not traverse its designated path, and in so doing will produce an output that is not the desired one. When this happens, the software is declared failed. It is possible that the presence of a bug prevents the software from producing any output. That is, the flawed logic could lead an input through an indefinite number of loops. Thus, implicit to the notion of software failure is that of a time interval within which an output should be produced. That is, associated with each input, there is an *allowable service time.*

We have said before that with hardware failures the random element is the time it takes for a dominant crack to propagate beyond a threshold. What are the sources of uncertainty with software failures? One source is the uncertainty about the presence and the location of a bug. Another is the type of input and the possibility of it encountering a bug. In either case, with the monitoring of software failures there are two types of random variables that can be conceived: binary and continuous. We first discuss, albeit briefly, the nature of the binary random variables.

Suppose that Y_i, $i = 1, 2, \ldots, k$, is a binary random variable taking the value 1 if the *i*th type of input to the software results in a correct output within its allowable service time; otherwise, Y_i is 0. The number of distinct input types is assumed to be k. Let p_i be the probability that $Y_i = 1$; thus the Y_is are the Bernoulli random variables of Section 2.2.2. If $p_i = p$, for all values of i, and if given p the Y_is are assumed to be independent, then $\sum Y_i$ has a binomial distribution, and the reliability of the software is simply p. In actuality, p is not known and the number of input types is conceptually infinite. Consequently, the sequence of random variables Y_i, $i = 1, 2, \ldots$, can be judged *exchangeable* (see Section 2.1.5), and if $\pi(p \mid \mathcal{H})$ describes our uncertainty about p, then $\pi(p \mid \mathcal{H})$ is a measure of the reliability of the software. Upon observing some of the Y_i's, $\pi(p \mid \mathcal{H})$ will be updated (via Bayes' Law) and this updated quantity will be a measure of the reliability of the software. The preceding two measures of reliability are naive; the assumption that $p_i = p$, for all values of i, ignores the possibility that some input types will be encountered more often than the others and that some may not be encountered at all. Approaches that improvise on this theme are outlined in Singpurwalla and Soyer (1996) who, following Chen and Singpurwalla (1996), propose a hierarchical model for the p_is.

The second type of random variable used for describing the software failure process pertains to the time between software failures. It is motivated by the notion that the arrival times to the software of the different input types are random. As before, those inputs that traverse through their designated paths will produce the desired outputs. Those that do not, due to bugs, will produce faulty outputs. To assess the software's reliability, we observe $T_1, T_2, \ldots$, the times (in CPU units) between software failures; we apologize to the reader for the change in notation from that used in Section 2.3. With this conceptualization, even though the failure of software is not generated stochastically, the detection of flaws is stochastic, and the result is that there is an underlying random process that governs the failure of software.

Most of the well-known models for software reliability are centered around the interfailure times $T_1, T_2, \ldots$, or the point processes that they generate; see Singpurwalla and Wilson (1994). In what follows, we introduce and describe some of these models. Whereas the monitoring of time is very conventional in hardware reliability studies, we see several issues that arise when this convention is applied to software reliability. For one, monitoring the times between failures ignores the amount of time needed to process an input. Consequently, an input that is executed successfully, but which takes a long time to process will contribute more to the reliability than one which takes a small time to process. Second, also ignored is the fact that between two successive failure times there could be several successful iterations of inputs that are of the same type. In principle, there could be an interfailure time of infinite length and still the software could be riddled with flaws. Of course, one can argue that monitoring the interfailure times takes into account the frequency with which the different types of inputs occur and in so doing the assessed reliability tends to be more realistic than the one which assumes that all the input types occur with equal frequency. In view of these considerations, it appears that a meaningful approach

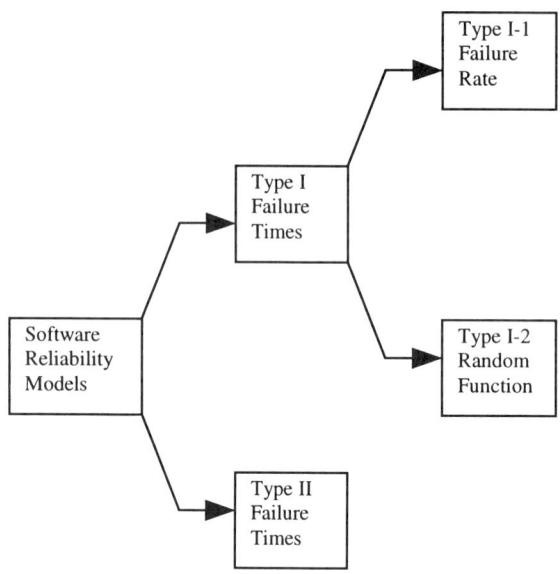

FIGURE 3.1. A Classification Scheme for Software Reliability Models.

to describe the software failures is via the scheme used to study *marked point process* [e.g., Arjas and Haara (1984)], wherein associated with each *interarrival* time, say W_i, of inputs, there is an indicator δ_i, with $\delta_i = 1$, if the ith input is successfully processed, and $\delta_i = 0$, otherwise; $i = 1, 2, \ldots$. Progress in this direction has been initiated but more development is needed. The point process approach to software reliability modeling was initiated by Goel and Okumoto (1979), and was followed up by Musa and Okumoto (1984), Langberg and Singpurwalla (1985), Miller (1986), Fakhre-Zakeri and Slud (1995), Kuo and Yang (1996), Chen and Singpurwalla (1997), and Slud (1997). Many of these authors have attempted to unify several of the existing software reliability models so that this topic can be studied under a common structure; see Section 3.5.

3.1.2 *Classification of Software Reliability Models*

Since many models based on the interfailure times $T_1, T_2, \ldots$, use similar modeling principles, the differences being only in the detailed application of a principle, it is possible to classify the models according to the principle used. Such a scheme adds structure to the disparate set of models and provides an

3.1 Background: The Failure of Software

explanation as to why certain modeling strategies were predominant at certain times. Our classification scheme (see Figure 3.1) follows that of Singpurwalla and Wilson (1994), and divides models into two broad types:

Type I : those that model the times between successive failures;

Type II : those that model the number of failures up to a given time.

Under Type I, the random variables $T_1, T_2, \ldots$, are modeled directly. This is often done by specifying the failure rate function for each random variable, $r_{T_i}(t \mid \mathcal{H})$, $i = 1, 2, \ldots$, and then invoking the exponentiation formula (2.30) to obtain $\mathcal{P}(T_i \geq t \mid \mathcal{H})$.

Type I models that use the failure rate as a modeling tool are said to be of Type I-1. Typically, $r_{T_i}(0 \mid \mathcal{H}) \leq r_{T_{i-1}}(0 \mid \mathcal{H})$, for $i = 1, 2, \ldots$, to reflect the fact that as software evolves over time, more bugs are discovered and fixed, and that each $r_{T_i}(t \mid \mathcal{H})$ is a nondecreasing function of t, for $t \geq 0$, to reflect the fact that between failures our opinion of the credibility of the software increases. In actuality, the fixing of bugs may introduce new ones in the code, so that the inequality given previously is not realistic. Nevertheless, many of the proposed Type I-1 models that we review here reflect such a feature; see Figures 3.2 to 3.4.

Another approach to modeling the times between successive failures is to describe each T_i as a random function of the previous T_is. Models that describe the T_is in this manner are said to be of Type I-2. A simple example of this is a time series model of the *random coefficient autoregressive* type wherein we may postulate that $T_i = \rho T_{i-1} + \epsilon$, where $\rho \geq 0$ is an unknown constant, and ϵ is a random disturbance term having mean zero. With $\rho > 1$, the successive failure times would tend to increase, indicating that the software is becoming more reliable with aging, whereas with $\rho < 1$, the opposite is to be true. In general, a Type I-2 model will have the feature that

$$T_i = \mathcal{F}(T_1, \ldots, T_{i-1}, \epsilon), \tag{3.1}$$

for some random function $\mathcal{F}$, or a known function having random coefficients.

With Type II models, we do not propose a model for the interfailure times T_i; rather, we propose a counting process model (see Section 2.3) for $N(t)$, the number of times the software fails in an interval $[0, t]$. The earliest and perhaps best known models of the Type II kind are those which assume that $N(t)$ is described by a Poisson process whose mean value function is based on assumptions about how the software experiences failure. The more recent contributions to Type II models do not suffer from the independent increments restriction (see Section 2.3.1) of Poisson process models.

3. Models for Measuring Software Reliability

We close this section with the remark that, in principle, a model of either type defines a model of the other. Specifically, for a sequence of inter-failure times $T_1, T_2, \ldots$, for which a Type I model has been proposed, there is an implicit Type II model [cf. Kuo and Yang (1996), Chen and Singpurwalla (1997)], because

$$N(t) = \max\{n \mid \sum_{i=1}^{n} T_i \leq t\}, \quad (3.2)$$

and conversely, for a Type II model there is a Type I model, because with $T_0 = 0$, and $i = 2, 3, \ldots$,

$$T_i = \inf\{t \mid N(t) = i\} - T_{i-1}. \quad (3.3)$$

3.2 Models Based on the Concatenated Failure Rate Function

3.2.1 The Failure Rate of Software

In Section 2.4.4 we introduced the notion of a concatenated failure rate function and have used it as a proxy for the failure rate of software. The earliest models in software reliability were based on specific forms of the concatenated failure rate function which, we recall from (2.34), is a side-by-side placement of the failure rates of conditional distributions. From a subjective point of view one is free to specify any general form for the failure rate function, and indeed this has been the basis of many such proposals, each motivated by a view of the software development process. We start with one of the earliest, and perhaps the most widely discussed of such models. All the models discussed in this section are of Type I-1.

3.2.2 The Model of Jelinski and Moranda (1972)

According to Iannino, Musa, and Okumoto (1987), a model for describing software failures proposed by Hudson (1967), and based on the structure of "birth and death processes," predates all known models. However, it is the model by Jelinski and Moranda (1972) that appears to be the first one to be widely known and used; also, it has formed the basis on which several other models have been developed. Jelinski and Moranda assume that the software contains an unknown number, say N, of bugs and that each time the software fails, a bug is detected and corrected. Furthermore, the failure rate of T_i is proportional to $N - i + 1$, the number of bugs remaining in the code; that is, for some constant $\Lambda > 0$, and software failure times $0 \equiv S_0 \leq S_1 \leq \ldots \leq S_i, i = 1, 2, \ldots,$

3.2 Models Based on the Concatenated Failure Rate Function 73

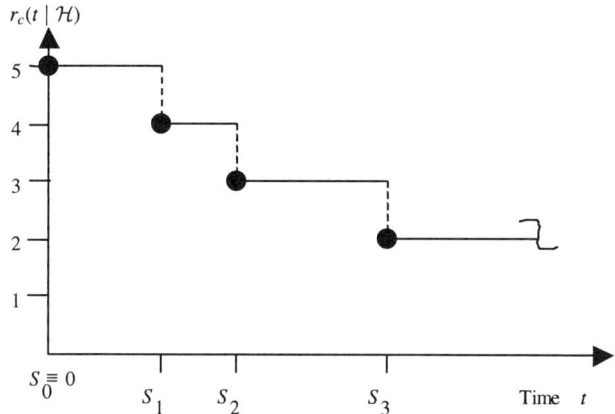

FIGURE 3.2. The Concatenated Failure Rate Function of the Model by Jelinski and Moranda.

$$r_{T_i | S_{i-1}}(t - S_{i-1} \mid N, \Lambda) = \Lambda (N - i + 1), \text{ for } t \geq S_{i-1}. \tag{3.4}$$

In Figure 3.2 we show a plot of the concatenated failure rate function for the model of Jelinski and Moranda, with $N = 5$ and $\Lambda = 1$. Since each failure leads to the removal of precisely one bug, the failure rate drops by a constant amount Λ. Since the right-hand side of (3.4) is a constant, it follows from the exponentiation formula that the conditional distribution of T_i given S_{i-1} is an exponential distribution with a mean of $(\Lambda(N - i + 1))^{-1}$; that is, for $t \geq S_{i-1}$,

$$\mathcal{P}(T_i \geq t \mid S_{i-1}, N, \Lambda) = e^{-\Lambda(N-i+1)t}. \tag{3.5}$$

The assumptions underlying the model of Jelinski–Moranda are: a perfect detection and repair of bugs, and a type of constant relationship between the number of bugs and the failure rate. This model is also known as a *de-eutrophication model,* because the process of removing bugs from software is analogous to the removal of pollutants from rivers and lakes. Both assumptions are unrealistic; perfect repair does not always occur, and each bug cannot be assumed to contribute the same amount to the failure rate. Some bugs may be benign and some may never be encountered.

Langberg and Singpurwalla (1985) provide an alternative *nonbug counting* perspective of the software failure process which also results in (3.5). They assume that there are N^* distinct input types to the program of which $N \leq N^*$ results in the inability of the program to perform its desired function. Conceptually, N^* is assumed to be infinite and N is assumed unknown. The N^* inputs arrive at the software as a homogeneous Poisson process with intensity λ.

74   3. Models for Measuring Software Reliability

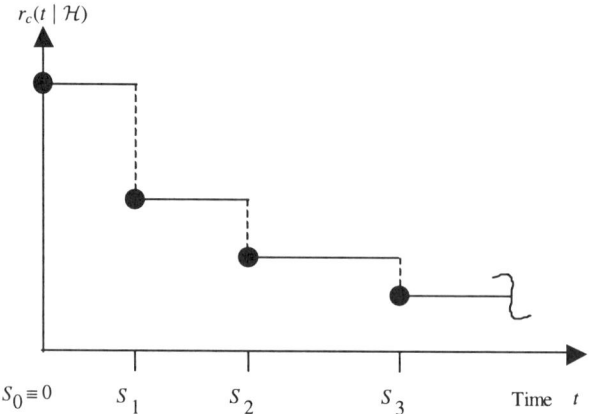

**FIGURE 3.3. The Concatenated Failure Rate Function
of the Model by Moranda.**

For any specified mission time, say t, there is a constant (but unknown) probability that the software will not encounter an input type that results in its failure; this probability depends on both N^* and N. Using a shock model type argument of reliability theory [see Barlow and Proschan (1975), p.128], Langberg and Singpurwalla show that (3.5) holds with $\Lambda = \lambda N/N^*$.

Despite its limitations, Jelinski and Moranda's *bug counting model* is important in software reliability for several reasons. Historically speaking, it established a line of thinking vis-à-vis its depiction of the concatenated failure rate function, and, as described in the following, stimulated the development of several other models. In fact, many subsequent models are generalizations of this model. Secondly, it appears to be ubiquitous in the sense that no matter how we attempt to look at the software failure process (see Section 3.5.3) the model always reappears as a special case. Indeed, the model by Jelinski and Moranda is as fundamental to software reliability as the exponential distribution is to hardware reliability. In hardware reliability the exponential distribution has served a foundational role. This is despite the fact that its practical application is subject to questioning; however, deviations from exponentiality serve as useful guidelines giving the distribution a benchmark status [cf. Barlow and Proschan (1975)]. Similarly, the model by Jelinski and Moranda plays a benchmark role with respect to software reliability.

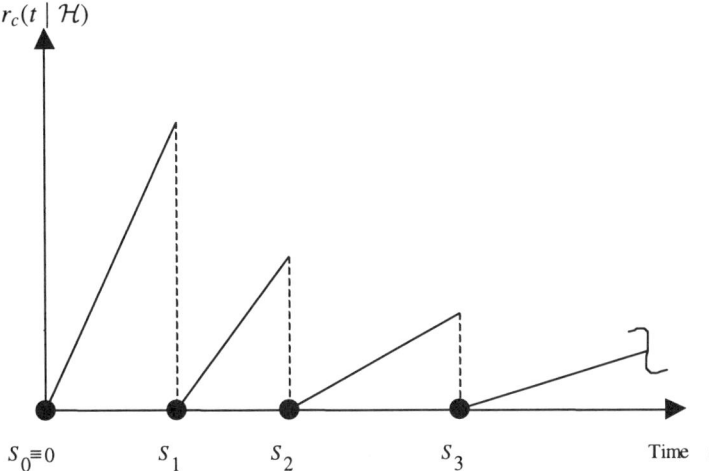

FIGURE 3.4. The Concatenated Failure Rate Function of the Model by Schick and Wolverton.

3.2.3 Extensions and Generalizations of the Model by Jelinski and Moranda

To address the concern that in Jelinski and Moranda's model every bug contributes equally to the failure rate, Moranda (1975) proposed a modification in which bugs that appear early are viewed as contributing more to the failure rate than those that appear later. Accordingly, the segments of the concatenated failure rate form a geometrically decreasing sequence, so that for constants $D > 0$, and $0 < k < 1$,

$$r_{T_i|S_{i-1}}(t - S_{i-1} \mid D, k) = D\, k^{i-1}, \text{ for } t \geq S_{i-1}; \qquad (3.6)$$

see Figure 3.3 which shows the concatenated failure rate function with $D = 8$ and $k = 0.5$.

Subsequent to Moranda's modification is the generalization of (3.5) by Goel and Okumoto (1978), who introduced a parameter p in (3.4) to address the criticism of perfect repair; p is the probability that a bug is successfully detected and repaired. Accordingly, (3.4) becomes

$$r_{T_i|S_{i-1}}(t - S_{i-1} \mid N, \Lambda) = \Lambda\, [N - p(i-1)], \text{ for } t \geq S_{i-1}. \qquad (3.7)$$

The model by Jelinski and Moranda is a special case of the preceding when $p = 1$; in (3.7), the assumption that no new bugs are introduced during debugging continues to hold.

A significant departure from the preceding line of thinking is due to Schick and Wolverton (1978), who assumed that the conditional failure rate of T_i is proportional to both the number of remaining bugs in the software and the elapsed time since last failure. That is,

$$r_{T_i|S_{i-1}}(t - S_{i-1} \mid N, \Lambda) = \Lambda (N - i - 1)(t - S_{i-1}), \text{ for } t \geq S_{i-1}. \quad (3.8)$$

Figure 3.4 is a plot of the concatenated failure rate function associated with (3.8). Note that each segment of this plot commences at zero and increases linearly with time. From a subjective point of view, this feature can be given an interesting interpretation. It says that every time a failure is encountered and the bugs corrected, our opinion of the software is so greatly enhanced that its failure rate drops to zero. However, as time elapses and we do not experience any failure, we become apprehensive about encountering one, and so our failure rate increases until we experience a failure at which time the failure rate drops to zero. Verify, using the exponentiation formula, that the distribution of T_i, given S_{i-1}, is the *Rayleigh*, which is a Weibull distribution with shape 2. It is important to bear in mind that in (3.4) through (3.8) the S_is are not the actually observed failure times; rather, they represent supposed failure times.

3.2.4 Hierarchical Bayesian Reliability Growth Models

In the models of Sections 3.2.2 and 3.2.3, the stochastic behavior of the times between observed failures is described in terms of the (unobservable) number of bugs in the software. This is why such models have been referred to as "bug counting models." Since the relationship between the number of bugs and the frequency of failure is tenuous, models that are devoid of such considerations have been proposed. One such model is the one due to Langberg and Singpurwalla (1985) mentioned at the end of Section 3.2.2; another model is due to Mazzuchi and Soyer (1988). Here the $\Lambda(N - i + 1)$ of (3.4) and (3.5) is replaced by an unknown parameter Λ_i, and a prior distribution, $\pi_{\Lambda_i}(\lambda \mid \bullet)$, assigned to Λ_i. Specifically, Λ_i is assumed to have a gamma distribution [see (2.15)] with shape parameter α and scale parameter $\psi(i)$, where $\psi(i)$ is a monotonically decreasing function of i. The function $\psi(i)$ is supposed to reflect the quality of the programming effort. A particular form is $\psi(i) = \beta_0 + \beta_1 i$; this form ensures that $\alpha/(\psi(i))$, the expected value of Λ_i, decreases in i. Consequently, for $i = 2, 3, \ldots$, the Λ_is will form a *stochastically decreasing* sequence; that is, for any $\lambda \geq 0$, $\mathcal{P}(\Lambda_i \leq \lambda) \geq \mathcal{P}(\Lambda_{i-1} \leq \lambda)$. Because

$$\pi_{\Lambda_i}(\lambda \mid \alpha, \psi(i)) = \frac{(\psi(i))^\alpha}{\Gamma(\alpha)} \lambda^{\alpha-1} e^{-\psi(i)\lambda}, \quad (3.9)$$

and since $\Lambda(N - i + 1)$ of (3.5) is replaced by Λ_i, it can be verified that subsequent to the $(i - 1)$th failure, the reliability of the software for a mission of duration t has a *Pareto distribution* of the form

$$\mathcal{P}(T_i \geq t \mid S_{i\text{-}1}, \alpha, \psi(i)) = [\frac{\psi(i)}{\psi(i)+t}]^\alpha, \text{ for } t \geq S_{i\text{-}1}. \qquad (3.10)$$

Both α and the parameters of $\psi(i)$ are treated as unknown. The prior on α is chosen to be a uniform on $[0, \omega]$, with $\omega > 0$ a constant, and the prior on β_1 a gamma independent of α; the prior on β_0 (given β_1) is supposed to be a *shifted gamma*, with β_1 being the extent of the shift. For specific details, see Section 4.4. An initial version of this model was proposed by Littlewood and Verall (1973). The model has been extended by Kuo and Yang (1995) who take $\psi(i)$ to be a polynomial of degree k, and by Soyer (1992) who lets Λ_i have expectation αi^β. Under Soyer's scheme positive (negative) values of β suggest an improvement or *growth (decay)* of reliability from one stage of testing to the other.

3.3   Models Based on Failure Counts

In the same paper where Moranda proposed his de-eutrophication model, he also proposed the very first of a Type II model [cf. Moranda (1975)]. Recall that in Type II models we look at $N(t)$, the number of failures to time t, rather than the interfailure times $T_1, T_2, \ldots$. Under such models the reliability of the software for a mission of duration t is simply $\mathcal{P}(N(t) = 0 \mid \mathcal{H})$. Moranda's motivation for considering models for $N(t)$ was that often data on software failures did not give times between failures; rather they gave the number of failures in fixed time intervals. For the ith interval Moranda assumed that $N(t)$ was a homogeneous Poisson process of intensity λk^{i-1}, with constants $\lambda > 0$ and $0 < k < 1$. This model reflects the lingering influence of the kind of thinking used in Type I models; we have here a sequence of decreasing intensity functions instead of a sequence of failure rates, one for each interval.

3.3.1   Time Dependent Error Detection Models

The model by Goel and Okumoto (1979) was the first Type II model to break free from the idea of describing $N(t)$ by a sequence of homogeneous Poisson processes. Instead, $N(t)$ is described by a single nonhomogeneous Poisson process with a mean value $\Lambda(t)$ and intensity $\lambda(t)$; see Section 2.3.1. These authors argued that $\Lambda(t)$ should be bounded because the expected number of failures over the life of the software is finite. Specifically, for a constant $a > 0$, $\Lambda(0) = 0$, and $\lim_{t \to \infty} \Lambda(t) = a$. Furthermore, the expected number of failures in an interval of time $(t, t + \Delta t)$ is assumed proportional to the product of the expected number of undetected failures times the length of the interval. That is, for a constant $b > 0$, known as the *fault detection rate*,

$$\Lambda(t + \Delta t) - \Lambda(t) = b(a - \Lambda(t))\Delta t + o(\Delta t). \tag{3.11}$$

Dividing (3.11) by Δt and letting $\Delta t \to 0$, we have a differential equation which for the boundary conditions on $\Lambda(\bullet)$ has the unique solution: $\Lambda(t) = a(1 - e^{-bt})$, or $\lambda(t) = ((d\Lambda(t))/dt) = abe^{-bt}$. Thus

$$\mathcal{P}(N(t) = n \mid a, b) = \frac{(\Lambda(t))^n}{n!} e^{-\Lambda(t)},$$

and the reliability of the software for a mission of duration t, starting at time 0, is

$$\mathcal{P}(N(t) = 0 \mid a, b) = e^{-a(1-e^{-bt})}. \tag{3.12}$$

Given S_i, the time of the ith failure, we can also obtain the distribution of T_{i+1} as

$$\mathcal{P}(T_{i+1} \geq t \mid a, b, S_i = s) = \exp(-a(e^{-bs} - e^{-b(s+t)})). \tag{3.13}$$

The model of Goel and Okumoto was the first of many nonhomogeneous Poisson process models that have been proposed, each based on different assumptions about the detection of failures. It has some noteworthy differences from the Type I-1 models that precede it. First, the total number of potential failures is assumed to be infinite so that the number of observed failures is a random variable having a Poisson distribution, as opposed to a fixed (but unknown) number of bugs N that had been previously assumed. Second, Equation (3.13) implies that the interfailure times are dependent, whereas in the Type I-1 models they were assumed independent. Both these differences appear to be sensible improvements of a description of software failure.

Experience has shown that the rate at which failures in software are observed increases initially and then decreases. To accommodate such phenomena, Goel (1985) proposed an intensity function of the type

$$\lambda(t) = a \cdot b \cdot c \cdot t^{c-1} e^{-bt^c}, \text{ for positive constants } a, b, \text{ and } c.$$

However, it is the proposal by Musa and Okumoto (1984), who postulate a relationship between the intensity function and the mean value function of a Poisson process, that has gained popularity with users. Specifically, for positive constants λ and θ,

$$\lambda(t) = \lambda e^{-\theta \Lambda(t)}; \tag{3.14}$$

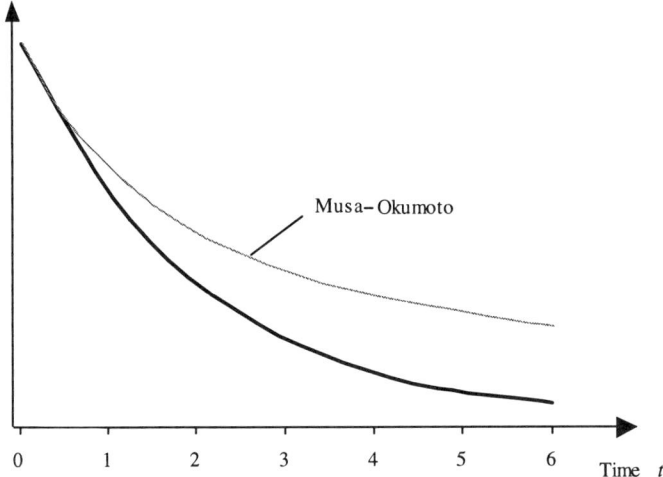

FIGURE 3.5. Intensity Functions of the Goel–Okumoto and the Musa–Okumoto Models.

that is, the rate at which failures occur exponentially decreases with the expected number of failures. Since $\lambda(t)$ is the derivative of $\Lambda(t)$, and if $\Lambda(0) \equiv 0$, then the differential equation (3.14) is solved to obtain:

$$\lambda(t) = \frac{\lambda}{\lambda\theta t + 1}, \text{ and} \tag{3.15}$$

$$\Lambda(t) = \frac{\ln(\lambda\theta t + 1)}{\theta}. \tag{3.16}$$

Figure 3.5 compares the intensity functions of the models proposed by Goel and Okumoto (1979) and by Musa and Okumoto (1984). The main difference is in the tails, wherein the intensity function of the latter decays more slowly. Under (3.14), the reliability for a mission of duration t is

$$\mathcal{P}(N(t) = 0 \mid \lambda, \theta) = (\lambda\theta t + 1)^{-1/\theta}, \tag{3.17}$$

and the analogue of (3.13) is

$$\mathcal{P}(T_{i+1} \geq t \mid \lambda, \theta, S_i = s) = \left(\frac{\lambda\theta s + 1}{\lambda\theta(s+1)+1}\right)^{1/\theta}. \tag{3.18}$$

In Chapter 4, we describe how expert opinion and failure data can be used for predicting future lifetimes using the model (3.16), referred to by Musa and

Okumoto as the *logarithmic Poisson execution time model*. There we also discuss inferential issues pertaining to the model of Jelinski and Moranda.

3.4 Models Based on Times Between Failures

3.4.1 The Random Coefficient Autoregressive Process Model

This model, introduced by Singpurwalla and Soyer (1985), describes the relationship between the successive interfailure times $T_1, T_2, \ldots, T_i, \ldots, i = 1, 2, \ldots$, via a *power law* of the form

$$T_i = (T_{i-1})^{\theta_i}, \qquad (3.19)$$

where T_0 is the time to first failure, and the θ_is are a sequence of unknown constants. If the T_is are scaled so that they are greater than one, then values of θ_i greater (less) than one suggest an increasing (decreasing) sequence of T_is; thus a stage-by-stage growth or decay in reliability can be described.

To account for uncertainty about the relationship (3.19), an error term δ_i is introduced so that

$$T_i = (T_{i-1})^{\theta_i} \times \delta_i, \qquad (3.20)$$

and an assumption made that the T_is and the δ_is have a lognormal distribution (see Section 2.2.2), with the latter having parameters 0 and σ_1^2; in other words, $\delta_i \sim \Lambda(0, \sigma_1)$. By taking logarithms in (3.20) we have what is known as a *linear model* in logarithms of the interfailure times; specifically,

$$\log T_i = \theta_i \log T_{i-1} + \log \delta_i,$$
$$= \theta_i \log T_{i-1} + \epsilon_i, \qquad (3.21)$$

if $\epsilon_i = \log \delta_i$.

The assumption that the T_is and the δ_is are lognormal implies that their logarithms ϵ_i have a Gaussian distribution (see Section 2.2.2) with ϵ_i having mean zero and variance σ_1^2; recall that the T_is have been scaled so that they are greater than 1. The linear model (3.21) is one of the most well-known time series models; with θ_i unknown, it is known as a *random coefficient autoregressive process of order 1* [henceforth RCAP(1)]. The model generalizes so that T_i depends on $k > 1$ previous T_is. Also, the variance of ϵ_i could change with i.

A final specification for this model pertains to the treatment of θ_is, and the authors propose several alternatives. One is to make the θ_is exchangeable (see Section 2.1.5); a way for doing this is to assume that each θ_i has a Gaussian distribution with mean λ and variance σ_2^2, with λ itself having a Gaussian distribution with mean μ and variance σ_3^2. Under this scheme, the quantities σ_1^2, σ_2^2, σ_3^2, and μ need to be specified; strategies for doing this are given by Singpurwalla and Soyer (1985). As an alternative to exchangeability, we may describe the θ_is by an autoregressive process of order 1, so that

$$\theta_i = \alpha\, \theta_{i-1} + \omega_i, \tag{3.22}$$

where α is a constant and $\omega_i \sim \mathcal{N}(0, W_i^2)$, with W_i^2 specified. When α is specified, along with the W_i^2 and σ_1^2, (3.21) and (3.22) constitute what is called a *Gaussian Kalman filter model*, for which there exists extensive literature; an expository description is in Meinhold and Singpurwalla (1983b). When α is unknown, its uncertainty is described by a distribution, and the preceding equations define an *adaptive Gaussian Kalman filter model;* Singpurwalla and Soyer (1992) discuss such models and their merits for analyzing software failure data.

3.4.2 A Non-Gaussian Kalman Filter Model

A Kalman filter model is specified by two equations, an *observation equation*—(3.21) in our case—which describes how the observables evolve as a function of time, and a *system equation*—(3.22) in our case—which describes how unknown coefficients in the observation equation evolve with time. The Kalman filter models of the previous section were called Gaussian, because the unknown quantities were assumed to have Gaussian distributions. *Non-Gaussian Kalman filter models* are those in which the underlying distributions are not Gaussian. The tradition of assuming that the error terms of linear models have Gaussian distributions dates back to Gauss who argued that since measurement errors tend to be symmetric about a mean the adoption of DeMoivre's (Gaussian) distribution is reasonable. However, failure data, be they for hardware, software, or biological entities, tend to be highly skewed; consequently the assumption of Gaussian distributions comes into question. Furthermore, in observing failure data it is difficult to conceptualize the notion of observational errors caused by instrumental inaccuracies. With Kalman filter models, or for that matter any general linear model, the main advantage of using Gaussian distributions is computational tractability; this advantage has diminished with modern computing. Because of these considerations, the routine use of Gaussian Kalman filter models, even on logarithms of the observed failure times, needs to be re-examined.

Bather (1965) introduced the idea of "invariant conditional distributions" and discussed their properties. This work provided Chen and Singpurwalla (1994) with the necessary framework for developing a *non-Gaussian Kalman filter model* for tracking software failure data. Here, instead of assuming that the T_is are lognormally distributed, it is assumed that they have a gamma distribution with a scale parameter θ_i which evolves according to a beta distribution; both the gamma and the beta distributions were discussed in Section 2.2.2. Specifically, for known constants C, ω_i, σ_i, and ν_i, such that $\sigma_{i-1} + \omega_i = \sigma_i + \nu_i$, $i = 2, 3, \ldots$, we have the observation equation as

$$(T_i \mid \theta_i, \omega_i) \sim \mathcal{G}(\theta_i, \omega_i), \tag{3.23}$$

and the system equation as

$$\theta_i = \frac{\theta_{i-1}}{C_i} \epsilon_i, \tag{3.24}$$

with ϵ_i having a beta distribution with parameters σ_{i-1} and ν_{i-1}. The initial value θ_0, required for starting the iterative process of Kalman filter models, is assumed to have a gamma distribution with scale parameter $\sigma_0 + \nu_0$ and a shape parameter u_0. Note that (3.24) is the analogue of (3.22) except that in the former the error term ϵ_i relating the θ_i's is multiplicative whereas in the latter it is additive.

In Section 4.5, where we discuss statistical inference using the preceding models, more insights about their hierarchical structure are given.

3.5 Unification of Software Reliability Models

In Section 3.1 we classified software reliability models according to the modeling strategy used to define them. The first few models were almost exclusively of Type I-1 (such as that of Jelinski and Moranda); then, in the late seventies, Type II models began to gain popularity. More recently, Type I-2 models appear to be coming into their own. To an outside observer, it would appear that all of these models are motivated by seemingly unrelated arguments. Indeed, even among software engineers, the topic of reliability modeling has been the subject of active debates and discussion; see, for example, Tausworthe and Lyu (1996). Software engineers have been too eager to come up with new models and to compare the predictive performance of the various competing models. In fact there even exist so-called "expert systems" devoted to selecting a software reliability model. The software industry would like a universal model that is equipped to accommodate as many nuances of the software reliability evolution formula as is possible. Whereas the search for an ideal model continues to be a futile exercise, at least for the immediate future, the possibility of viewing most of the available models from a unifying perspective appears to be at hand. The advantage of unification is the availability of a common structure under which the problem of reliability growth or decay can be studied.

Our classification scheme can be thought of as a step towards model unification, in the sense that a set of models becomes a special case of a more general model. Thus, all the Type II models that we have discussed are special cases of the nonhomogeneous Poisson process, whereas all Type I-1 models are special cases of a general model that models failure times as random variables with differing failure rates. Can we take this unification further? Is there a sense in which a larger group of models can be unified as being special cases of a general model? If so, what would this general model be?

The issue of unification has also arisen because of the contrast in the state of the art of software reliability and classical hardware reliability where only a few models, notably the Weibull, play a dominant role. A unifying perspective on the many software reliability models can hopefully simplify the task of model selection that a user faces.

3.5.1 Unification via the Bayesian Paradigm

One of the earliest attempts at unifying the then prevailing software reliability models was by Langberg and Singpurwalla (1985). By specifying prior distributions on the parameters N and Λ of the model by Jelinski and Moranda [see (3.5)], it was shown that the models by Goel and Okumoto and by Littlewood and Verall (1973) arise as special cases. Specifically, if

Case 1. The prior distribution on N is a Poisson with mean θ, and if Λ is degenerate at λ [i.e., $\mathcal{P}(\Lambda = \lambda \mid \mathcal{H}) = 1$],

then $N(t)$, the number of bugs discovered up to time t, is a nonhomogeneous Poisson process with a mean value function $\theta(1 - e^{-\lambda t})$; see Theorem 3.3 of Langberg and Singpurwalla. This is precisely the model of Goel and Okumoto. Also, if

Case 2. The prior distribution of N is degenerate at some n, and if Λ has a gamma distribution with a scale parameter $\psi(i)$,

then T_i will have the Pareto distribution (3.10). Also the sequence of T_is is *stochastically decreasing*; that is, for all $i = 1, 2, \ldots$, and any $a \geq 0$, $\mathcal{P}(T_i \geq a) \leq \mathcal{P}(T_{i+1} \geq a)$; see Theorem 3.4 of Langberg and Singpurwalla (1985). This is precisely the idea behind Littlewood and Verall's version of the hierarchical Bayes reliability growth model of Mazzuchi and Soyer (1988).

The fact that the model of Goel and Okumoto and that of Mazzuchi and Soyer can be derived as generalizations of the Jelinski and Moranda model is interesting from several perspectives. Jelinski and Moranda's model is the most widely known, and both the other models can be viewed as attempts to improve on it by moving away from its assumptions. Given this, the unifying result that has just been described is perhaps surprising. Secondly, note that the unified models are of both Type I and Type II. That models of both types can be so easily thought of as special cases of a more general model suggests that there is less fundamental difference between the two types than there appears at first sight. This is indeed the case, but to appreciate how closely the two types are related we need to revisit our discussion of point process models.

3.5.2 Unification via Self-Exciting Point Process Models

In Section 2.3.3 we have argued that Type II models, being based on the nonhomogeneous Poisson process, are special cases of the self-exciting point process with memory $m = -\infty$. If it can be shown that Type I-1 models can also be represented as self-exciting point processes, then much progress towards unification can be made. That such is indeed the case can be seen if we regard the concatenation points of the concatenated failure rate function of a Type I-1 model as events in a point process. By assuming that only one failure occurs at each failure time, the number of bugs discovered up to a time t will evolve as a point process. But is the resulting process a Poisson process? In general, the answer is no. First, many models make the sensible assumption of a finite, albeit unknown, number of bugs in the software. Second, the concatenated failure rate shows us that the evolution of the process is a function of the number of bugs that have already been discovered. Both these features violate the independent increments property of Poisson process models. However, they are not at variance with the postulates of self-exciting point processes; indeed, the second feature is a defining characteristic of the SEPP. The following theorem formally sets this down.

Theorem 3.1 [Chen and Singpurwalla (1997)]. *Under conditional orderliness, the sequence of failure times $S_1, S_2, \ldots,$ generated by a concatenation of inter-failure times T_i, having failure rates $r_i(\bullet)$, $i = 1, 2, \ldots,$ are described by a self-exciting point process whose intensity function is the concatenated failure rate function*

$$r_c(t \mid \mathcal{H}_t) = r_{T_i \mid S_{i-1}}(t - S_{i-1} \mid \mathcal{H}_t), \quad S_{i-1} \leq t < S_i,$$

with $\mathcal{H}_t$ denoting the history of the process up to time t.

Theorem 3.1 presupposes conditional orderliness. This implies that there are sequences of failure times for which conditional orderliness fails to hold. The following theorem gives sufficient conditions for the property to hold.

Theorem 3.2 [Chen and Sinpurwalla (1997)]. *Consider the set-up of Theorem 3.1. Suppose that the probability density function of T_i at t exists; let it be denoted by $f_{T_i}(t \mid \mathcal{H}_{S_{i-1}})$. Then, if there exists an $h > 0$ and an M, $0 < M < \infty$, such that*

$$f_{T_i}(t - S_{i-1} \mid \mathcal{H}_{S_{i-1}}) \leq M, \text{ for all } t \in [S_{i-1}, S_{i-1} + h],$$

and for all possible histories $\mathcal{H}_{S_{i-1}}$, then the counting process $\{N(t); t > 0\}$ has the conditional orderliness property. N(t) is the number of failures up to time t.

Now it is always so that for any random variable T_i having a probability density f_{T_i} and failure rate r_{T_i}, $f_{T_i}(t \mid \bullet) \leq r_{T_i}(t \mid \bullet)$, for all $t \geq 0$. Thus a convenient way of using Theorem 3.2 to verify conditional orderliness is to see if $r_{T_i}(t - S_{i-1} \mid \mathcal{H}_{S_{i-1}}) \leq M < \infty$, for all $t \in [S_{i-1}, S_{i-1} + h]$, that is, to see if the failure rate is bounded.

Clearly, all the Type I-1 models discussed in Section 3.2 have bounded failure rates and can therefore be viewed as self-exciting point processes. For example, in the model of Jelinski and Moranda, if N and Λ are assumed finite, then

$$r_{T_i \mid S_{i-1}}(t - S_{i-1} \mid N, \Lambda) = \Lambda(N - i + 1) \leq M,$$

for some $M < \infty$, and similarly, in the model by Schick and Wolverton

$$r_{T_i \mid S_{i-1}}(t - S_{i-1} \mid N, \Lambda) = \Lambda(N - i + 1)(t - S_{i-1}) \leq M.$$

Observe that the model by Jelinski and Moranda is of memory $m = 0$, whereas those of Schick and Wolverton and the hierarchical Bayes model of Mazzuchi and Soyer are of memory $m = 1$. Intuitively, it would appear that the greater the memory of the process the more refined is our ability to describe the phenomenon that generates the events under study. Consequently, in Section 3.6 we introduce a model for software failures whose underlying self-exciting point process is of memory $m > 2$.

Thus far we have said nothing about the other Type I models that are not specified via the failure rate. In Type I-2 models, a stochastic relationship between the consecutive failure times was given; for example, with the random coefficient autoregressive process (see Section 3.4.1) we postulated the relationship $T_i = \delta_i T_{i-1}^{\theta_i}$, with T_i having a lognormal distribution, whereas in Section 3.4.2 T_i had a gamma distribution. For both these examples, we can verify that the density functions of the interfailure times satisfy the conditions of Theorem 3.2, and thus these models can be viewed as members of the self-exciting point process family; furthermore, they are of memory $m = 2$. As a consequence of the preceding, we state the main result of this section.

All the software reliability models discussed in Sections 3.2–3.4 are special cases of a self-exciting point process model having memory $m \leq 2$.

It is important to note that not all software reliability models that have been proposed are special cases of self-exciting point processes. Whenever failures occur in clusters [see e.g., Crow and Singpurwalla (1984) and Sahinoglu (1992)], the conditional orderliness property fails and the underlying models cannot be viewed as members of the self-exciting point process family.

An advantage of the preceding unification is a common structure under which the problem of software reliability can be addressed. We can now think of other self-exciting point processes as potential models with the aim of making the intensity function better reflect our opinion of the evolution of the software's reliability; see Section 3.6.

3.5.3 Other Approaches to Unification

Whereas unifying software reliability models by viewing them as special cases of self-exciting point processes appears to be broadly encompassing, it is not the only way in which this issue has been addressed. For example, Koch and Spreij (1983) published work that investigated unification using the martingale theory for point processes. A contribution was subsequently made in a similar vein by van Pul (1993). Also conceptually important is the work of Fakhre-Zakeri and Slud (1995), and of Slud (1997), who use the idea of a *mixture model* in which a point process with an intensity function that depends on unobservable variables is considered. By specifying these unobservables in different ways, both randomly and deterministically, Fakhre–Zakeri and Slud (1995) obtain the ubiquitous model of Jelinski and Moranda, the time dependent error detection models of Section 3.3.1, and also a model by Dalal and Mallows (1988) that we have not discussed here.

More recently, Kuo and Yang (1996) have presented an elegant development on the relationship between the models in the Type I-1 category and those in the Type II category via the perspective of "order statistics" and "record values." These notions have played an important role in applied statistics and in probability theory, and it behooves us to gain an appreciation of their essential features. We start with an introduction to the former followed by its relevance to the models in the Type I-1 and the Type II categories.

The General Order Statistics Models

Suppose that $X_1, \ldots, X_n$ is a collection of random variables. Let $F(x \mid \underline{\theta}_i)$ be a probability model for X_i, $i = 1, 2, \ldots, n$, where $\underline{\theta}_i$ is a vector of parameters; that is, $F(x \mid \underline{\theta}_i) = \mathcal{P}(X_i \leq x \mid \underline{\theta}_i)$. Let $\bar{F}(x \mid \underline{\theta}_i) = 1 - F(x \mid \underline{\theta}_i)$, and suppose that $F(x \mid \underline{\theta}_i)$ is absolutely continuous so that the probability density $f(x \mid \underline{\theta}_i)$ exists for all x. Suppose that given $\underline{\theta}_i$, the X_is are judged independent. Thus if $\underline{\theta}_i = \theta$, $i = 1, 2, \ldots, n$, then the X_is are independent and identically distributed.

We now order the X_is from the smallest to the largest values, and denote the ordered values via the inequalities $-\infty < X_{(1)} < X_{(2)} < \ldots < X_{(n)} < +\infty$. The motivation for ordering comes from many applications; examples are hydrology, strength of materials, reliability, and life testing. For example, if X_i denotes the lifelength of the ith component, then $X_{(1)}$ is the smallest lifetime and $X_{(n)}$ the largest lifetime. If the n-component system is a series (parallel redundant) system, then $X_{(1)}$ ($X_{(n)}$) would be the time to failure of the system; the reliability of the series system would be $\mathcal{P}(X_{(1)} \geq x \mid \bullet)$. In statistical terminology, $X_{(1)}\{X_{(i)}\}$ [$X_{(n)}$] is known as the *smallest* {*i*th} (*largest*) *order statistic* in a sample of size n, and interest generally centers around the distribution of the ith order statistic, $i = 1, \ldots, n$. Especially, we may be interested in assessing

$$\mathcal{P}(X_{(1)} \geq x \mid \bullet) = \mathcal{P}(X_{(1)} \geq x, \ldots, X_{(n)} \geq x \mid \underline{\theta}_1, \underline{\theta}_2, \ldots, \underline{\theta}_n)$$

$$= \prod_{i=1}^{n} \bar{F}(x \mid \underline{\theta}_i),$$

since given the $\underline{\theta}_i$s, the X_is are assumed to be independent. Similarly, we have

$$\mathcal{P}(X_{(n)} \leq x \mid \bullet) = \mathcal{P}(X_{(1)} \leq x, \ldots, X_{(n)} \leq x \mid \underline{\theta}_1, \ldots, \underline{\theta}_n)$$

$$= \prod_{i=1}^{n} F(x \mid \underline{\theta}_i).$$

For most cases of practical interest $\underline{\theta}_i = \underline{\theta}$, for all values of i, and now $X_{(i)}$ is known as the ith order statistic in a sample of size n from $F(x \mid \underline{\theta})$. When such is the case, the preceding expressions for $X_{(1)}$ and $X_{(n)}$ simplify as $(\bar{F}(x \mid \underline{\theta}))^n$ and $(F(x \mid \underline{\theta}))^n$, respectively. Furthermore, using the binomial distribution (see Section 2.2.2) we can verify that

$$\mathcal{P}(X_{(i)} \leq x \mid \underline{\theta}) = \sum_{j=i}^{n} \binom{n}{j} (F(x \mid \underline{\theta}))^j (\bar{F}(x \mid \underline{\theta}))^{n-j}, \qquad (3.25)$$

and if $f_{X_{(i)}}(x \mid \underline{\theta})$ denotes the probability density of $X_{(i)}$ at x, were we to know $\underline{\theta}$, then

$$f_{X_{(i)}}(x \mid \underline{\theta}) = \frac{n!}{(i-1)!(n-i)!} (F(x \mid \underline{\theta}))^{i-1} f(x \mid \underline{\theta}) (\bar{F}(x \mid \underline{\theta}))^{n-i}. \qquad (3.26)$$

The preceding line of thinking can be extended in such a way so that if $f_{X_{(1)},\ldots,X_{(k)}}(x_1, \ldots, x_k \mid \underline{\theta}, n)$ denotes the joint probability density of $X_{(1)}, \ldots, X_{(k)}$, the smallest k order statistics out of a sample of size n, at $x_1, \ldots, x_k$, respectively, then

$$f_{X_{(1)},\ldots,X_{(k)}}(x_1, \ldots, x_k \mid \underline{\theta}, n) = \frac{n!}{(n-k)!} \prod_{i=1}^{k} f(x_i \mid \underline{\theta}) (\bar{F}(x_k \mid \underline{\theta}))^{n-k}. \qquad (3.27)$$

The lefthand side of (3.27) gives us what is known as the *joint distribution of the first k order statistics* in a sample of size n from $F(x \mid \underline{\theta})$. Verify that when $k = n$, we get the important result

$$f_{X_{(1)},\ldots, X_{(k)}}(x_1, \ldots, x_k \mid \underline{\theta}, n) = n! \prod_{i=1}^{n} f(x_i \mid \underline{\theta}),$$

whose import is that order statistics formed from independent random variables are dependent; the act of ordering destroys independence.

Armed with these preliminaries we can now describe how order statistics play a role with respect to software reliability models in the Type I-1 category. We start by noting (see Figures 3.2 to 3.4) that since the times to software failure $0 \equiv S_0 \leq S_1 \leq \ldots \leq S_i \leq \ldots$, are ordered, they constitute a natural framework for an order statistics type analysis. We start by asking if there is a common distribution that generates these order statistics and if so what would it be? It turns out that the answer depends on the assumed probability model for the interfailure times $T_1, T_2, \ldots$, that generate the ordered failure epochs S_i, $i = 1, 2, \ldots$. Conversely, given an $F(x \mid \underline{\theta})$ and having specified an n, the joint distribution of the first k out of n order statistics from $F(x \mid \underline{\theta})$ prescribes failure models for the k interfailure times $T_1, \ldots, T_k$.

As an example of the preceding, suppose that $\bar{F}(x \mid \underline{\theta}) = \bar{F}(x \mid \Lambda) = e^{-\Lambda x}$, an exponential distribution with scale Λ. Let n be specified as N. Then, given N and Λ, the joint distribution of the first k out of n order statistics from $\bar{F}(x \mid \Lambda)$ is, from (3.27), of the form

$$f_{X_{(1)},\ldots, X_{(k)}}(x_1, \ldots, x_k \mid \Lambda, n) = \frac{N!}{(N-k)!} \prod_{i=1}^{k} \Lambda e^{-\Lambda x_i} (e^{-\Lambda x_k})^{N-k}.$$

But this is precisely the joint distribution of $0 \equiv S_0 < S_1 < \ldots < S_k$, when the T_is are independent and each T_i has an exponential distribution with a scale parameter $\Lambda(N - i + 1)$, $i = 1, 2, \ldots, k$, the form specified by Jelinski and Moranda. To verify this claim we use the fact that $S_1 \equiv T_1$, $S_2 = T_1 + T_2, \ldots$, $S_i = T_1 + T_2 + \ldots + T_i$. Different forms for $F(x \mid \underline{\theta})$, say the Pareto, the Weibull, the gamma, and so on, will lead to different probability models for the interfailure times, and models constructed via the preceding mechanism have been referred to by Raftery (1987) as the *general order statistics models*, abbreviated GOS. With $\bar{F}(x \mid \Lambda) = e^{-\Lambda x}$, the resulting model is called the *exponential order statistics model*, abbreviated EOS; this terminology is due to Miller (1986), whose work predates that of Raftery and Kuo and Yang. The EOS model has also been considered by Ross (1985b) and by Kaufman (1996). Kaufman's work is noteworthy because he makes some fascinating connections between software reliability modeling and "successive sampling," that is, sampling without replacement that is proportional to size [cf. Gordon (1983), Scholz (1986), and Andreatta and Kaufman (1986)]. Estimation (see Section 4.2.2) under successive sampling schemes is described by Nair and Wang (1989) and by Bickel, Nair, and Wang (1992).

Once the model by Jelinski and Moranda is interpreted in the light of an EOS framework, the road for relating GOS models with some (but not all) models in the Type II category is paved. The signal for this connection comes from the work of Langberg and Singpurwalla (1985) who show that the nonhomogeneous Poisson process model of Goel and Okumoto (1979) is a consequence of assuming a Poisson distribution for the parameter N of the model by Jelinski and Moranda. Specifically, one can prove the following theorem.

Theorem 3.3 [cf. Kuo and Yang (1996)]. *Suppose that failure epochs are described by a GOS model with a distribution function $F(t \mid \underline{\theta})$ and a parameter N. Let $N(t)$, $t > 0$, denote the number of epochs in time $[0, t]$. Then $\{N(t); t > 0\}$ can be described by a nonhomogeneous Poisson process with mean value function $\mu F(t \mid \underline{\theta})$ if N has a Poisson distribution with parameter μ.*

As a special case of Theorem 3.3, if $F(t \mid \underline{\theta}) = 1 - e^{-\Lambda t}$, then the resulting Poisson process has a mean value function $\mu(1 - e^{-\Lambda t})$, which is the mean value function used by Goel and Okumoto (1979); see Section 3.3.1. There is a drawback to the limiting behavior of this mean value function and consequently to the essence of the result of Theorem 3.3. We note that for any choice of $F(t \mid \underline{\theta})$, $\lim_{t \to \infty} \mu F(t \mid \underline{\theta}) < \infty$, suggesting that the mean value function of the resultant nonhomogeneous Poisson process is bounded. This means that GOS models cannot be used in those situations wherein new faults get introduced during the debugging process. In the model by Musa and Okumoto (1984) the mean value function [see (3.16)] $((\ln(\lambda \theta t + 1))/\theta) \to \infty$, as $t \to \infty$, and thus the GOS model is unable to accommodate those models in the Type II category for which the mean value function is unbounded. It is for this reason that point process models generated by "record value statistics" have been explored, and this matter is taken up next. But before we close our discussion of GOS models another noteworthy feature of such models needs to be mentioned. This pertains to the distribution of T_1, the first interfailure time, or the time to occurrence of the first epoch in a GOS model. We note that

$$\mathcal{P}(T_1 > t \mid \mu, F(t \mid \underline{\theta})) \, \mathcal{P}(N(t) = 0 \mid \mu, F(t \mid \underline{\theta})) = e^{-\mu F(t \mid \underline{\theta})}.$$

For any $\mu < \infty$, $\lim_{t \to \infty} e^{-\mu F(t \mid \underline{\theta})}$ is not zero; that is, the distribution function of T_1 is *defective*. The implication of this result is that unless $\mu \uparrow \infty$, there is a nonzero probability that the software will never experience any failure. The smaller the μ, the larger is this probability. Other special cases of Theorem 3.3

are a model by Goel (1983) for which $F(t \mid \underline{\theta}) = 1 - e^{-\beta t^\alpha}$, a model by Ohba and Yamada [see Yamada and Osaki (1984)] for which

$$F(t \mid \underline{\theta}) = 1 - (1 + \beta t)e^{-\beta t},$$

and a generalized order statistics model by Achcar, Dey, and Niverthy (1998) in which $F(t \mid \underline{\theta}) = I_k(\beta t^\alpha)$, where $I_k(s)$ is an incomplete gamma integral.

The Record Value Statistics Models

Suppose that $X_1, X_2, \ldots, X_n$ is a collection of independent and identically distributed random variables, with $F(x \mid \underline{\theta})$ as a probability model for each X_i, $i = 1, \ldots, n$. Suppose that $F(x \mid \underline{\theta})$ is absolutely continuous so that the probability density $f(x \mid \underline{\theta})$ exists for all x. We define the sequence of *record values* $\{Z_n\}$, $n \geq 1$ and *record times* R_k, $k \geq 1$, as follows.

$$R_1 = 1,$$
$$R_k = \min\{i : i > R_{k-1}, X_i > X_{R_{k-1}}\}, \text{ for } k \geq 2, \text{ and}$$
$$Z_k = X_{R_k}, \text{ for } k \geq 1.$$

An example best illustrates the preceding construction. Suppose that $X_1 = 4$, $X_2 = 1$, $X_3 = 7$, $X_4 = 5$, $X_5 = 9$, $X_6 = 3$, $X_7 = 13$, $X_8 = 6$, $X_9 = 18$, $X_{10} = 14$, and $X_{11} = 15$. Then, the *record pairs* (R_k, Z_k) are: (1, 4), (3, 7), (5, 9), (7, 13), and (9, 18). That is, a record value is the largest value that we have observed in the process of traversing from X_1 to X_n, one step at a time, and the record time is the index associated with a record value. Even though with $n \to \infty$, R_k will tend to get rare, the sequence of record values can be shown to be infinite. Since the record values constitute an increasing sequence, they can be viewed as epochs of the occurrence of an event over time; that is, they can be modeled as a point process. Thus it is meaningful to regard the epochs of software failure as record values from some underlying distribution that we are free to specify. The following theorem is beautiful; it gives us an interesting property of the point process generated by record values.

Theorem 3.4 [Dwass (1964)]. *Suppose that the epochs of failure are described as the record values generated by a collection of independent and identically distributed random variables having a common distribution $F(t \mid \underline{\theta}) = 1 - \bar{F}(t \mid \underline{\theta})$. Let $N(t)$ denote the number of epochs in time $[0, t]$. Then $\{N(t); t > 0\}$ can be described by a nonhomogeneous Poisson process with mean value function $\ln(1/(\bar{F}(t \mid \underline{\theta})))$, and intensity function $(f(t \mid \underline{\theta}))/(\bar{F}(t \mid \underline{\theta}))$, where $f(t \mid \underline{\theta})$ is the probability density at t, if it exists.*

Since the failure rate of $\bar{F}(t \mid \underline{\theta})$ is $(f(t \mid \underline{\theta}))/(\bar{F}(t \mid \underline{\theta}))$, the record value statistics provide an interesting relationship between the intensity function of a point process and the failure rate.

It is easy to see that if $\bar{F}(t \mid \underline{\theta}) = \alpha/(\alpha + t)$, a Pareto distribution with parameter $\alpha = 1/\lambda$, for $t \geq 0$, then the mean value function of the process is (3.16), the form specified by Musa and Okumoto (1984) with their $\theta = 1$. Besides the model by Musa and Okumoto, there are other models for describing the growth in reliability of engineering systems, all having the property that the mean value function of the underlying point process is not bounded [cf. Duane (1964), Cox and Lewis (1966)]. The record value statistics approach is an elegant way of looking at all of these in a comprehensive manner.

Before closing this section, it is important to note that both the order statistics and the record value statistics perspectives result in Poisson processes, which as we have said before possess the independent increments property. This is a disadvantage, but one that can be overcome by a more general model that is described next.

3.6 An Adaptive Concatenated Failure Rate Model

We have seen that all the models introduced in Sections 3.2 through 3.4 are special cases of self-exciting point processes with a memory of at most two. Specifically, all the models in the Type II category, being based on the postulates of the Poisson process, have memory $m = -\infty$, and possess the independent increments property. This latter feature may not be appropriate in the context of software testing. The models in the Type I-1 category are of memory $m = 0$ (for the model of Jelinski and Moranda) or of memory $m = 1$ (for the models of Schick and Wolverton and the hierarchical Bayes model of Mazzuchi and Soyer). The models in the Type I-2 category are of memory $m = 2$, but can be easily extended to have a memory $m = k$; all that we need to do is consider a kth order autoregressive process with random coefficients. We have said before that models with large memories tend to be more refined than those with smaller memories vis-à-vis their predictive capabilities. Thus it is desirable to introduce models that are conceptually of infinite memory, where the notion of infinite memory is akin to the notion of *invertibility* in time series analysis; see Box and Jenkins (1976), p. 50. One way to achieve this objective would be to extend the random coefficient autoregressive process model to all its previous terms. Whereas such an approach would indeed provide for good predictability, it would suffer from the criticism of a lack of interpretive features. A model such as that by Schick and Wolverton is attractive because it attempts to incorporate some of the more pragmatic aspects of the software testing and the bug discovery phenomena. Time series models, such as the random coefficient autoregressive processes are often viewed by practicing engineers as "black-box" models; they are purely mechanistic.

92   3. Models for Measuring Software Reliability

The concatenated failure rate model that we present in the following is guided by the preceding considerations. It is motivated by ideas that are analogous to those of Schick and Wolverton but has the added feature of adaptivity. A consequence of adaptivity is improved predictivity. The model has two parameters and possesses characteristics that are intuitively appealing and which generalize those of other models. This model was introduced by Al-Mutairi, Chen, and Singpurwalla (1998); also see Singpurwalla (1998b).

3.6.1   The Model and Its Motivation

In keeping with the notation of Section 3.2, we let $0 \equiv S_0 \leq S_1 \leq \ldots \leq S_i$, $i = 1, 2, \ldots$, denote the software failure times (in CPU units) and $T_1, T_2, \ldots$, the interfailure times; that is, $T_i = S_i - S_{i-1}$. Also, let $r_{T_i}(t)$ denote the failure rate function of the distribution function of T_i. We have stated before, in Section 3.2.3, that from a subjective point of view, the functional form of $r_{T_i}(t)$ can be given an interpretation that reflects a software engineer's judgments about the stochastic behavior of each T_i. Such judgments, although personal, should capture the engineer's knowledge about the software failure, its bug elimination process, previous data, and the experimental control under which the software is tested. For the model that is proposed here, we assume that for each version of the software's code, one's intuition is that small time intervals between successive failures should result in a judgment of poor reliability, and vice versa, for large time intervals. However, upon the occurrence of failure any judgment of enhanced reliability should be replaced by a judgment of enhanced unreliability; that is, the judgment of unreliability should take a sharp upward jump. The more frequent the failures the higher the upward jump in unreliability should be and vice versa. These characteristics parallel the sample path of a "shot noise process," once the failure rate is identified as being analogous to stress. The shot noise process is a stochastic process that is popular in engineering and physics; see, for example, Cox and Isham (1980), p. 135. The process consists of two parts, a "shot process" and a "stress process." The shot process is a point process that generates the epochs of events, whereas the stress process generates a function of time that takes jumps of random size at each shot and which decreases deterministically between the adjacent jumps. The stress function is, like the concatenated failure rate function, a random function because both the jump sizes and their locations are random; also, in most applications its value at any time t depends on the history of the process to time t. Accordingly, we propose that given the parameters k and b, and conditional on $S_{n-1} = s_{n-1}$, $n = 2, 3, \ldots$,

$$r_{T_n}(t \mid S_{n-1}, k, b) = \frac{1}{\frac{t-S_{n-1}}{k} + \frac{S_{n-1}}{(n-1)b}}, \quad t \geq S_{n-1}. \qquad (3.28)$$

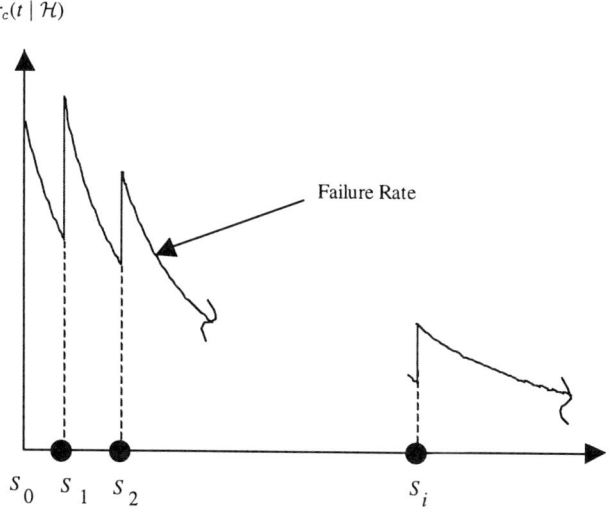

FIGURE 3.6. The Concatenated Failure Rate Function as a Sample Path of a Shot Noise Process.

Verify, that for any n, $r_{T_n}(t \mid \bullet)$ is decreasing in t, and that at $t = S_{n-1}$, it is proportional to $(n-1)/S_{n-1}$, which is a proxy for the "failure intensity" until the $(n-1)$th failure. Furthermore $r_{T_n}(t \mid \bullet)$ increases as $(n-1)/S_{n-1}$ increases. See Figure 3.6 which is a plot of the concatenated failure rate function defined by (3.28).

Verify that the plot of Figure 3.6 displays the following characteristics that capture our subjective views about the credibility of the software.

- (a) Frequent failures should result in a judgment of poor reliability. This suggests that the failure rate should take a large upward jump; see S_{i-1} and S_i.

- (b) When the software experiences no failure, our opinion of its reliability is enhanced. This suggests that its failure rate should decrease.

- (c) Upon the occurrence of failure, our opinion of the reliability is on the poor side. This suggests that the failure rate should take an upward jump.

(d) Large interfailure times correspond to small jump sizes and viceversa; see S_2 and S_{i-1}.

3.6.2 *Properties of the Model and Interpretation of Model Parameters*

The model proposed in the previous section possesses several attractive features. To see these, we replace $(t - S_{n-1})$ of (3.28) by τ, with $\tau \geq 0$, and then verify, via the exponentiation formula (2.30), that

$$\mathcal{P}(t_{n+1} \geq \tau \mid S_n, k, b) = \left(\frac{nb\tau}{kS_n} + 1\right)^{-k}, \qquad (3.29)$$

and that the probability density of T_{n+1} at τ is

$$f_{T_{n+1}}(\tau \mid S_n, k, b) = \frac{nb}{S_n}\left(\frac{nb\tau}{kS_n} + 1\right)^{-(k+1)}. \qquad (3.30)$$

Thus the conditional (given S_n, k, b) mean and variance of T_{n+1} are

$$E(t_{n+1} \mid S_n, b, k) = \frac{k}{(k-1)nb} S_n, \quad \text{and}$$

$$V(t_{n+1} \mid S_n, b, k) = \frac{k^3}{(k-1)^2(k-2)}\left(\frac{S_n}{nb}\right)^2, \quad \text{respectively.}$$

Clearly, we need $k \geq 2$, for the predictive mean and variance to exist. Also, since $E[T_{n+1} \mid S_n, \bullet] = \infty$, T_{n+1} has a decreasing failure rate.

The model construction (3.28) suggests that there is a growth in reliability in going from stage n to stage $(n+1)$, if given b and k,

$$r_{T_{n+1}}(\tau \mid S_n = s_n) < r_{T_{n+1}}(\tau \mid S_{n-1} = s_{n-1}), \text{ for } \tau \geq 0.$$

A consequence of the preceding is that t_{n+1}, a realization of T_{n+1}, should satisfy the inequality

$$t_{n+1} > \frac{S_{n+1}}{(n+1)}. \qquad (3.31)$$

But the import of (3.31) is that the time between the nth and the $(n+1)$th failure is greater than the average of all past failures. This implies that our model reflects the feature of having a memory. Furthermore, we can show that

$$\frac{E[T_{n+1}]}{E[T_n]} > (\leq) 1 \Leftrightarrow b < (\geq) \frac{k}{k-1};$$

this suggests that, on the average, there is growth in reliability if and only if $b < k/(k-1)$. The preceding gives us a criterion for choosing model parameters or for specifying prior distributions for them; see Section 4.7. Let $\eta(n) \overset{\text{def}}{=} ((\mathrm{E}(t_{n+1}) - T_n))/\mathrm{E}(t_n))$ be the *relative growth in reliability* at stage n. Then we are able to show that

$$\eta(n) = \frac{k-(k-1)b}{(k-1)b} \frac{1}{n},$$

Now if $b < k/(k-1)$, then $\eta(n) \downarrow n$. This implies that under the assumption of reliability growth, the bugs detected and eliminated during the early phases of testing contribute more to reliability growth than those detected later. Thus $b < k/(k-1)$ reflects the feature that the greatest improvement in a software's reliability occurs during the early stages of testing.

Finally, suppose that $D_n = nb/S_n$ has been recorded as d_n, and we are at the $(n+1)$th stage of testing. If at time w, w measured from s_n, failure has not occurred, then we can show that

$$\mathrm{E}(t_{n+1} - w \mid T_{n+1} > w) = \frac{k}{k-1} w + \frac{k}{(k-1)d_n}$$

$$= \textit{mean residual life} \ (\mathrm{MRL}).$$

Thus the longer the elapsed time since last failure, the longer the expected time to next failure. The MRL is a linear function of w with coefficient $k/(k-1)$. This helps us pin down the range of values for k; see Section 4.7.

3.7 Chapter Summary

The focus of this chapter is on models for software reliability, their classification, and their unification. The models were introduced in the chronological context in which they were developed. They were classified into two broad types, those that model the successive times between failures, and those that model the number of failures up to a given time. The former category was further subdivided into two classes, those that use the concatenated failure rate function as a modeling device, and those that model the interfailure times. Representative members of the first of the preceding two classes are the famous model by Jelinski and Moranda, and its extensions via a hierarchical Bayes scheme. Representative members of the second of these classes are the random coefficient autoregressive process and the Gaussian and the non-Gaussian Kalman filter models. Of the models that describe the number of failures up to a given point in time, those based on the Poisson point process, such as the models by Goel and Okumoto, and Musa and Okumoto were discussed.

The search for an omnibus model to describe software failures leads us to the topic of model unification. Unification provides a common structure under which the general problem of assessing reliability growth or decay, and the prediction of failure times can be accomplished. Unification was discussed from several perspectives, such as the Bayesian paradigm which involves specifying prior distributions for model parameters, the order statistics perspective wherein the successive times to failure are described as order statistics from independent but not identically distributed random variables, and from the point of view of record values which note the times at which the successive records get broken. Unification was also achieved via the perspective of looking at software failures as points in a self-exciting point process. Such processes need not possess the independent increments property of Poisson processes, and are therefore more general than those that do. It was argued that practically all of the proposed models for software reliability are special cases of a general point process model, namely, the self-exciting point process.

The chapter concludes with the introduction of a new model for software failures, a model that combines the attractive features of many of the previously proposed models. The model is based on a concatenation of several failure rate functions, the behavior of each function being determined by the past history. The model embodies the defining features of a self-exciting point process and captures a software engineer's overall judgments about the failure process.

Exercises for Chapter 3

1. **Hierarchical Bayesian Reliability Growth Models.** Consider the Bayesian reliability growth model, with each T_i exponentially distributed with failure rate Λ_i, and Λ_i assumed to have a gamma distribution with shape parameter α and scale parameter $\psi(i)$. Verify, using the laws of probability, that the reliability function for T_i is given by the Pareto distribution

   $$\mathcal{P}(T_i \geq t \mid S_{i-1}, \alpha, \psi(i)) = \left[\frac{\psi(i)}{\psi(i)+t}\right]^\alpha, \ t \geq S_{i-1}.$$

   Show that the probability density of the preceding Pareto distribution at t is of the form

   $$\frac{\alpha(\psi(i))^\alpha}{(t+\psi(i))^{\alpha+1}}$$

2. **Derivation of the Goel–Okumoto Model.** As stated in this chapter, the model by Goel and Okumoto is a nonhomogeneous Poisson process with mean value function $\Lambda(t)$, where it is assumed that

   $$\Lambda(t + \delta t) - \Lambda(t) = b(a - \Lambda(t)) \, \delta t + \mathrm{o}(\delta t).$$

   (a) By dividing both sides of the preceding equation by δt and letting $\delta t \to 0$, show that $\lambda(t) = b(a - \Lambda(t))$, where $\lambda(t)$ is the derivative of $\Lambda(t)$.

   (b) Verify that $\Lambda(t) = a(1 - \mathrm{e}^{-bt})$ satisfies the preceding equation.

3. **Model of Musa and Okumoto.** Check that the intensity function of the model by Musa and Okumoto,

   $$\lambda(t) = \frac{\lambda}{\lambda \theta t + 1},$$

   satisfies the relationship

   $$\lambda(t) = \lambda \mathrm{e}^{-\theta \Lambda(t)}.$$

4. **Statistical Analysis of Software Failure Data.** In the random coefficient autoregressive process model of order one each T_i depends on its previous value T_{i-1}. How would you generalize the model so that each T_i depends on its k previous values $T_{i-1}, T_{i-2}, \ldots, T_{i-k}$. Is there more than one way to generalize the model?

3. Models for Measuring Software Reliability

5. **General Order Statistics Models.** Let $0 \equiv S_0 \leq S_1 \leq S_2 \leq \cdots \leq S_N$ be the failure times of a piece of software that contains N bugs. Assume that these failure times are the order statistics from a sample of N independent realizations of a Weibull distribution with parameters α and β; that is,

$$P(X_i \geq x \mid \alpha, \beta) = \exp(-\alpha x^\beta), \quad i = 1, \ldots, N,$$

and $S_i = X_{(i)}$. Derive the joint density of $(S_1, \ldots, S_k)$, for $k \leq N$.

6. **The Adaptive Concatenated Failure Rate Model.** This model assumes that the failure rate for the nth time to failure, given the time of the $(n-1)$th failure s_{n-1}, is:

$$r_{T_n}(t \mid s_{n-1}, k, b) = \frac{1}{\frac{t - s_{n-1}}{k} + \frac{s_{n-1}}{(n-1)b}}, \quad t \geq s_{n-1}.$$

(a) Verify that

    (i) r_{T_n} is a decreasing function of t;
    (ii) at $t = s_{n-1}$, r_{T_n} is proportional to the inverse of the average time between failures up to s_{n-1}, that is, $(n-1)/s_{n-1}$.

(b) Let $\tau = t - s_n$ be the time since the nth failure. Using the exponentiation formula, verify that the survival function of T_{n+1}, given s_n, k, and b, is

$$P(T_{n+1} \geq \tau \mid s_n, k, b) = \left(\frac{nb}{ks_n} \tau + 1 \right)^{-(k+1)}.$$

(c) The expected value of T_{n+1}, given s_n, is $ks_n/((k-1)nb)$. Using the relationship between T_n and S_n, as well as the identity $E(S_{n+1}) = E_{S_n}(E(S_{n+1} \mid S_n))$, show that

$$E(S_{n+1}) = \left(1 + \frac{k}{nb(k-1)} \right) E(S_n).$$

7. **Recent Developments.** The adaptive concatenated failure rate model of Figure 3.6 reflects the disposition of an optimist (in the sense of Exercise 7 of Chapter 2).

(a) Using Figure 3.6 as a guide describe the disposition of a pessimist, and using this analogue develop results along the lines of those given in Section 3.6.2.

(b) In practice it is more likely that a software engineer starts off with a pessimistic disposition but then after encountering and correcting several initial bugs begins to become optimistic. Thus a more realistic depiction of the concatenated failure rate function is a combination of that given in Figure 3.6 with the one developed in Part (a). Propose such a concatenated function and discuss its development, delineating a mechanism that describes the gradual evolution of the change in disposition, from pessimism to optimism.

4
STATISTICAL ANALYSIS OF SOFTWARE FAILURE DATA

4.1 Background: The Role of Failure Data

In Chapters 2 and 3, we introduced several models for describing our uncertainties about the software failure process. These models involved unknown parameters, often denoted by Greek symbols. The parameters entered into the picture because of our invoking the law of the extension of conversation, as a way of simplifying the probability specification process. The parameters being unobservable, our uncertainty about them was described by a prior distribution. The prior distribution is specific to an individual and may vary from individual to individual.

Software failure data, if available, are assumed to provide additional information about the failure process. That is, the data enhance our appreciation of the underlying uncertainties. There are certain strategies through which data can be incorporated into the assessment process. The first is to simply make the data a part of the background $\mathcal{H}$, and then to reassess the relevant uncertainty in the light of this expanded $\mathcal{H}$. There is nothing in the calculus of probability that forbids us from using this strategy, as long as our assessments remain coherent. However, ensuring coherence is not easy to do, and so this strategy is difficult to implement. The second, and the more commonly used strategy, is to use the data for an enhanced appreciation of the unknown parameters. This is done through Bayes' Law whereby the prior distribution gets updated—via the data—to what is known as the *posterior distribution*; see the exercises of Chapter 2. The process of going from the prior distribution to the posterior distribution is known as *Bayesian inference*. There may be, of course, broader interpretations as to what constitutes Bayesian inference, but for the present purposes the preceding seems adequate. Section 4.2 describes the appropriate machinery for making the transition from the prior to the posterior, and then using the latter for assessing uncertainties about the phenomena of interest.

At this point in time it is useful to mention the role of data in non-Bayesian, or what is known as *frequentist inference*. If our interpretation of probability is a relative frequency, then the probability model becomes an objective entity in which the parameters take fixed (but unknown) values. Under these circumstances a prior distribution is not assigned to the parameters, and so the matter of updating it does not arise. When such is the case, the role of the data is to provide a vehicle for *estimating* the unknown parameters; estimation involves the tasks of obtaining a single number, called a *point estimate*, or an interval, called an *interval estimate*, that covers the true (but unknown) value of the parameter. An overview of one of the most commonly used frequentist procedures, the "method of maximum likelihood" is given in Section 4.2. Frequentist inference includes estimation as well as testing hypotheses about the parameters. Here, notions such as "unbiasedness," "efficiency," "confidence limits," "significance levels," "Type I and II errors," and the like, come into play. Since our interpretation of probability is not in terms of a relative frequency, the preceding notions are not germane to us. This does not mean to say that frequentist inference has not been used in the context of software failure data. On the contrary, much of the inferential work in software reliability has been frequentist; see, for example, Musa, Iannino, and Okumoto (1987) for an overview. What distinguishes the material here from much of what has been written is our interpretation of probability, and the ensuing Bayesian inference which is its consequence.

Thus to summarize, irrespective of whether inference is Bayesian or frequentist, a key role played by the data is the information that they provide about the unknown parameters in probability models. There are of course other roles that the data can play, a common one being *model selection*, but this too stems from the theme that the data facilitate an enhanced appreciation of the model parameters. Model selection has become a central problem in software reliability because of the huge number of models that have been proposed—over one hundred by the latest count. In the frequentist paradigm, model selection is formally done via "goodness-of-fit" testing [cf. Box and Jenkins (1976), for a general flavor of this topic], whereas in the Bayesian paradigm it is done via *Bayes factors* and *prequential prediction*; see Section 4.6. The main idea underlying these approaches is an investigation of how well a proposed model describes the data. In actual practice models are often selected because of their simplicity or their familiarity to the analyst. Often, the type of data that are available will also help us to choose a model. For example, if the data consist of times between software failures, then a Type I model (see Chapter 3) will be selected; if the data consist of the number of bugs discovered at certain times, then a Type II model will be selected. Finally, a question arises as to whether model selection should precede inference. In principle, model selection should precede inference, because the latter is conducted within the framework of the former. However, model evaluation requires that inference be performed first, and thus model selection and inference are iterative procedures conducted in a step-by-step fashion [cf. Box (1980)].

The material in this chapter pertains to a use of Bayesian approaches for inference, prediction, and model selection. Prior distributions being central to the Bayesian paradigm, a section has been devoted to their discussion. The general plan of this chapter is to introduce a theme, such as Bayesian inference, and to follow it up with an application involving one or more models of Chapter 3. Thus a discussion on elicitation of prior distributions is followed up by an application involving the logarithmic Poisson model of Musa and Okumoto (1984). As a consequence inference procedures for the models of Chapter 3 are not discussed in the same order in which the models are introduced.

4.2 Bayesian Inference, Predictive Distributions, and Maximization of Likelihood

In this section we give an overview of Bayesian inference, and hypothesis testing using Bayes factors. We also introduce and discuss the notion of prequential prediction, which in the context of assessing software reliability models plays a natural role. We start by recalling (see Section 2.2) that for any unknown quantity X, the law of total probability and the assumption of conditional independence result in the relationship

$$\mathcal{P}(X = x \mid \mathcal{H}) = \sum_{\theta} \mathcal{P}(X = x \mid \theta) \, \mathcal{P}(\theta \mid \mathcal{H}) \qquad (4.1)$$

for any parameter θ taking discrete values.

Suppose now, that in addition to $\mathcal{H}$, we have at our disposal the realizations of n random quantities, $X_1, \ldots, X_n$ that are judged exchangeable (see Section 2.1.5) with X. Let x_i denote the realization of X_i, $i = 1, \ldots, n$, and let $\underline{x} = (x_1, \ldots, x_n)$. How should we revise $\mathcal{P}(X = x \mid \mathcal{H})$ in the light of this added information? That is, how should we update the $\mathcal{P}(X = x \mid \mathcal{H})$ of Equation (4.1) to $\mathcal{P}(X = x \mid \underline{x}, \mathcal{H})$? In the context of software failure, X could be the time to failure of the current version of the software, and $X_1, \ldots, X_n$, the times to failure of its n previous versions. The assumption that $X, X_1, \ldots, X_n$ is an exchangeable sequence is crucial. Intuitively, it says that the X_is provide us with information about X. As stated in Section 2.1.5, exchangeability is a subjective judgment which to some may not be meaningful in a particular application.

4.2.1 *Bayesian Inference and Prediction*

To address the question posed, we start with the proposition $\mathcal{P}(X = x \mid X_1, \ldots, X_n, \mathcal{H})$. Using the law of total probability, together with the assumption of conditional independence, an analogue to Equation (4.1) can be written as

$$\mathcal{P}(X = x \mid X_1, \ldots, X_n, \mathcal{H}) = \sum_\theta \mathcal{P}(X = x \mid \theta, X_1, \ldots, X_n) \times$$
$$\mathcal{P}(\theta \mid X_1, \ldots, X_n, \mathcal{H}). \quad (4.2)$$

We then invoke Bayes' Law to obtain

$$\mathcal{P}(\theta \mid X_1, \ldots, X_n, \mathcal{H}) \propto \mathcal{P}(X_1, \ldots, X_n \mid \theta, \mathcal{H}) \mathcal{P}(\theta \mid \mathcal{H})$$
$$= \mathcal{P}(X_1 \mid X_2, \ldots, X_n, \theta, \mathcal{H}) \times \mathcal{P}(X_2 \mid X_3, \ldots, X_n, \theta, \mathcal{H}) \times$$
$$\cdots \times \mathcal{P}(X_n \mid \theta, \mathcal{H}) \times \mathcal{P}(\theta \mid \mathcal{H}). \quad (4.3)$$

A consequence of the judgment of exchangeability of the sequence $X, X_1, \ldots, X_n$ is a result, due to de Finetti (1937), which says that given θ, X is independent of $X_1, \ldots, X_n$, and that X_1 is independent of $X_2, \ldots, X_n$ and $\mathcal{H}$, and so on. Consequently, we may write Equation (4.2) as

$$\mathcal{P}(X = x \mid X_1, \ldots, X_n, \mathcal{H}) \propto \sum_\theta \mathcal{P}(X = x \mid \theta) \times$$
$$\prod_{i=1}^n \mathcal{P}(X_i = x_i \mid \theta) \mathcal{P}(\theta \mid \mathcal{H}), \quad (4.4)$$

and Equation (4.3) as

$$\mathcal{P}(\theta \mid X_1, \ldots, X_n, \mathcal{H}) \propto \prod_{i=1}^n \mathcal{P}(X_i = x_i \mid \theta) \mathcal{P}(\theta \mid \mathcal{H}). \quad (4.5)$$

Equations (4.4) and (4.5) provide the probabilistic foundations for a Bayesian approach to prediction (about X) and inference (about θ). The logic for this assertion is the premise that the preceding equations prescribe how we will assess our uncertainty about X and θ, in the light of $\mathcal{H}$, and *were we to know* $X_1, \ldots, X_n$. Consequently, when $X_1, \ldots, X_n$ are *actually* observed as $x_1, \ldots, x_n$, respectively, we are obliged to do what we said we would do, and thus our use of Equations (4.4) and (4.5) as the basis for prediction and inference. However, there is a caveat. When X_i is observed as x_i, the entity $\mathcal{P}(X_i = x_i \mid \theta)$ is no longer a probability; rather, it is a likelihood of θ for a fixed value of x_i; see Section 2.1.4. Accordingly, the product $\prod_i \mathcal{P}(X_i = x_i \mid \theta)$, when viewed as a function of

θ, for the fixed values $x_1, \ldots, x_n$, is a likelihood function of θ; it is denoted $\mathcal{L}(\theta; \underline{x})$.

Thus to summarize, with $X_1, \ldots, X_n$ observed as $x_1, \ldots, x_n$, respectively, the posterior distribution of θ, now denoted as $\mathcal{P}(\theta \mid \underline{x}, \mathcal{H})$, is obtained via Equation (4.5) as

$$\mathcal{P}(\theta \mid \underline{x}, \mathcal{H}) \propto \mathcal{L}(\theta; \underline{x}) \, \mathcal{P}(\theta \mid \mathcal{H}), \qquad (4.6)$$

and the *predictive distribution* of X, $\mathcal{P}(X = x \mid \underline{x}, \mathcal{H})$, is obtained via Equation (4.4) as

$$\mathcal{P}(X = x \mid \underline{x}, \mathcal{H}) \propto \sum_{\theta} \mathcal{P}(X = x \mid \theta) \, \mathcal{P}(\theta \mid \underline{x}, \mathcal{H}). \qquad (4.7)$$

In Equations (4.6) and (4.7), the constant of proportionality ensures that the left-hand sides are legitimate probabilities. If θ is assumed to be continuous, then the summation sign on the right-hand side of Equation (4.7) will be replaced by an integral, and now $\mathcal{P}(\theta \mid \mathcal{H})$ is a probability density function.

The foregoing material gives the bare essentials of Bayesian inference and prediction. Whereas the logical underpinnings of this approach are relatively straightforward, its implementation poses some difficulties. Besides model specification, specification of the prior is an onerous task, and often the computation of the posterior and the predictive distributions can be demanding. More details on these issues plus related matters can be found in the books by Berger (1985), Bernardo and Smith (1994), and Lee (1989), which is an introductory, but an otherwise comprehensive treatment of the subject.

4.2.2 The Method of Maximum Likelihood

The predominant mode of inference in software engineering has been the method of maximum likelihood and confidence interval estimation. There could be several reasons behind this choice. One is familiarity with the method and its widespread use; the second could be a desire for being "objective." The third could be ease of application: one does not have to specify a prior distribution on θ. Irrespective of the reasons, the method of maximum likelihood is employed so often by software engineers that a few words about the rationale behind this frequentist procedure are in order.

There are many views as to what constitutes a frequentist procedure. The one that appeals to us is based on the notion that the prior distribution of the unknown parameters of a probability model is a degenerate one, and that inference pertains to learning about this degenerate value using data alone. Thus frequentist procedures mandate the availability of data for inference and prediction. The method of maximum likelihood is one such procedure. It is based on the premise that for any given datum, $\underline{x} = (x_1, \ldots, x_n)$, some values of a parameter θ are more likely than the others. The *maximum likelihood estimate* of θ, say $\widehat{\theta}$, is that value of θ which maximizes the likelihood function $\mathcal{L}(\theta; \underline{x})$; that is, $\widehat{\theta}$ is the most likely value of θ. Thus $\widehat{\theta}$ is a point estimate of θ; it is based on the datum $\underline{x}$ alone, and is independent of the analyst's background information $\mathcal{H}$.

Point estimates on their own give no clue about the degree of uncertainty associated with the estimate. In order to gauge this uncertainty a subtle philosophical principle involving the long-range performance of an estimation *procedure* is invoked, and the datum $\underline{x}$ is used to obtain an interval called a *confidence interval*. Associated with a confidence interval is a number between zero and one (both inclusive), called the *confidence level*, say α. Usually, α is chosen to be between 0.90 and 1.0. The interpretation of a confidence interval is tricky. Contrary to what many users believe, a confidence interval with a confidence coefficient α does not imply that the probability that the interval contains the true value of the unknown parameter is α. Rather, the coefficient α represents the proportion of times that intervals, such as the one based on $\underline{x}$, the datum at hand, would contain the unknown parameter.

Implicit in this interpretation is the recognition that data other than the observed $\underline{x}$ could be obtained were another sample of size n to be taken. The idea of repeated sampling also enables one to judge the quality of point estimates (like the maximum likelihood estimate) via notions such as *unbiasedness, efficiency, consistency, uniqueness*, and the like. These notions do not appear in the context of Bayesian inference. Once an estimate such as, say $\widehat{\theta}$ is obtained and its quality evaluated, it can be plugged into the probability model for purposes of prediction. Hogg and Craig (1978) give a good account of frequentist inference at an intermediate level; a gentler introduction to the topic is Chatfield (1983). Thus to summarize, the method of maximum likelihood mandates the availability of failure data, and the quality of the estimate is gauged by the long-term performance of the procedure. Confidence limits that usually accompany maximum likelihood estimates do not convey a sense of coverage probabilities in the usual sense; they too reflect long-term performance based on a repeated application of the confidence limit construction.

4.2.3 Application: Inference and Prediction Using Jelinski and Moranda's Model

As an illustration of how the methodology of the previous section has been applied for inferential problems in software reliability, we consider two sets of data. The first set is shown in Table 4.1; it consists of 136 successive times (in seconds) between software failure. These data are taken from Musa (1975); the entries are to be read across rows. The second set of data is given later, in Table 4.2. Let us suppose that the failure process generating these data can be meaningfully described by the model of Jelinski and Moranda (1972); see

4.2 Bayesian Inference, ..., and Maximization of Likelihood

Table 4.1. Successive Times Between Software Failure [data from Musa (1975)]

3	30	113	81	115	9	2	91	112	15
138	50	77	24	108	88	670	120	26	114
325	55	242	68	422	180	10	1146	600	15
36	4	0	8	227	65	176	58	457	300
97	263	452	255	197	193	6	79	816	1351
148	21	233	134	357	193	236	31	369	748
0	232	330	365	1222	543	10	16	529	379
44	129	810	290	300	529	281	160	828	1011
445	396	1755	1064	1783	860	983	707	33	868
724	2323	2930	1461	843	12	261	1800	865	1435
30	143	109	0	3110	1247	943	700	875	245
729	1897	447	386	446	122	990	948	1082	22
75	482	5509	100	10	1071	371	790	6150	3321
1045	648	5485	1160	1864	4116				

Section 3.2.2. We have no scientific basis for this supposition; it is made for illustrative purposes only.

An analysis of these data has been conducted by Meinhold and Singpurwalla (1983a). They show that under the model of Equation (3.5), $\hat{N}$, the maximum likelihood estimator of N, as a function of the sample size k, fails to provide meaningful answers. For example, when $k = 6$, $\hat{N} = 11$, whereas when $k = 7$, $\hat{N}$ is infinite; for $k = 8$, $\hat{N}$ becomes finite again, as $\hat{N} = 27$ (see Table 2 of the preceding reference). This erratic behavior of the maximum likelihood estimator is also true if the interfailure times were generated by a simulation of Equation (3.5) [cf. Forman and Singpurwalla (1977)]. It may be claimed that the initial impetus for considering Bayesian approaches in software reliability has arisen from experiences like this; the motivation was pragmatic, rather than philosophical.

A Bayesian analysis of the preceding data using the Jelinski–Moranda model calls for the specification of prior distributions for N and Λ. A discussion about choosing prior distributions is given in the section that follows, but for now we use the choices made by Meinhold and Singpurwalla (1983a). Specifically, the prior distribution for N is a Poisson with mean θ, and the prior for Λ is a gamma with a scale μ and a shape α, *independent* of the distribution of N (see Section 2.2.2). With this choice of priors, and the k observed interfailure times $(t_1, \ldots, t_k) = \underline{t}^{(k)}$, it can be shown (left as an exercise for the reader) that:

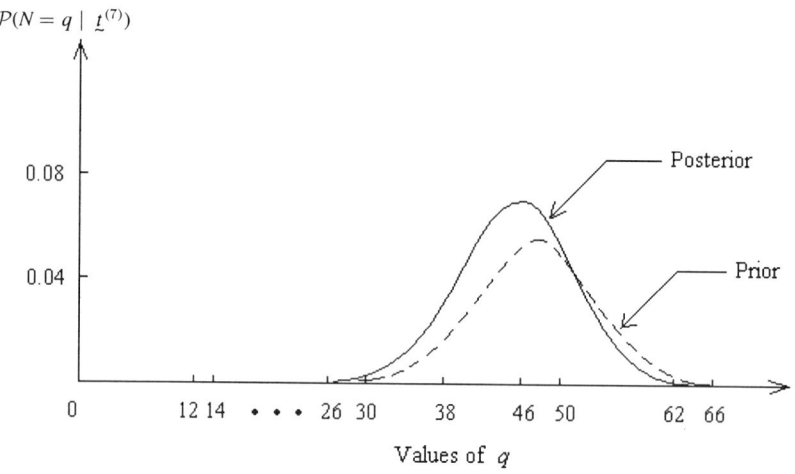

FIGURE 4.1a. Plot of the Prior and Posterior ($k = 7$) Probabilities of N.

(i)   the posterior probability that $N = q$, $q \geq k$ is

$$\mathcal{P}(N = q \mid \underline{t}^{(k)}) \propto \exp(-\theta)\, \theta^{q(q-k)!} \left\{ \mu + \sum_{j=1}^{k}(q-j+1)t_j \right\}^{-(\alpha+k)};$$

(ii)   the posterior density of Λ, given that $N = q$, is a gamma with scale parameter $\mu + \sum_{j=1}^{k}(q-j+1)t_j$, and a shape parameter $(\alpha + k)$;

(iii)   the joint posterior distribution of N and Λ, at q and λ, respectively, is of the form

$$\mathcal{P}(N = q, \Lambda = \lambda \mid \underline{t}^{(k)}) = \frac{(\lambda^{\alpha+k-1})\exp[-\lambda\{\mu+\sum_{j=1}^{k}(q-j+1)t_j\}]}{\Gamma(\alpha+k)} \times$$

$$\left\{ \mu + \sum_{j=1}^{k}(q-j+1)t_j \right\}^{(\alpha+k)} \times \frac{q!}{(q-k)!}$$

$$\times \frac{\left\{\mu+\sum_{j=1}^{k}(q-j+1)t_j\right\}^{-(\alpha+k)} \mathcal{P}\{N=q\}}{\sum_{r=k}^{\infty}\frac{r!}{(r-k)!}\left\{\mu+\sum_{j=1}^{k}(r-j+1)t_j\right\}^{-(\alpha+k)} \mathcal{P}\{N=r\}},$$

where $\mathcal{P}(N = q) = (e^{-\theta}\, \theta^q)/q!$.

It is useful to note that even though N and Λ were a priori independent, once the data $\underline{t}^{(k)}$ are at hand, they are a posteriori dependent, as the preceding expression shows. This is to be expected because posterior inference for both parameters is based on the same set of data.

Figure 4.1a) shows a plot of the posterior probability $\mathcal{P}(N = q \mid \underline{t}^{(k)})$ when $k = 7$, for $q = 8, 9, \ldots$. Recall that when $k = 7$, the maximum likelihood estimator of N was infinite; the likelihood function was flat. The values chosen for the parameters of the prior distributions were $\theta = 50$, $\mu = 1$, and $\alpha = 2$. For purposes of comparison Figure 4.1a) also shows the prior probability of N. We observe that the flatness of the likelihood has not interfered with our ability to perform inference in the Bayesian framework. Rather, the paucity of information in the sample has resulted in a posterior that does not differ dramatically from the prior. This again points out the importance of the knowledge that the experimenter puts into the problem. The posterior probability of N is sensitive to the choice of the parameter θ.

The second set of data, given in Table 4.2 is taken from Jelinski and Moranda (1972). It pertains to a large military software system, called the Naval Tactical Data System (NTDS), which consists of 38 distinct modules. The module selected here is Module-A, and Table 4.2 shows 34 times (in days) between software failure, split into four phases of the development process: production, testing, user experience, and further testing. The interfailure times are denoted by t_i, $i = 1, 2, \ldots, 34$, and the S_i denote the cumulatives of the interfailure times; that is, $S_i = \sum_{j=1}^{i} t_j$. For the purposes of this section, we assume that these data can be described by Jelinski and Moranda's model. Later on, in Section 4.4.1, we consider alternate models.

For a Bayesian analysis of these data, the mean of the Poisson prior on N was chosen to be 50, and the scale (shape) parameter of the gamma prior on Λ was taken to be $\mu = 0.5$ ($\alpha = 0.01$). Thus the prior mean of Λ is 0.02. Using this prior, plus the first 31 interfailure times, the posterior distribution of $(N - 31)$ was calculated. The formula for $\mathcal{P}(N = q \mid \underline{t}^{(31)})$, given before, was used. A plot of this posterior distribution is shown in the top part of Figure 4.1b). The mean of this posterior distribution is 1.3, which accords well with the observed failures in the "user experience phase."

To obtain the predictive distribution of the time to next failure T_{k+1}, with $k = 31$, we use the fact that T_{k+1} has density at t of the form

$$f_{T_{k+1}}(t \mid \underline{t}^{(k)}, \theta, \mu, \alpha) = \sum_{j=k}^{\infty} \int_0^{\infty} \lambda(j-k) e^{-\lambda(j-k)t} \times$$

$$\mathcal{P}(N = j, \Lambda = \lambda \mid \underline{t}^{(k)}) \, d\lambda . \tag{4.8}$$

Table 4.2. Interfailure Times of the NTDS
[Data from Jelinski and Moranda (1972)]

Error No. (i)	t_i	s_i	Error No. (i)	t_i	s_i
Production Phase			*Production Phase (cont'd)*		
1	9	9	20	1	105
2	12	21	21	11	116
3	11	32	22	33	149
4	4	36	23	7	156
5	7	43	24	91	247
6	2	45	25	2	249
7	5	50	26	1	250
8	8	58	*Test Phase 1*		
9	5	63	27	87	337
10	7	70	28	47	384
11	1	71	29	12	396
12	6	77	30	9	405
13	1	78	31	135	540
14	9	87	*User Experience Phase*		
15	4	91	32	258	798
16	1	92	*Test Phase 2*		
17	3	95	33	16	814
18	3	98	34	35	849
19	6	104			

The preceding expression when solved numerically yields the predictive density shown in the bottom part of Figure 4.1b).

The upper 95th percentile of this density is 285 days; it accords well with the observed 258 days of Table 4.2. As an alternative, see Appendix A.3.1 on Gibbs sampling.

4.2.4 Application: Inference and Prediction Under an Error Detection Model

In Section 3.3.1 we introduced a Type II model by Goel and Okumoto (1979), called a "time dependent error detection model." An attractive feature of this model is that it lends itself nicely to a closed form Bayesian analysis, as the following development shows.

4.2 Bayesian Inference , . . . , and Maximization of Likelihood 111

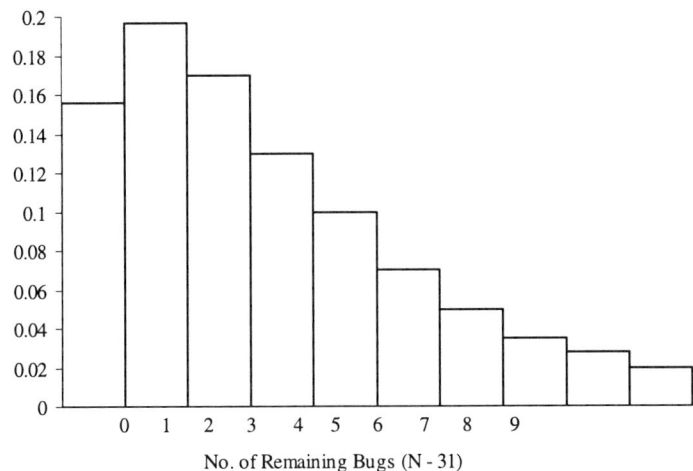

No. of Remaining Bugs (N - 31)

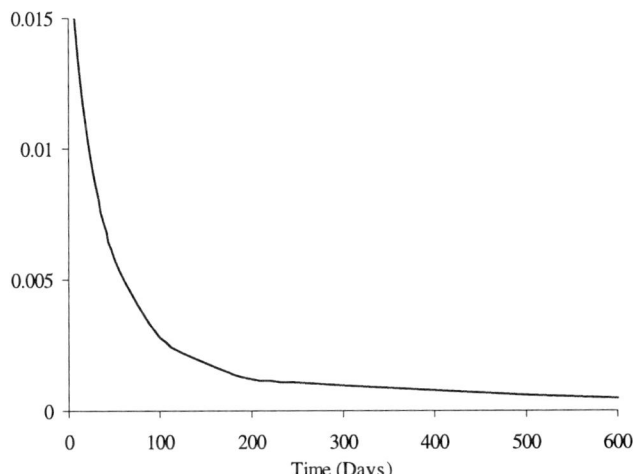

Time (Days)

FIGURE 4.1b. Bayesian Analysis of the NTDS Data Using Jelinski and Moranda's Model.

4. Statistical Analysis of Software Failure Data

Suppose that a piece of software is tested for T units of time, and that n failures at times $0 < S_1 < S_2 < \cdots < S_n \leq T$ are observed. Recall that the mean value of the nonhomogeneous Poisson process that supposedly generates these failures is $\Lambda(t) = a(1 - e^{-bt})$, where a and b are unknown parameters. McDaid and Wilson (1999) propose a Bayesian analysis of the foregoing process by assuming independent gamma priors on a and b. Specifically, given the quantities λ, τ, α, and μ (see Section 6.6.2), the joint prior density at a and b is of the form

$$\pi(a, b \mid \lambda, \tau, \alpha, \mu) = \left(\frac{\lambda^\tau}{\Gamma(\tau)} a^{\tau-1} e^{-\lambda a} \right) \left(\frac{\alpha^\mu}{\Gamma(\mu)} b^{\mu-1} e^{-\alpha b} \right).$$

It is easy to show (details left as an exercise for the reader) that the joint posterior of a and b, given n, T, and $S_1, \ldots, S_n$, is proportional to the quantity

(i) $\qquad a^{n+\tau-1} \, b^{n+\alpha-1} \, e^{-(1+\lambda)a} \, e^{-(\mu+S_n)b} \, e^{ae^{-bT}} \, ;$

the normalizing constant is $(K)^{-1}$, where

$$K = \Gamma(n + \tau) \int_0^\infty e^{-(\mu+S_n)b} b^{n+\alpha-1} (\lambda + 1 - e^{-bT})^{n+\tau} \, db.$$

It is noteworthy that the posterior distribution given previously depends only on n, T, and S_n, the last observed time of failure. The calculation of the various other quantities of interest is also straightforward. For example, if $N^*(T)$ denotes the number of failures that will be observed subsequent to time T, then, for $i = 0, 1, 2, \ldots,$

(ii) $\qquad \mathcal{P}(N^*(T) = i \mid n, S_n, T) \propto \dfrac{\Gamma(n+\tau+i)}{i! \, (1+\lambda)^i \, (\mu+iT+S_n)^{n+\alpha}},$

where the constant of proportionality is

$$\left(\sum_{j=0}^\infty \frac{\Gamma(n+\tau+j)}{j! \, (1+\lambda)^j \, (\mu+jT+S_n)^{n+\alpha}} \right)^{-1}.$$

Finally, if S_T denotes the time to next failure, as measured from T, then for $t \geq T$, the predictive distribution of S_T is

(iii) $\quad \mathcal{P}(S_T \geq t \mid n, S_n, T) = \mathcal{P}(N(T+t) - N(T) = 0 \mid n, S_n, T)$

$$= \frac{\Gamma(n+\tau)}{K} \int_0^\infty \frac{b^{n+\alpha-1} e^{-(\mu+S_n)b}}{(\lambda+1-e^{-(T+t)b})^{n+\tau}} \, db \,;$$

this quantity will have to be numerically evaluated.

4.3 Specification of Prior Distributions

The specification of prior distributions has been a roadblock for the application of Bayesian techniques ever since the days of Bayes and Laplace. However, recent advances in elicitation methodologies and computational methods have done much to ease this roadblock. When specifying a prior distribution, certain issues need to be addressed: the first is a choice of the family of distributions to use, and the second is a choice of the parameters of the chosen family; such parameters are called *hyperparameters*.

With regards to the first issue, the choice has sometimes been guided by mathematical tractability, under the disguise of what are known as "natural conjugate priors." Much of the early literature on Bayesian inference subscribed to this form of analysis [cf. Raiffa and Schlaifer (1961)]. Indeed, the priors used in Section 4.2.3 are natural conjugate priors. With natural conjugate priors, the choice of hyperparameters has been based on approximations like matching moments with beliefs [cf. Martz and Waller (1982), p. 222]. Another favorite approach for prior specification, and one that is gaining current popularity, is based on the philosophy espoused by Jeffreys (1961). According to Jeffreys, priors should be selected by convention, as a "standard of reference;" this is in keeping with the use of standards in other scientific settings [cf. Kass and Wasserman (1996)] (see Section 4.3.1 for an overview). Since the dominant philosophical foundation for Bayesian inference is subjectivism, the use of a natural conjugate, and standard of reference priors, is foundationally objectionable. Priors that are induced via a subjective elicitation of future observables, such as those discussed in Tierney and Kadane (1986), are in keeping with the subjectivistic foundations, and are therefore worthy of serious consideration. In Section 4.3.4 we describe an approach for constructing a subjectively elicited prior for the parameters of the model by Musa and Okumoto [see Equation (3.18)]. In Sections 4.3.2 and 4.3.3, we give an overview of the ideas underlying the material of Section 4.3.4.

There are two other matters about prior distributions that need to be mentioned: hierarchically constructed priors and sensitivity. A hierarchically constructed prior is one wherein a prior distribution is assigned to the hyperparameters of a prior distribution. Such priors are called *hierarchical priors*, and the hierarchical construction need not be limited to a single stage. That is, we may assign prior distributions on the hyperparameters at each stage of the prior construction. Probability models that involve hierarchically constructed priors are known as *hierarchical models*. Such models have proved to be very useful for addressing many problems in science and engineering, the "Kalman Filter Model" [cf. Meinhold and Singpurwalla (1983b)] being a prime example. In the context of software failure, the models introduced in Sections 3.2.4, 3.4.1, and 3.4.2 are examples of hierarchical models. Section 4.4 on "Inference and Predictions Based on Hierarchical Models" shows how such models provide meaningful descriptions of software failure data. In principle, hierarchical models also serve as a foundation for what are known as "empirical Bayes methods," because they are, in fact, a consequence of a repeated application of the law of total probability [cf. Singpurwalla (1989a)].

Our final comment about prior distributions pertains to sensitivity. Irrespective of how the prior is chosen, an investigation of the sensitivity of the posterior distribution to changes in the prior distribution and its hyperparameters is an important feature of any Bayesian analysis. Often, the posterior is insensitive to small changes in the prior, especially when the amount of data is large; in such cases we need not be too concerned about the imprecisions in our priors, if any. On the other hand, if our investigations reveal that small variations in a particular prior have large effects on the posteriors, then more careful attention should be paid to assessing the prior. Alternatively we may want to present a family of posterior distributions generated by a large class of priors so that a potential user of the results may make decisions in cognizance of the alternate possibilities that are revealed by the analyses.

4.3.1   Standard of Reference—Noninformative Priors

Jeffreys' notion of using priors that are a standard of reference has found appeal with many investigators who hold the view that analysts should say as little as possible about the parameters in question; this enables the data to speak for themselves. Supporting this position are those who maintain that often an analyst has no relevant experience to specify a prior, and that subjective elicitation in multiparameter problems is next to impossible. Priors that are developed to react to these points of view are called *noninformative priors*. Bernardo (1997), who claims that "noninformative priors do not exist," touches on these and related issues, from both a historical as well as a mathematical perspective.

A simple strategy for constructing priors that (supposedly) convey little information about a parameter, say θ, is to let the prior density function be flat over all allowable values of θ. If θ can take values only in a finite range, say $[a, b]$, then the obvious noninformative prior is the uniform density on $[a, b]$; that is,

$$\pi(\theta \mid a, b) = \begin{cases} (b-a)^{-1}, & a \leq \theta \leq b \\ 0, & \text{otherwise.} \end{cases} \tag{4.9}$$

When θ takes values over an infinite range, then a limiting form of noninformative prior would be the uniform, with constant weight given to all possible values. Unfortunately, this is not a probability density, as the integral is infinite, and for this reason is known as an *improper prior*. Although not a legitimate density, it turns out that the posterior distribution, calculated from Bayes' formula, may be a proper density. So, in terms of computing posterior distributions, it may be possible to work with this type of prior distribution.

There are, however, some objections to using improper priors. First, the posterior distribution is not guaranteed to be proper; it may be improper, in which case one cannot calculate posterior means or sensibly find marginal distributions. Second, there would always appear to be at least some prior information on θ, even if it is just some fantastically large bound on its possible values; in such cases a uniform prior density can be used. There is also a more fundamental problem with assigning an equal weight to all values of θ. For example, suppose that our prior on θ is of the form given by Equation (4.9), but that our problem is parameterized in such a way that inference needs to be made about $\psi = \theta^2$. Then, it can be shown, using the calculus of probability, that our prior on ψ is of the form

$$\pi^*(\psi \mid a, b) = 0.5 \, \psi^{-0.5} \, \pi(\psi \mid a, b). \tag{4.10}$$

But Equation (4.10) suggests that a uniform prior on θ (chosen to reflect an absence of knowledge about θ) results in a prior for ψ that is proportional to $\psi^{-0.5}$. This is contrary to intuition; thus we cannot choose uniform priors for both θ and ψ at the same time.

The preceding type of scenarios has motivated a lot of research into finding noninformative priors that are invariant under transformations. This kind of work was initiated by Jeffreys, and has been continued by, among others, Jaynes (1968), Zellner (1971), (1977), and Bernardo (1979). It has also spawned a variety of new ideas; the one that has seen many applications in physics and engineering is the principle of "maximum entropy priors" [see Jaynes (1983), and Good (1983)].

4.3.2 Subjective Priors Based on Elicitation of Specialist Knowledge

The subjective specification of prior distributions often entails, in addition to the background knowledge $\mathcal{H}$, the use of information that an analyst, say $\mathcal{A}$, elicits from users and subject matter specialists, called *experts*. The term expert is generic, and could include the information provided by mathematical and engineering models, simulation algorithms, empirical experience, and the like. How does an analyst incorporate expert information into the background knowledge $\mathcal{H}$ that $\mathcal{A}$ has, to arrive at prior distributions for parameters? Also, since most experts are subject matter specialists, the information they provide is about observable entities, not parameters, which to them are Greek symbols concocted by analysts. How should $\mathcal{A}$ induce prior distributions on parameters from information about observables?

The foregoing problems have been addressed by many. The general plan was first proposed by Morris (1974, 1977), and subsequently improved upon by French (1980), Tversky, Lindley, and Brown (1979), and Lindley (1983). Application to problems in reliability has been considered by Lindley and Singpurwalla (1986a), Singpurwalla (1988b), and Singpurwalla and Song (1988).

For purposes of discussion, suppose that interest centers around an unknown quantity, say X, and we (the analyst $\mathcal{A}$), possess background information $\mathcal{H}$ about X. Let $\mathcal{P}(X \mid \mathcal{H})$ denote our uncertainty about X in the light of $\mathcal{H}$. To obtain an enhanced appreciation of X, we consult an expert, say $\mathcal{E}$, who provides us with an assessment of X in terms of two quantities m and s, where m represents $\mathcal{E}$'s best guess about X, and s a measure of $\mathcal{E}$'s uncertainty about m. Note that whereas X often denotes some observable quantity, it could in principle also be an unknown parameter. $\mathcal{A}$'s problem therefore is to assess $\mathcal{P}(X \mid m, s, \mathcal{H})$; this is $\mathcal{A}$'s uncertainty about X in the light of m, s, and $\mathcal{H}$. By Bayes' Law

$$\mathcal{P}(X = x \mid m, s, \mathcal{H}) \propto \mathcal{L}(X = x; m, s, \mathcal{H}) \, \mathcal{P}(X = x \mid \mathcal{H}),$$

where $\mathcal{L}(X = x; m, s, \mathcal{H})$ is $\mathcal{A}$'s likelihood that $\mathcal{E}$ will declare the values m and s, were $X = x$. The likelihood reflects $\mathcal{A}$'s opinion of the expertise of $\mathcal{E}$, and may be better expressed through additional coefficients that are introduced by $\mathcal{A}$. For example, if $\mathcal{A}$ is of the opinion that $\mathcal{E}$ tends to overestimate or underestimate the location of $\mathcal{E}$'s distribution for X, then m is actually the location of $\alpha + \beta x$; the case $\alpha = 0$, $\beta = 1$ corresponds to $\mathcal{A}$'s view that $\mathcal{E}$ is unbiased. If in $\mathcal{A}$'s view, $\mathcal{E}$ tends to underestimate the standard deviation of $\mathcal{E}$'s distribution for X, then $\mathcal{A}$ modulates s to γs, with $\gamma > 1$; if $\mathcal{E}$ tends to overestimate the standard deviation, then $\gamma < 1$. Some further simplification in the specification of the likelihood occurs if in $\mathcal{A}$'s opinion, $\mathcal{E}$'s declared value s is independent of the value of X. If such be the case, then $\mathcal{A}$ may reflect the expertise and the attitudes of $\mathcal{E}$, via the normal (Gaussian) form

$$\mathcal{L}(X = x; m, s, \mathcal{H}) \propto \exp\left[-\tfrac{1}{2}\left(\tfrac{m - (\alpha + \beta x)}{\gamma s}\right)^2\right], \qquad (4.11)$$

where the *tuning coefficients* α, β, and γ are chosen by $\mathcal{A}$ to reflect $\mathcal{A}$'s view of the biases and the assertiveness of $\mathcal{E}$. The choice $\alpha = 0$, $\beta = \gamma = 1$, reflects $\mathcal{A}$'s willingness to accept the values m and s without any modification (tuning).

4.3 Specification of Prior Distributions 117

It is often the case that analysts who consult experts are unwilling to impose their own views about X in a manner that will greatly distort the expert's inputs beyond that which is done through the tuning parameters. If such be the case, then the analyst's prior $\mathcal{P}(X = x \mid \mathcal{H})$ will tend to be flat over the range of values x where the likelihood is appreciable. Consequently, $\mathcal{A}$'s posterior probability density for X at the point x is of the form

$$f(x \mid m, s, \mathcal{H}) \propto \exp\left[-\tfrac{1}{2}\left(\tfrac{m-(\alpha+\beta x)}{\gamma s}\right)^2\right], \qquad (4.12)$$

with the constant of proportionality chosen to make the preceding quantity integrate to one. This posterior density represents $\mathcal{A}$'s assessment of the uncertainty of X in the light of $\mathcal{E}$'s inputs and $\mathcal{A}$'s views about the expertise and attitudes of the expert. If $\mathcal{A}$ chooses to incorporate $\mathcal{A}$'s own views about X, then the right-hand side of Equation (4.12) must be multiplied by the probability density of X in the light of $\mathcal{H}$ alone; the latter is a proxy for $\mathcal{P}(X = x \mid \mathcal{H})$.

Thus to summarize, the crux of the plan for incorporating expert inputs into an analysis is to view such inputs as data, and then to invoke Bayes' Law using as the likelihood a model for the expertise of the expert. The attitudes of the expert, as perceived by the analyst, get reflected in the likelihood via the tuning coefficients.

4.3.3 Extensions of the Elicitation Model

There are several possible directions in which the model of Section 4.3.2 can be extended, the most natural one being the case of several experts, say $\mathcal{E}_1, \ldots, \mathcal{E}_k$, $k \geq 2$. Now $\mathcal{A}$ has to contend with the quantities $(m_1, s_1), \ldots, (m_k, s_k)$ and the corresponding tuning coefficients $(\alpha_i, \beta_i, \gamma_i)$, $i = 1, \ldots, k$. The principle is the same except that in writing $\mathcal{L}(X = x; (m_i, s_i), i = 1, \ldots, k, \mathcal{H})$, the likelihood, $\mathcal{A}$ has to consider possible correlations between the expert announcements. The treatment of this possibility has been considered by Lindley (1983) in a general context, and by Lindley and Singpurwalla (1986) in the context of reliability.

Another generalization of the elicitation model is motivated by the difficulty in specifying the tuning coefficients α, β, and γ. One approach for easing this difficulty is to gather information about $\mathcal{E}$'s previous announcements (m_i, s_i), $i = 1, \ldots, n$, and to relate them to x_i, the revealed values of X. Once the (m_i, s_i) and the corresponding x_i are at hand, we may invoke Bayes' Law, with a flat (vague) prior on α, β, and γ, to obtain the posterior distribution

$$\mathcal{P}(\alpha, \beta, \gamma \mid (m_i, s_i), x_i, i = 1, \ldots, n, \mathcal{H})$$

$$\propto \gamma^{-n} \times \exp\left[-\tfrac{1}{2}\left(\tfrac{m_i-(\alpha+\beta x_i)}{\gamma s_i}\right)^2\right]. \qquad (4.13)$$

The mode of this posterior provides us with suitable values of α, β, and γ for use in future elicitations.

Finally, the matter of inducing prior distributions on unknown parameters using the elicited distribution of observables $\mathcal{P}(X = x \mid m, s, \mathcal{H})$ remains to be settled. This is generally a straightforward matter if a simple relationship between the observable X and a parameter θ can be established. For example, with exponentially distributed lifetimes, the mean time to failure is θ, and so the X of Section 2.2.2 is now the mean lifelength. Consequently, $\mathcal{E}$ will therefore be asked to provide assessments for the mean lifelength. Often there is a simple relationship between the median and the parameters; see, for example, Singpurwalla (1988b). In such cases expert elicitation about the median is sought. Psychological studies have shown that experts are more at ease assessing medians and other percentiles than the mean.

In the next section we describe an application of the foregoing general methodology to a commonly used model for describing the software failure process.

4.3.4 Example: Eliciting Priors for the Logarithmic-Poisson Model

Recall (see Section 2.3), that the Poisson process is completely determined by $\Lambda(t)$, its mean value function. The logarithmic-Poisson execution time model for describing software failures, introduced by Musa and Okumoto (1984), takes for $\Lambda(t)$ the functional form $\ln(\lambda \theta t + 1)/\theta$, where λ and θ are parameters; see Equation (3.16). In this section we describe how the elicitation techniques of the previous two sections, plus some empirical experience reported by software engineers, can be used to assess the priors on λ and θ.

Since $\Lambda(t)$ represents the expected number of software failures encountered by time t (see Section 2.3.1), $\Lambda(t)$ is an observable, and thus it is meaningful to elicit expert opinion on $\Lambda(t)$ rather than on θ and λ. The latter quantities lack an intuitive import. Accordingly, if two time points T_1 and T_2, $T_1 \leq T_2$, are chosen and expert opinion in terms of a measure of location and scale, say m_i and s_i, elicited for $\Lambda(T_i)$, $i = 1, 2$, then a prior on λ and θ can be induced from the fact (verification left as an exercise for the reader) that:

$$\frac{e^{\Lambda(T_1)\theta} - 1}{e^{\Lambda(T_2)\theta} - 1} = \frac{T_1}{T_2}, \quad \text{and} \quad \lambda = \frac{e^{\Lambda(T_1)\theta} - 1}{\theta T_1}. \tag{4.14}$$

The preceding will yield a solution for $\theta > 0$, and $\lambda > 0$, if and only if $0 \leq \Lambda(T_1) \leq \Lambda(T_2) \leq T_2((\Lambda(T_1))/T_1)$.

The simplifying assumptions that pertain to the joint distribution of $\Lambda(T_1)$ and $\Lambda(T_2)$, given (m_1, s_1) and (m_2, s_2), are in the same spirit as those given in Section 4.3.2 with suitable modifications to account for the fact that $\Lambda(T_1) \leq \Lambda(T_2)$. The motivation and details are in Campodónico and Singpurwalla (1994), (1995); the following is an overview of the essentials.

(i) The likelihood of $\Lambda(T_2)$, for fixed values of m_1, m_2, s_1, and s_2, is of the truncated normal shape, centered at $\alpha + \beta m_2$ with a scale γs_2; see Equation (4.13). The left truncation point is $m_1 + ks_1$, and the right truncation point is $m_1 T_2/T_1$; k is specified by the analyst.

(ii) The likelihood of $\Lambda(T_1)$, for a fixed value of m_1 and s_1 is also of the truncated normal shape centered at $\alpha + \beta m_1$ and a scale γs_2. The left truncation point is zero.

(iii) The likelihood of the difference $(\Lambda(T_2) - \Lambda(T_1))$ is truncated to the left at zero, and for a fixed value of $(s_2 - s_1)$, it is proportional to the quantity

$$(\Lambda(T_2) - \Lambda(T_1))^{-(1/2)} \exp\left[-\frac{1}{2} \frac{(s_2-s_1)^2}{(\Lambda(T_2)-\Lambda(T_1))^2}\right];$$

for a fixed value of s_1 it is of the form $(\Lambda(T_2) - \Lambda(T_1)) \times \exp(-s_1(\Lambda(T_2) - \Lambda(T_1)))$.

(iv) The joint prior on $\Lambda(T_1)$ and $\Lambda(T_2)$ is a constant over the range of values of $\Lambda(T_1)$ and $\Lambda(T_2)$ for which the likelihood is appreciable.

Under the preceding assumptions, the density of the joint posterior of $\Lambda(T_1)$ and $\Lambda(T_2)$, at the points λ_1 and λ_2, $0 < \lambda_1 < \lambda_2 < \lambda_1(T_2/T_1)$, is proportional to (the formidable looking expression)

$$\frac{\exp\left\{-\frac{1}{2}\left(\frac{m_2-\alpha-\beta\lambda_2}{\gamma s_2}\right)^2\right\}}{\left(\Phi\left(\frac{\frac{T_2 m_1}{T_1}-\alpha-\beta\lambda_2}{\gamma s_2}\right) - \Phi\left(\frac{m_1+ks_1-\alpha-\beta\lambda_2}{\gamma s_2}\right)\right)\gamma s_2} \times$$

$$\frac{\exp\left\{\frac{-s_1}{\lambda_2-\lambda_1} - \frac{1}{2}\left(\frac{m_1-\alpha-\beta\lambda_1}{\gamma s_1}\right)^2 - \frac{1}{2}\left(\frac{s_2-s_1}{\lambda_2-\lambda_1}\right)^2\right\}}{(\lambda_2-\lambda_1) \times \gamma s_1 \left(1-\Phi\left(\frac{-\alpha-\beta\lambda_1}{\gamma s_1}\right)\right) \times (\lambda_2-\lambda_1)\left(1-\Phi\left(\frac{-s_1}{\lambda_2-\lambda_1}\right)\right)}, \quad (4.15)$$

where $\Phi(x)$ is the cumulative distribution of the standard normal distribution [so $\Phi(x) = \int_{-\infty}^{x} (1/\sqrt{2\pi}) \exp(-u^2/2)\,du$].

This prior distribution, although complex, is easily manipulated numerically. Using the relationships given in Equation (4.13), it has been used to compute the joint distribution of (λ, θ), posterior distributions in light of data on the Poisson process, as well as various expectations and variances; see Campodónico and Singpurwalla (1994) (1995). A computer code for carrying out the needed calculations is described by Campodónico (1993).

The principal remaining issue is to discuss how the expert might in practice specify the various values: T_1, T_2, m_1, m_2, s_1, and s_2. First, T_1 and T_2 are chosen; recommended values for T_2 are the total time that testing is scheduled for, or some proportion of the total hours worked [Myers (1978) suggests one half]. Typically, T_1 will then be some reasonably small percentage of T_2 for which it is felt that a number of bugs will have been discovered; for example, if T_2 is chosen to be the scheduled testing time, and the expert thinks that 10% of bugs will be discovered in the first 1% of the test, then it is reasonable to define $T_1 = 0.01 T_2$.

The expert then specifies a mean and standard deviation for $\Lambda(T_2)$, denoted m_2 and s_2. If T_2 is the total testing time, we might set m_2 to be the total number of bugs expected in the code; Gaffney (1984) has suggested various empirical formulae that relate the length of code S to the number of bugs B:

- $B = 0.021 S$;
- $B = 4 + 0.0014 S^{4/3}$;
- $B = 4.2 + 0.0015 S^{4/3}$.

Given the rather ad hoc nature of these formulae, it is wise to set the standard deviation s_2 large, to reflect large uncertainty in the estimate of m_2.

Next is the specification of a mean and standard deviation on T_1. These may be simply specified as fixed proportions of m_2 and s_2, or alternatively the expert can use experience from previous tests; if m_2 and s_2 describe the total number of bugs and, on average, the expert knows 10% of bugs occur up to testing time T_1, then $m_1 = 0.1 m_2$ and $s_1 = 0.1 s_2$.

The final part of the specification is the tuning coefficients α, β, γ, and k. If there is no basis for assuming any bias by the analyst, we choose $\alpha = 0$, $\beta = 1$, $\gamma = 1$, and $k = 1$.

4.3.5 Application: Failure Prediction Using Logarithmic-Poisson Model

To illustrate the workings of the procedure described in the previous section, we consider some software failure data given by Goel (1985); these are given in Table 4.3. The data consist of the observed number of failures of a software system that was tested for 25 hours of CPU time. For purposes of illustration, we choose the logarithmic-Poisson model of Musa and Okumoto (1984) to analyze these data. The choice of this model has no basis other than the need for exposition. The standard approach for analyzing such data has been the method of maximum likelihood. However, as discussed by Campodónico and Singpurwalla (1994), this approach may lead one to difficulties, the main one

4.3 Specification of Prior Distributions

Table 4.3. Data on Software Failures During System Test

of testing	per CPU hour	# of Failures	of testing	per CPU hour	# of Failures
1	27	27	14	5	111
2	16	43	15	5	116
3	11	54	16	6	122
4	10	64	17	0	122
5	11	75	18	5	127
6	7	82	19	1	128
7	2	84	20	1	129
8	5	89	21	2	131
9	3	92	22	1	132
10	1	93	23	2	134
11	1	97	24	1	135
12	7	104	25	1	136
13	2	106			

being nonunique estimators when the data are such that only the total number of failures in the first interval of testing is available.

The data of Table 4.3 pertain to a system consisting of 21,700 object instructions. Thus we take 21,700 as our length of code S, and using the first of the three formulas of Gaffney (1984), choose $m_2 = 0.021 \times (21,700) \doteq 455$. Considering a long term for the debugging horizon, we take $T_2 = 250$ (CPU hours). Given the very general nature of our choices for m_2 and T_2, we choose $s_2 = 200$ to reflect a high degree of uncertainty in our specifications. Experience of software engineers suggests that, on average, about 10% of system failures occur during the first 1% of debugging time. Consequently, we choose $T_1 = 2.5$ and $m_1 = 45.5$. As a measure of uncertainty about our choice of m_1, we choose $s_1 = 4$, and as an alternative, $s_1 = 20$. Since we have no basis for tuning all of these selections, we choose $\alpha = 0, \beta = \gamma = k = 1$.

In Figure 4.2 we show plots comparing the cumulative number of failures that are actually observed during the first five intervals of testing and those predicted via a Bayesian analysis of the model with prior parameters $m_1 = 45.5$, $s_1 = 4$, $m_2 = 455$, and $s_2 = 200$, for $T_1 = 2.5$ and $T_2 = 250$. The predictions shown are *one-step-ahead* predictions. That is, the predicted cumulative failures at the end of the second interval of testing incorporate the data observed at the end of the first interval of testing, the predictions at the end of the third interval of testing incorporate the data observed at the end of the second interval of testing, and so on.

The plots of Figure 4.2 suggest that the approach described here provides good predictive capability vis-à-vis the chosen parameters. When the one-step-

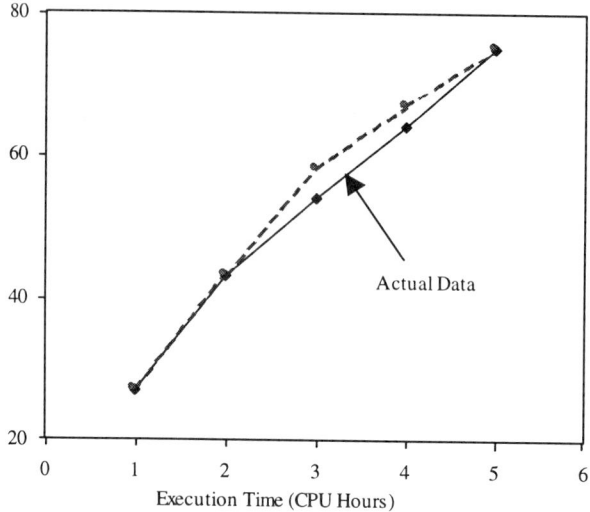

FIGURE 4.2. Comparison of Actual Versus One-Step-Ahead Predictions.

ahead prediction is extended to a horizon covering the 25 intervals of testing, a less promising picture appears. The predictions tend to overestimate the actual; see Figure 4.3. Note that the predicted values for the 25-interval horizon are based on the data up to and including the fifth interval of testing only. Presumably, the one-step-ahead predictions would be better, but in practice, it is the several steps ahead predictions that are useful.

In view of Figure 4.3, it appears desirable to explore the sensitivity of our analysis to the choice of prior parameters. Table 4.4 shows our selections for three other priors considered by us: Prior I is the selection previously described, and Prior II is identical to Prior I except that $s_1 = 20$ instead of 4. Prior IV uses the second formula of Gaffney (1984) to specify m_2, and Prior III uses the actual data from the first interval of testing to specify T_1 and m_1. Prior III is intended to reflect the feature of maximum likelihood estimation that would necessitate the use of some data for inference; this is in contrast to Bayesian inference which can be based on the prior alone.

In Table 4.5 we compare the one-step-ahead predictions based on the four priors of Table 4.4. Also given are the mean square errors (MSE) of the predictions over the five testing intervals. A comparison of the predictions based on the MSE suggests that Prior II appears to provide better predictivity than Prior I. A possible reason for this is that the higher uncertainty associated with

4.3 Specification of Prior Distributions   123

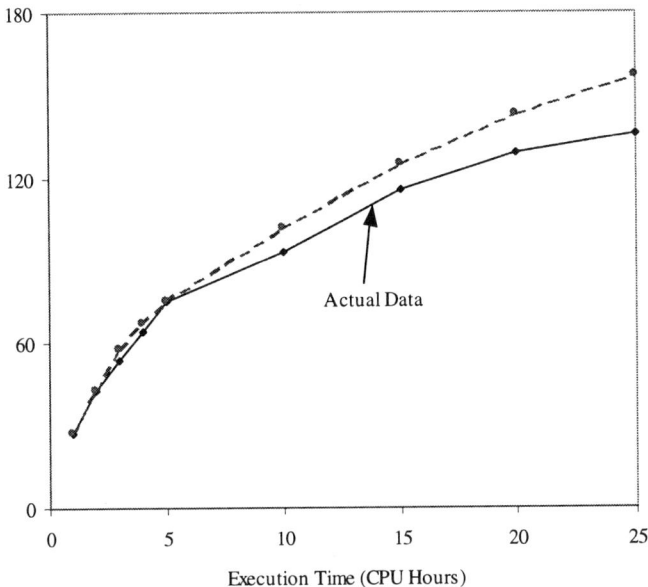

FIGURE 4.3. Comparison of Actual Versus Predicted Failures—
Forecast Horizon of 25.

Table 4.4. The Class of Priors Considered

	T_1	T_2	m_1	m_2	s_1	s_2	a	b	γ	k
Prior I	2.5	250	45.5	455	4	200	0	1	1	1
Prior II	2.5	250	45.5	455	20	200	0	1	1	1
Prior III	1	250	27	455	0.5	200	0	1	1	1
Prior IV	1	250	27	851	0.5	300	0	1	1	1

Prior II ($s_1 = 20$ instead of 4) better compensates any misspecifications in m_1. The MSE of Prior III is slightly smaller than that of Prior II because under Prior III, the predicted failures for the first interval of testing equal the observed failures.

124   4. Statistical Analysis of Software Failure Data

Table 4.5. Comparison of the One-Step-Ahead Predictions Under Different Priors

	CPU Hour Interval	Prior I	Prior II	Prior III	Prior IV	Observed Failures
1.	(0, 1]	22.8	25	27	27	27
2.	(1, 2]	16	18.7	18.3	22.1	16
3.	(2, 3]	12.6	13.3	13.8	20.6	11
4.	(3, 4]	9.5	9.7	10.1	15.6	10
5.	(4, 5]	8.6	8.5	8.1	12.8	11
	MSE	5.2	4.5	4.3	32.7	
	Based on 5 predictions					

4.4   Inference and Prediction Using a Hierarchical Model

In Section 3.2.4, we introduced a model for tracking the growth in reliability of software using a prior that was hierarchically constructed in two stages. In this section we discuss inferential aspects of this model using actual data on software failures. The model of Section 3.2.4 was proposed by Mazzuchi and Soyer (1988). The first step of model construction involves the specification

$$\mathcal{P}(T_i \geq t \mid \Lambda_i) = e^{-\Lambda_i t}, \qquad (4.16)$$

where the parameter Λ_i is such that the collection of Λ_is, $i = 1, 2, \ldots$, constitute a decreasing sequence. The prior distribution on Λ_i is a gamma with a scale parameter $\psi(i)$ and a shape parameter α; see Equation(3.9). Furthermore, $\psi(i)$ is reparameterized as $\psi(i) = \beta_0 + \beta_1 i$, and the predictive distribution of T_i, given $\psi(i)$ and α, is of the form [see Equation (3.10)]

$$\mathcal{P}(T_i \geq t \mid S_{i-1}, \alpha, \psi(i)) = \left[\frac{\beta_0 + \beta_1 i}{t + \beta_0 + \beta_1 i} \right]^{\alpha}, \; t \geq S_{i-1}. \qquad (4.17)$$

For the second stage of the hierarchy, the following prior structure is assumed for the hyperparameters

$$\pi(\alpha \mid \omega) = \omega^{-1}, \; 0 < \alpha < \omega;$$
$$\pi(\beta_1 \mid c, d) = \frac{d^c}{\Gamma(c)} \beta_1^{c-1} e^{-\beta_1 d}, \; \beta_1 > 0; \text{ and}$$

4.4 Inference and Prediction Using a Hierarchical Model

$$\pi(\beta_0 \mid \beta_1, a, b) = \frac{b^a}{\Gamma(a)} (\beta_0 - \beta_1)^{a-1} e^{-b(\beta_0 - \beta_1)}, \quad \beta_0 \geq \beta_1. \quad (4.18)$$

The shape parameter α is independent of both β_0 and β_1, but β_0 and β_1 are a priori dependent. The joint prior of α, β_0, and β_1 is obtained via the relationship

$$\pi(\alpha, \beta_0, \beta_1) = \pi(\alpha \mid \omega) \times \pi(\beta_1 \mid c, d) \times \pi(\beta_0 \mid \beta_1, a, b).$$

The foregoing prior distributions are more in the spirit of natural conjugate priors than priors based on elicitation.

Given k interfailure times $t_1, \ldots, t_k = \underline{t}^{(k)}$, the joint posterior of α, β_0, and β_1 is of the form (details left as an exercise for the reader):

$$\pi(\alpha, \beta_0, \beta_1 \mid \underline{t}^{(k)}) \propto \pi(\alpha, \beta_0, \beta_1) \prod_{i=1}^{k} \frac{\alpha(\beta_0 + \beta_1 i)^\alpha}{(t_i + \beta_0 + \beta_1 i)^{\alpha+1}}, \quad (4.19)$$

where the constant of proportionality is such that the preceding integrates to one. The posterior distribution (4.18), although not in closed form, is relatively straightforward to numerically compute.

In practice, interest may often center around the parameters Λ_i, $i = 1, 2, \ldots$. This is because Λ_i could be regarded as a proxy for the quantity $\Lambda(N - i + 1)$ in Jelinski and Moranda's model. Also, a decreasing sequence of Λ_is implies a growth in reliability, suggesting that the debugging process which is subsequent to every observed failure is producing desirable results.

Given $\underline{t}^{(k)}$, and conditional on α, β_0, and β_1, we can show (details left as an exercise for the reader) that the posterior density of Λ_i at λ_i is:

$$\mathcal{P}(\lambda_i \mid \alpha, \beta_0, \beta_1, \underline{t}^{(k)}) = \frac{\lambda_i^\alpha (t_i + \beta_0 + \beta_1 i)^{\alpha+1}}{\Gamma(\alpha+1)} e^{-\lambda_i (t_i + \beta_0 + \beta_1 i)}. \quad (4.20)$$

Consequently, the posterior density at λ_i given the data $\underline{t}^{(k)}$ alone, is of the form

$$\mathcal{P}(\lambda_i \mid \underline{t}^{(k)}) = \int_{(\alpha, \beta_0, \beta_1)} \mathcal{P}(\lambda_i \mid \alpha, \beta_0, \beta_1, \underline{t}^{(k)}) \pi(\alpha, \beta_0, \beta_1 \mid \underline{t}^{(k)}) d\alpha d\beta_0 d\beta_1. \quad (4.21)$$

The preceding integration will have to be done numerically. Mazzuchi and Soyer (1988) use an approximation, first suggested by Lindley (1980), valid for large values of k, to obtain $E(\Lambda_k \mid \underline{t}^{(k)})$ and $E(T_{k+1} \mid \underline{t}^{(k)})$, the mean of the posterior distribution of Λ_k, and the predictive distribution of T_{k+1}, respectively. Verify (left as an exercise for the reader) that the predictive density of T_{k+1}, at t, is given by

126   4. Statistical Analysis of Software Failure Data

$$\mathcal{P}(t \mid \underline{t}^{(k)}) = \int\limits_{(\alpha, \beta_0, \beta_1)} \frac{\alpha(\beta_0 + \beta_1 i)^\alpha}{(t + \beta_0 + \beta_1 i)^{\alpha+1}} \pi(\alpha, \beta_0, \beta_1 \mid \underline{t}^{(k)}) d\alpha d\beta_0 d\beta_1 \quad (4.22)$$

(see Exercise 1 of Chapter 3 for a hint). As an alternative, see Section A.3.2 of Appendix A for Gibbs sampling.

4.4.1   Application to NTDS Data: Assessing Reliability Growth

In Table 4.2 of Section 4.2.3, some software failure data from the NTDS system was analyzed using the Jelinski–Moranda model. The aim there was to assess the remaining number of bugs in the software. In this section we describe how the hierarchical model of Equations (4.16) and (4.18) can be used to see if the debugging process is effective; that is, it is improving the reliability of the software. One way of investigating this is to track the mean of the posterior distribution of Λ_i, $i = 1, 2, \ldots$. Alternatively, we may also monitor the behavior of the posterior distribution of β_1, and see if it reflects either a constant or an increasing central characteristic, such as the mean or the mode. Note that the posterior distribution of β_1 can be obtained from Equation (4.19) as

$$\pi(\beta_1 \mid \underline{t}^{(k)}) = \int\limits_{(\alpha, \beta_0)} \pi(\alpha, \beta_0, \beta_1 \mid \underline{t}^{(k)}) d\alpha d\beta_0 . \quad (4.23)$$

Equation (4.23) can be used to obtain $E(\beta_1 \mid \underline{t}^{(k)})$, the mean of $\pi(\beta_1 \mid \underline{t}^{(k)})$, or its mode $M(\beta_1 \mid \underline{t}^{(k)})$. The required computations will have to be done either numerically or by an approximation.

Mazzuchi and Soyer (1988) analyzed the data of Table 4.2, using the following values for the hyperparameters of Equations (4.18): $\omega = 500$; $a = 10$; $b = 0.1$; $c = 2$; $d = 0.25$. Using Equations (4.21)–(4.23) and Lindley's approximation, they calculated $E(\Lambda_i \mid \underline{t}^{(i)})$, $E(T_{i+1} \mid \underline{t}^{(i)})$, and $M(\beta_1 \mid \underline{t}^{(i)})$, for $i = 1, 2, \ldots, 26$, the production phase of the data. The values of the former two quantities are given in columns 3 and 4 of Table 4.6. A plot of $E(\Lambda_i \mid \underline{t}^{(i)})$ and $M(\beta_1 \mid \underline{t}^{(i)})$ is shown in Figure 4.4.

An examination of the upper plot of Figure 4.4 suggests that there has been an apparent growth in reliability during the initial stages of testing, followed by a modest decay for most of the middle stages of testing, and then an increase during the very last stages. The lower plot of Figure 4.4 suggests that the parameter β_1 is not relatively constant; rather, the downward drift in β_1 during the first 20 or so stages of testing confirms the decay in reliability during the middle stages of testing. The sharp upward drift in β_1 during the last stages of testing is a reflection of the growth in reliability during the final stages of testing. Our conclusion that the middle portion of the data is at odds with the structure of the model, namely, that the sequence of Λ_is be decreasing, suggests that the model should be weakened. Accordingly, Mazzuchi and Soyer do away with the

4.4 Inference and Prediction Using a Hierarchical Model

Table 4.6. Posterior and Predictive Means for NTDS Data

Error Number i	Actual Interfailure Times t_i	Means of Predictive Interfailure Time $E(T_{i+1}\|\underline{t}^{(k)})$	Means of the Posterior of Λ_i $E(\Lambda_i\|\underline{t}^{(k)})$
1	9.00	—	0.2215
2	12.00	9.75	0.1389
3	11.00	11.36	0.1197
4	4.00	11.77	0.1331
5	7.00	10.09	0.1265
6	2.00	9.87	0.1400
7	5.00	8.74	0.1375
8	8.00	8.45	0.1290
9	5.00	8.71	0.1318
10	7.00	8.50	0.1273
11	1.00	8.61	0.1392
12	6.00	7.92	0.1340
13	1.00	7.93	0.1449
14	9.00	7.35	0.1329
15	4.00	7.70	0.1383
16	1.00	7.50	0.1469
17	3.00	7.03	0.1488
18	3.00	6.78	0.1521
19	6.00	6.55	0.1483
20	1.00	6.61	0.1586
21	11.00	6.23	0.1425
22	33.00	6.68	0.1061
23	7.00	8.52	0.1173
24	91.00	8.57	0.0617
25	2.00	13.10	0.0847
26	1.00	12.66	0.0875

128   4. Statistical Analysis of Software Failure Data

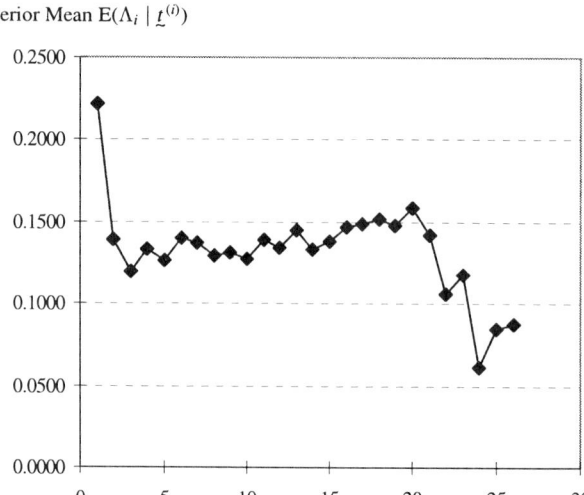

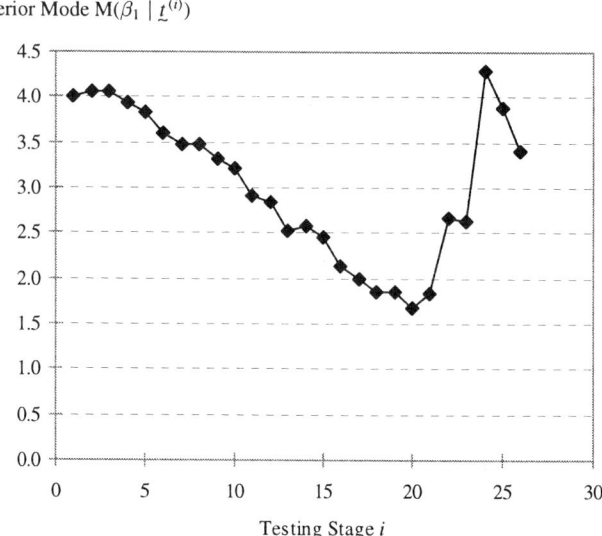

FIGURE 4.4. Plots of the Posterior Means and Modes of Λ_i and β_1.

parameter $\psi(i)$ and its reparameterization. Instead, they assume that the Λ_is have a common gamma distribution with shape (scale) $\alpha(\beta)$. They next assume that α has a uniform distribution over $(0, \omega)$, and β a gamma distribution, independent of the distribution of α. This scheme makes the Λ_is exchangeable; see Section 2.1.5. As regards inference, we note that the new model is indeed a special case of the hierarchical model, with $\pi(\beta_1 \mid \bullet)$ degenerate at 0, and $\beta = \beta_0$; see Equation (4.18). Mazzuchi and Soyer have shown that the new (weaker) model provides better predictivity of the NTDS data than the parent model, but only by a small margin; the overall conclusions about reliability growth do not change.

4.5 Inference and Predictions Using Dynamic Models

In Section 3.4 we introduced three models for describing the times between software failure; these models are classified in the Type I-2 category of Section 3.1.2. Whereas the model of the previous section was based on a two-stage hierarchical construction, the models of Section 3.4 are based on hierarchical constructions involving several stages. This is because of two reasons: the autoregressive construction underlying Equation (3.19) and because the underlying parameters of these models are assumed to evolve dynamically over time, thus the label "dynamic;" see Equation (3.22). In the control theory literature, such models play a dominant role, and are known there as *Kalman filter models*. The dynamic feature underlying the models enables them to be more responsive to changes in the process generating the data, and in doing so they are able to better track the data. This results in enhanced predictivity. The purpose of this section is to discuss inferential issues pertaining to such models, and to illustrate how they can be applied to data on software failures for prediction and for assessing the growth (or decay) in reliability.

We start with the two models of Section 3.4.1 wherein the interfailure times T_i, $i = 1, 2, \ldots$, bear a relationship with each other via a sequence of parameters θ_i as

$$\log T_i = \theta_i \log T_{i-1} + \epsilon_i ; \tag{4.24}$$

see Equation (3.21).

The error terms ϵ_i are independent and identically normally (Gaussian) distributed, with a mean 0 and variance σ_1^2; $\epsilon_i \sim \mathcal{N}(0, \sigma_1^2)$. Recall, from Section 3.4.1, that the T_is are to be scaled (if necessary), so that they are all greater than one.

For the sequence of parameters θ_i, two models were proposed. The first is a two-stage hierarchical construction that makes the θ_is an exchangeable sequence. Specifically, conditional on λ, the θ_is are assumed to be independent and identically normally distributed with mean λ and variance σ_2^2. Furthermore, λ itself is normally distributed with mean μ and variance σ_3^2. Thus, for $i = 1, 2, \ldots$

$$\theta_i \sim \mathcal{N}(\lambda, \sigma_2^2), \text{ and}$$

$$\lambda \sim \mathcal{N}(\mu, \sigma_3^2). \tag{4.25}$$

The hyperparameters σ_1^2, σ_2^2, σ_3^2, and μ are to be specified by the user. Equations (4.24) and (4.25) constitute what is referred to as the *exchangeable model*. It is important to note that even though the prior construction involves only the two stages of Equation (4.25), the exchangeable model as a whole involves multiple stages. This is so because of the "autoregressive" nature of Equation (4.24); the multistage hierarchy is in the "observation equation" of the model.

As an alternative to Equation (4.25), we may assume that the θ_is also constitute an autoregressive process of order one, so that for some parameter α, and $i = 1, 2, \ldots$,

$$\theta_i = \alpha \theta_{i-1} + \omega_i, \tag{4.26}$$

where the ω_is are independent, and $\omega_i \sim \mathcal{N}(0, W_i^2)$. A uniform prior over (a, b), with a and b specified, is then assigned to α. Equations (4.24) and (4.26) constitute what is referred to as the *adaptive Kalman filter model*. Here the hierarchical feature is inherent in *both* the *observation equation* (4.24), and the *system equation* (4.26). In all the cases mentioned, the size of the hierarchy increases with i, $i = 1, 2, \ldots$, imparting an increasing memory to the process.

The non-Gaussian Kalman filter model of Section 3.4.2 does not require a scaling of the T_is, and assumes the following as observation and system equations, respectively,

$$(T_i \mid \theta_i, \omega_i) \sim \mathcal{G}(\theta_i, \pi_i), \text{ and} \tag{4.27}$$

$$\theta_i = \frac{\theta_{i-1}}{C_i} \epsilon_i. \tag{4.28}$$

The ϵ_is are assumed independent, and each ϵ_i has a beta distribution with parameters σ_{i-1} and ν_{i-1}. The hyperparameters ω_i, C_i, σ_i, and ν_i are assumed known and must satisfy the constraint $\sigma_{i-1} + \omega_i = \sigma_i + \nu_i$, $i = 2, 3, \ldots$. The initial (starting) value θ_0 is assumed to have a gamma distribution with scale parameter $\sigma_0 + \nu_0$, and shape parameter u_0, also assumed known. As written previously, the model consists of an excessive number of parameters that a user needs to specify; this is not practical. A simplification would be to let $C_i = C$, $\omega_i = \omega$, $\sigma_i = \sigma$, and $\nu_i = \nu$. The hierarchical nature of this model is due to the dynamic feature in the system equation (4.28).

In Sections 4.5.1 through 4.5.3 we discuss inferential aspects of the three models described previously, and then apply our procedures to a common set of data on software failures. This facilitates a comparison of the inferential and predictive capabilities of the three models. The actual data are given in column 2 of Table 4.7; they have been taken from Musa (1979), who has labeled them as "System 40 Data." The data consist of 100 interfailure times of a software system (comprised of 180,000 object instructions) which was undergoing conventional testing (as opposed to testing under an operational profile). There was an overlap between integration and system testing, and the earlier stages of testing were conducted with only a part of the system present (personal communication with Musa). A plot of entries in column 2 of Table 4.7 is shown in Figure 4.5. The large fluctuations towards the end of the data could be attributed to the introduction of the missing part of the system. To gain a better appreciation of the variability in the interfailure times, we plot their logarithms; these are shown in Figure 4.6.

The analysis of these data has proved to be challenging because of the absence of a discernible trend and the presence of wild fluctuations. Can we use the data to infer whether the debugging process that is subsequent to every failure is producing an improvement in reliability? Can we use the data to make meaningful predictions of the next time to failure? Do the models proposed here provide meaningful descriptions of the process that generates the data? If so, which of these models provides the best description? We propose to address these and related questions that may be germane to a software engineer's interests. For a general discussion on a paradigm for modeling reliability growth, see Singpurwalla (1998a).

4.5.1   Inference for the Random Coefficient Exchangeable Model

If we let $Y_i = \log_e T_i$, $i = 1, 2, \ldots$, then Equation (4.24) can be written as $Y_i = \theta_i Y_{i-1} + \epsilon_i$, $i = 1, 2, \ldots$; this is an autoregressive process of order one, with a random coefficient θ_i. As was mentioned in Section 3.4.1, θ_i provides information about the growth or decay in reliability at stage i, and since $\theta_i \sim \mathcal{N}(\lambda, \sigma_2^2)$, λ provides information about the *overall* growth or decay in reliability. If y_i denotes the realization of Y_i, then given the n interfailure times $y_1, \ldots, y_n$, interest centers around an assessment of θ_i and λ, given $\underline{y}^{(n)} = (y_1, \ldots, y_n)$. Interest also centers around the predictive distribution of Y_{n+1}.

An agreeable feature of the exchangeable model is that the relevant posterior and predictive distributions can be obtained in closed form. Specifically, the posterior distribution of λ, given the data $\underline{y}^{(n)}$, is of the form

$$(\lambda \mid \underline{y}^{(n)}, \bullet) \sim \mathcal{N}(m_n, s_n^2), \qquad (4.29)$$

where m_n and s_n^2 can be iteratively obtained as

132   4. Statistical Analysis of Software Failure Data

$$m_n = \frac{s_{n-1}^2 \, y_n \, y_{n-1} + m_{n-1} \, r_n}{s_{n-1}^2 \, y_{n-1}^2 + r_n},$$

$$s_n^2 = \frac{s_{n-1}^2 \, r_n}{s_{n-1}^2 \, y_{n-1}^2 + r_n}, \quad \text{and}$$

$$r_n = \sigma_2^2 \, y_{n-1}^2 + \sigma_1^2; \tag{4.30}$$

$m_0 = \mu$ and $s_0^2 = \sigma_3^2$ are the starting values of the iterative process.

Analogously, the posterior distribution of θ_i, given $\underline{y}^{(i)}$, $i = 1, 2, \ldots, n$, is of the form

$$(\theta_i \mid \underline{y}^{(i)}, \bullet) \sim \mathcal{N}(\widehat{\theta}_i, \Sigma_i^2), \tag{4.31}$$

where

$$\widehat{\theta}_i = \frac{\sigma_1^2 \, m_i + \sigma_2^2 \, y_i y_{i-1}}{r_i}, \quad \text{and}$$

$$\Sigma_i^2 = \frac{\sigma_1^2 \, (\sigma_1^2 \, s_i^2 + \sigma_2^2 \, r_i)}{r_i^2}. \tag{4.32}$$

Finally, the predictive distribution of Y_{n+1}, given $\underline{y}^{(n)}$ is specified via the relationship

$$(Y_{n+1} \mid \underline{y}^{(n)}, \bullet) \sim \mathcal{N}(m_n y_n, y_n^2 \, s_n^2 + r_{n+1}). \tag{4.33}$$

The details leading us to Equations (4.29) through (4.33) are relatively straightforward; they are based on elementary properties of Gaussian distributions. An interested reader may wish to develop them directly, or may consult Soyer (1985) to fill in the appropriate gaps.

Column 3 of Table 4.7 shows the logarithms (to base e) of the inter-failure times given in column 2, and column 4 gives the means of the one-step-ahead predictive distributions, that is, the quantities $m_i y_i$, $i = 1, 2, \ldots$, of Equation (4.33). In computing the entries of column 4, the following values of the hyperparameters were used: $\sigma_1^2 = \sigma_2^2 = 1$, $\sigma_3^2 = 0.25$, and $\mu = 1$. Figures 4.7 and 4.8 show plots of $\widehat{\theta}_i$ and m_i, the means of the posterior distributions of θ_i and λ, respectively, for $i = 1, 2, \ldots, 100$. Figure 4.7 reveals the lack of a consistent pattern of growth in reliability from one stage of testing to the other. Figure 4.8 shows that, overall, there is a very modest growth in reliability. Will an analysis of these data using the adaptive Kalman filter model, or the non-Gaussian model, reveal conclusions different from the preceding? We explore this matter in the following sections.

Table 4.7. Actual and Predicted Interfailure Times for System 40 Data [from Musa (1979)]

Failure Number	Observed Interfailure Times T_i	$Y_i = \ln(T_i)$	One-Step-Ahead Predicted Times Y_i or T_i Using:				
			Exchangeable Model (Y_i)	Adaptive Model (Y_i)	Non-Gaussian Kalman Filter Model (T_i)	Concatenated Failure Rate Model (T_i)
1	14390	9.574288800	—	—	—	12544.81
2	9000	9.104979856	10.80810	—	22058.89	9262.62
3	2880	7.965545573	10.00950	7.2985	23924.20	8130.37
4	5700	8.648221454	8.50250	5.9005	20255.50	11899.77
5	21800	9.989665249	9.250900	8.4024	25632.41	16959.86
6	26800	10.196157166	10.77910	10.6448	48750.13	54075.92
7	113540	11.639910477	10.94350	9.5254	88110.80	87617.82
8	112137	11.627476617	12.56530	12.4189	244956.63	61459.67
9	660	6.492239835	12.47370	10.7737	290003.82	46412.19
10	2700	7.901007052	6.70740	3.2575	110755.53	46080.66
11	28793	10.267887581	8.26530	8.8024	65888.62	37728.04
12	2173	7.683863980	10.91380	12.5537	46894.56	31922.12
13	7263	8.890548246	8.01610	5.2074	21388.95	29557.00
14	10865	9.293301893	9.33400	9.4330	16430.52	26816.24
15	4230	8.349957272	9.75450	8.9503	16066.28	25008.51
16	8460	9.043104453	8.69790	7.8826	11269.56	24830.08
17	14865	9.606764735	9.43860	9.0624	13728.46	23353.44
18	11844	9.379576689	10.03100	9.4851	19914.57	22963.70
19	5361	8.586905804	9.76250	8.5066	19629.38	

4.5 Inference and Predictions Using Dynamic Models

134   4. Statistical Analysis of Software Failure Data

Table 4.7. Continued

Failure Number	Observed Interfailure Times T_i	$Y_i = \ln(T_i)$	One-Step-Ahead Predicted Times Y_i or T_i Using:				Concatenated Failure Rate Model(T_i)
			Exchangeable Model (Y_i)	Adaptive Model (Y_i)	Kalman Filter Model (T_i)	Non-Gaussian Model (T_i)	
20	6553	8.787678239	8.89640	7.3107	13669.69	21381.85
21	6499	8.779403598	9.09980	8.3976	12744.48	19919.09
22	3124	8.046869511	9.07840	8.2151	11043.04	19904.28
23	51323	10.845894274	8.28470	6.9043	8872.73	21034.62
24	17010	9.741556685	11.29370	13.8507	51652.02	21053.48
25	1890	7.544332108	10.09390	8.1986	38217.43	19836.23
26	5400	8.594154233	7.74890	5.4473	17364.48	19565.52
27	62312	11.039909303	8.85920	9.1674	14942.98	21296.41
28	24826	10.119646770	11.47060	13.4389	64362.37	21987.84
29	26335	10.178654132	10.47560	8.7164	53669.84	21919.71
30	363	5.894402834	10.52760	9.6213	45508.51	21607.46
31	13989	9.546026585	6.01740	3.1938	21629.86	21327.26
32	15058	9.619664690	9.90590	14.3890	22453.09	20352.53
33	32377	10.385203573	9.97430	9.0268	24529.41	21329.76
34	41362	10.630117864	10.78000	10.4125	42373.38	22446.54
35	4160	8.333270353	11.03020	10.1203	56608.08	21457.42
36	82040	11.314962212	8.59260	6.0592	31062.48	23777.19
37	13189	9.487138428	11.75920	14.3135	89922.38	23509.34

4.5 Inference and Predictions Using Dynamic Models

Table 4.7. Continued

Failure Number	Observed Interfailure Times T_i	$Y_i = \ln(T_i)$	One-Step-Ahead Predicted Times Y_i or T_i Using:			Concatenated Failure Rate Model(T_i)
			Exchangeable Model (Y_i)	Adaptive Model (Y_i)	Non-Gaussian Kalman Filter Model (T_i)	
38	3426	8.139148679	9.81290	7.3783	52177.42	22419.51
39	5833	8.671286727	8.38440	6.4558	24836.98	22316.72
40	640	6.461468176	8.93980	8.5647	14455.37	21382.76
41	640	6.461468176	6.61960	4.4709	5929.40	20742.64
42	2880	7.965545573	6.61620	5.9634	3246.33	19928.91
43	110	4.700480366	8.19190	9.1343	3563.77	19301.98
44	22080	10.002427500	4.79040	2.5915	1699.38	19494.67
45	60654	11.012940864	10.41770	19.3881	19390.86	20754.35
46	52163	10.862128710	11.48350	11.0904	59924.55	21656.86
47	12546	9.437157169	11.31400	9.7274	69002.42	21525.54
48	784	6.664409020	9.79770	7.4402	37764.68	21065.59
49	10193	9.229456489	6.87640	4.2686	16920.65	20569.62
50	7841	8.967121656	9.58380	11.5589	15401.10	20237.79
51	31365	10.353447901	9.30040	7.9377	13336.63	20585.99
52	24313	10.098766466	10.76030	10.8762	32765.46	20751.92
53	298890	12.607830892	10.48520	8.9795	35197.25	27352.23
54	1280	7.154615357	13.13550	14.3821	267522.66	26741.66
55	22099	10.003287638	7.39520	3.6967	114274.11	26546.88

136 4. Statistical Analysis of Software Failure Data

Table 4.7. Continued

Failure Number	Observed Interfailure Times T_i	$Y_i = \ln(T_i)$	One-Step-Ahead Predicted Times Y_i or T_i Using:				
			Exchangeable Model (Y_i)	Adaptive Model (Y_i)	Non-Gaussian Kalman Filter Model (T_i)		Concatenated Failure Rate Model(T_i)
56	19150	9.860057995	10.40110	12.6137	65709.57		26151.42
57	2611	7.867488569	10.24330	8.8370	41151.91		25627.82
58	39170	10.575666427	8.14210	5.6897	20532.78		25984.09
59	55794	10.929421616	10.99740	12.8783	42022.25		27061.26
60	42632	10.660360424	11.36410	10.2797	64982.47		27803.60
61	267600	12.497248607	11.07360	9.4502	65464.91		34390.11
62	87074	11.374513611	133.00740	13.3468	254521.68		37547.57
63	149606	11.915760451	11.81620	9.4337	182153.15		41571.66
64	14400	9.574983486	12.38010	11.3776	234574.95		40917.79
65	34560	10.450452223	9.91470	7.0159	96365.12		41995.24
66	39600	10.586584397	10.82970	10.3937	69572.82		42686.65
67	334395	12.720078208	10.96730	9.8073	65522.52		53025.81
68	296105	12.598469400	13.20730	14.0040	316144.10		62481.41
69	177355	12.085908653	13.07230	11.4408	390860.27		67672.63
70	214622	12.276633620	12.52770	10.6270	329747.45		74798.96
71	156400	11.960172107	12.72190	11.4427	330965.95		83266.74
72	166800	12.024550769	12.38400	10.7036	280318.57		87694.62
73	10800	9.287301413	12.44590	11.1151	254832.05		85745.95

4.5 Inference and Predictions Using Dynamic Models

Table 4.7. Continued

Failure Number	Observed Interfailure Times T_i	$Y_i = \ln(T_i)$	One-Step-Ahead Predicted Times Y_i or T_i Using:			
			Exchangeable Model (Y_i)	Adaptive Model (Y_i)	Non-Gaussian Kalman Filter Model (T_i)	Concatenated Failure Rate Model (T_i)
74	267000	12.495003937	9.58090	6.5952	123497.40	92155.66
75	2098833	14.556892034	12.94030	15.4497	291172.69	158802.07
76	694080	13.450342507	15.09996	15.6550	1935391.90	177181.62
77	7680	8.946374826	13.80440	11.2419	1351849.40	176851.52
78	6269667	15.651233801	9.22570	5.5249	629965.61	249640.89
79	2948700	14.896874953	15.35580	22.4572	6591260.15	351333.28
80	187200	12.139932843	15.46920	13.8009	3667195.44	352901.75
81	18000	9.798127037	12.57370	9.0674	1677352.01	339966.29
82	178200	12.090661794	10.12180	7.2440	731019.82	343132.37
83	487800	13.097660765	12.51840	13.6803	477826.72	358802.80
84	639200	13.367972673	13.56830	13.0569	642735.98	376739.63
85	334560	12.720571515	13.84590	12.5528	813990.24	386330.09
86	1468800	14.199956299	13.16330	11.1382	651416.14	442939.91
87	86720	11.370439817	14.69870	14.5875	1555295.16	427873.68
88	199200	12.202064624	11.74690	8.3918	729633.94	434482.23
89	215200	12.279323107	12.61140	12.0519	482663.53	433168.49
90	86400	11.366742955	12.68760	11.4003	381221.35	426616.78
91	88640	11.392338502	11.73160	9.7078	235176.38	408154.91

Table 4.7. Continued

Failure Number	Observed Interfailure Times T_i	$Y_i = \ln(T_i)$	One-Step-Ahead Predicted Times Y_i or T_i Using:			
			Exchangeable Model (Y_i)	Adaptive Model (Y_i)	Non-Gaussian Kalman Filter Model (T_i)	Concatenated Failure Rate Model(T_i)
92	1814400	14.411265393	11.75450	10.5370	182481.61	474234.22
93	4160	8.333270353	14.90450	16.8667	1634926.93	456991.32
94	3200	8.070906089	8.57910	4.4545	678584.40	448554.73
95	199200	12.202064624	8.30400	7.1640	287712.72	432261.69
96	356160	12.783135347	12.61370	16.0111	298391.67	444069.66
97	518400	13.158502424	13.21610	12.3861	431492.47	462669.64
98	345600	12.753037316	13.60360	12.5033	626867.43	471487.66
99	31360	10.353288474	13.17630	11.4137	551257.58	462108.57
100	265600	12.489746697	10.67460	7.7609	266743.05	468026.58

4.5 Inference and Predictions Using Dynamic Models   139

Observed Interfailure Times

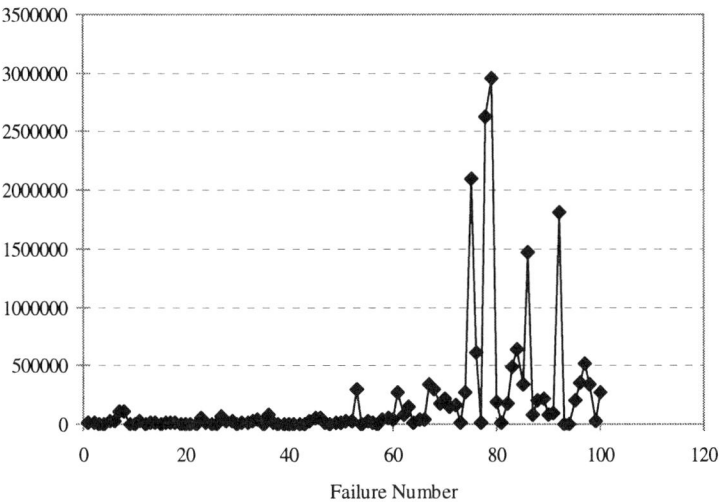

FIGURE 4.5. Plot of Interfailure Times—System 40 Data.

ln(Observed Interfailure Times)

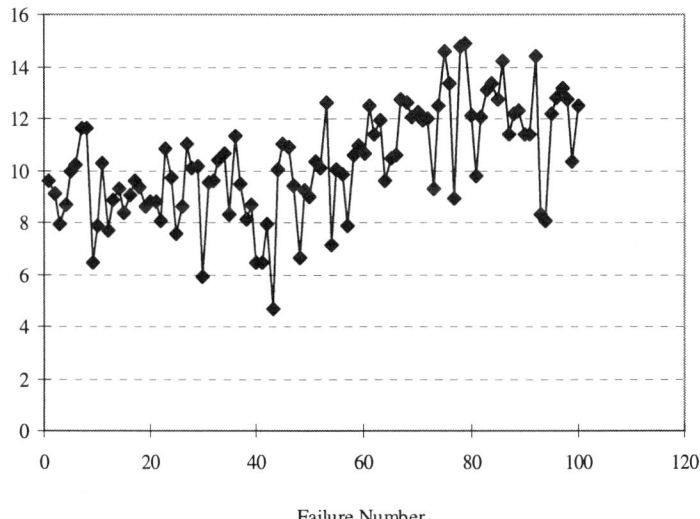

**FIGURE 4.6. Plot of Interfailure Times on a Logarithmic Scale
—System 40 Data.**

140   4. Statistical Analysis of Software Failure Data

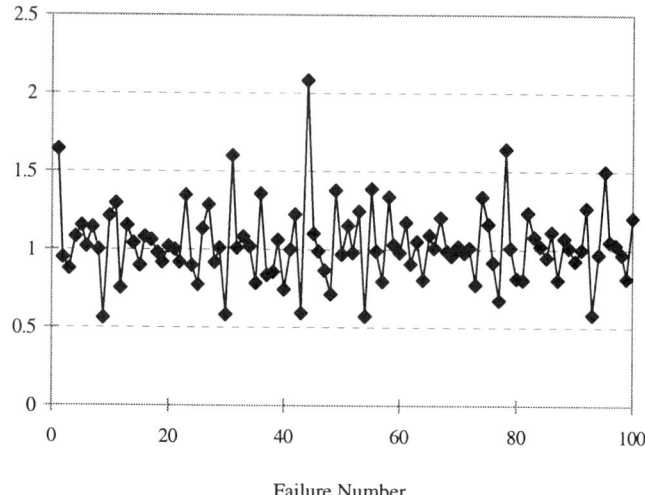

FIGURE 4.7. A Plot of the Posterior Means of θ_i Versus Failure Number for System 40 Data Using the Exchangeable Model.

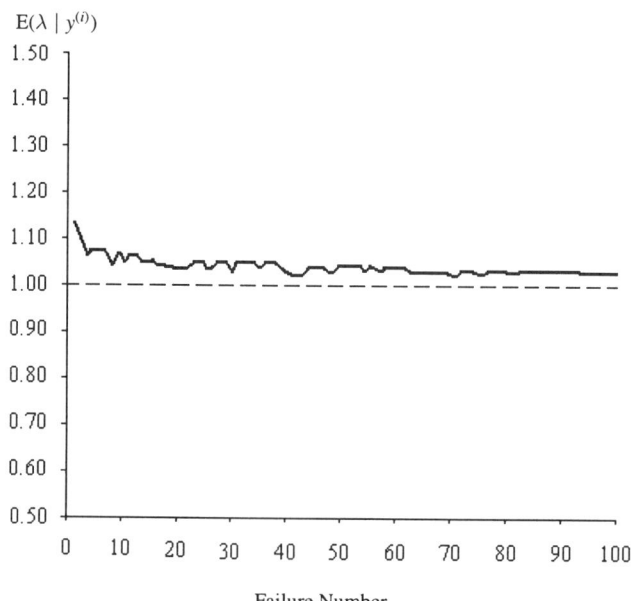

FIGURE 4.8. A Plot of the Posterior Mean of λ for System 40 Data

4.5.2 Inference for the Adaptive Kalman Filter Model

The adaptive Kalman filter model as prescribed by Equations (4.24) and (4.26) imposes a dependence structure on the θ_is that is stronger than the one prescribed by Equation (4.25) of the exchangeable model. This is so because a value of $\alpha > 1$ suggests an increasing sequence of θ_is, and this in turn implies a steady growth in reliability; the opposite is true for $\alpha < 1$. As to whether such a specific structure is justified is a matter of an analyst's judgment. Following the setup of Section 4.5.1, we let $Y_i = \log_e T_i$, $i = 1, 2, \ldots$, and assume that n interfailure times $\underline{y}^{(n)} = (y_1, \ldots, y_n)$ have been observed. Were α to be specified (i.e., assumed known), then the posterior and the predictive distributions of θ_i and Y_{n+1}, respectively, are Gaussian. Specifically, for $i = 1, \ldots, n$

$$(\theta_i \mid \underline{y}^{(i)}, \alpha, \bullet) \sim \mathcal{N}(\widehat{\theta}_i, \mathbf{\Sigma}_i^2), \text{ and} \qquad (4.34)$$

$$(Y_{i+1} \mid \underline{y}^{(i)}, \alpha, \bullet) \sim \mathcal{N}(\alpha y_i \widehat{\theta}_i, y_i^2 \, r_{i+1} + \sigma_1^2), \text{ where} \qquad (4.35)$$

$$\widehat{\theta}_i = \frac{\alpha \sigma_1^2 \widehat{\theta}_{i-1} + r_i \, y_i y_{i-1}}{y_{i-1}^2 \, r_i + \sigma_1^2},$$

$$\mathbf{\Sigma}_i^2 = \frac{r_i \, \sigma_1^2}{y_{i-1}^2 \, r_i + \sigma_1^2}, \text{ and}$$

$$r_i = \alpha^2 \, \mathbf{\Sigma}_{i-1}^2 + W_i^2, \qquad (4.36)$$

with the starting values $\theta_0 = \widehat{\theta}_0$, and $\mathbf{\Sigma}_0^2$ specified in advance.

The given closed form results are no longer valid when α cannot be specified. When such is the case, one possibility is to run the prescribed model for different values of α, and choose that selection which provides the best predictivity. A formal approach, however, is to assign a prior distribution on α, and then to approximate the ensuing results either via a simulation (see Section A.3.3), or via a scheme such as the one suggested by Lindley (1980). Singpurwalla and Soyer (1985) have done the latter assuming a uniform prior for α over [-2, +2]; the details are too cumbersome to reproduce here. However, their results on the means of the one-step-ahead predictive distributions of Y_i, $i = 3, 4, \ldots, 100$, are given in column 5 of Table 4.7, and plots of the means of the posterior distributions of θ_i are given in Figure 4.9. A visual comparison of Figures 4.7 and 4.9 does not reveal any noticeable differences between the two plots. The entries in columns 4 and 5 of Table 4.7 enable us to compare the predictive abilities of the exchangeable model versus the adaptive Kalman filter model. However, this can be formally done; see Section 4.6.3, where it is argued

142   4. Statistical Analysis of Software Failure Data

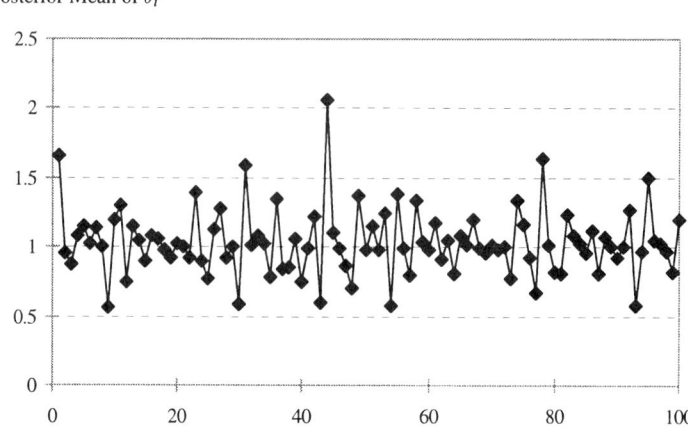

FIGURE 4.9. Posterior Means of θ_i Versus Failure Number for System 40 Data Using Adaptive Model.

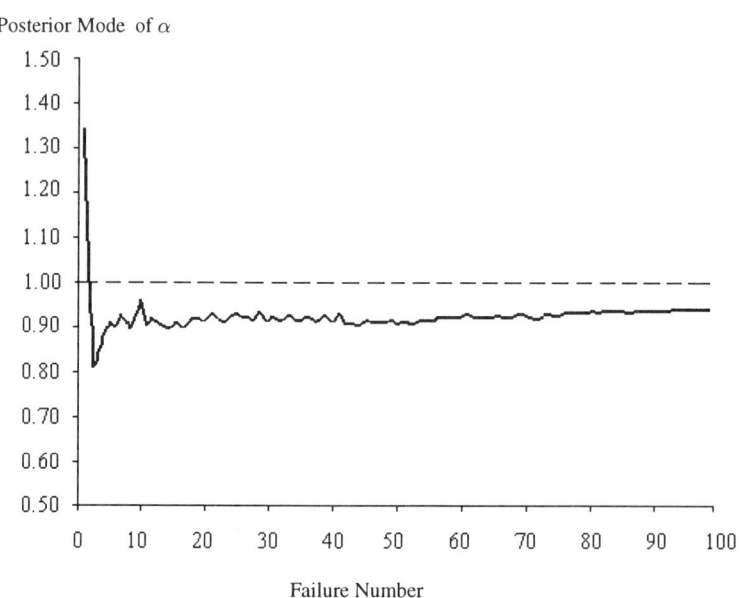

FIGURE 4.10. Posterior Mode of α Versus Failure Number for System 40 Data Using Adaptive Model.

that the exchangeable model provides better predictivity than the adaptive model.

The modal values of the posterior distribution of α, as a function of i, $i = 1, 2, \ldots, 100$, are shown in Figure 4.10. It suggests that the most likely value of α is (almost) always less than one. This means that there is an absence of a steady growth in reliability, a conclusion that is in mild contrast to that given by the exchangeable model. Recall that the latter suggested a modest overall growth in reliability. Could it be that the assumption of dependence in the parameters through autoregression is too strong for these data?

4.5.3   Inference for the Non-Gaussian Kalman Filter Model

Even though the model specified by Equations (4.27) and (4.28) is non-Gaussian, closed form results for the posterior and predictive distributions can be produced, provided that C is known. Specifically, suppose that n interfailure times $t_1, \ldots, t_n$, are observed; then given $\underline{t}^{(n)} = (t_1, \ldots, t_n)$, the posterior distribution of θ_i, $i = 1, 2, \ldots, n$, is of the form

$$(\theta_i \mid \underline{t}^{(i)}, C, \bullet) \sim \mathcal{G}(u_i, \sigma_{i-1} + \nu_i), \qquad (4.37)$$

where $u_i = Cu_{i-1} + t_i$; recall that u_0 is the shape parameter of the gamma distribution of θ_0. It is also shown that

$$(\theta_{n+1} \mid \underline{t}^{(i)}, C, \bullet) \sim \mathcal{G}(Cu_i, \sigma_i). \qquad (4.38)$$

Similarly, given $\underline{t}^{(i)}$, the predictive distribution of T_{i+1} has a density at t of the form

$$\mathcal{P}(t \mid \underline{t}^{(i)}, C, \bullet) \propto \frac{t^{\omega_{(i+1)}-1}}{(Cu_i+t)^{\sigma_i+\omega_{i+1}}}. \qquad (4.39)$$

When $\omega_i = 1$, the observation equation is governed by an exponential distribution, and Equation (4.39) is a Pareto density. The development of Equations (4.37) to (4.39) are left as an exercise for the reader; they can, however, be found in Chen and Singpurwalla (1994).

Assessing Reliability Growth

Consider a special case of the foregoing model, with $\omega_i = \nu_i = \sigma_i = 2$ for all values of i. Verify that the mean of the predictive distribution of T_{i+1}, conditional on C, is

$$E(T_{i+1} \mid \underline{t}^{(i)}, \bullet) = 2C \sum_{j=0}^{i+1} C^j t_{(i+1-j)}. \qquad (4.40)$$

Clearly, the value of C is crucial for determining whether the times between failure are expected to increase or decrease. Specifically $C > 1$ would suggest a strong growth in reliability, whereas C close to zero would imply the reverse. Intermediate values of C would indicate a growth or decay, depending on the values of t_i; see Appendix B of Chen and Singpurwalla (1994). Thus to assess whether the software is experiencing a growth or decay in reliability, it is necessary to make inferences about C. Accordingly, we assign a uniform on (0,1) as a prior distribution on C. If we have prior notions about growth or decay in reliability, a prior such as a beta may be entertained. Unfortunately, allowing C to be unknown destroys the closed form nature of the predictive and the posterior distributions. One way to overcome this difficulty is via a Markov Chain Monte Carlo simulation of the inferential mechanism; see Section A.3.4 of Appendix A. Alternatively, we may discretize the uniform distribution of C at k points so that

$$\mathcal{P}(C = \tfrac{j}{k-1}) = \tfrac{1}{k}, \ j = 0, 1, \ldots, (k-1),$$

and given $\underline{t}^{(i)}$, $i = 1, 2, \ldots, n$, compute its posterior distribution

$$\mathcal{P}\left(C = \tfrac{j}{k-1} \mid \underline{t}^{(i)}\right), \quad \text{where}$$

$$\mathcal{P}\left(C = \tfrac{j}{k-1} \mid \underline{t}^{(i)}\right) \propto \mathcal{P}(t \mid \underline{t}^{(i-1)}, \bullet) \, \mathcal{P}\left(C = \tfrac{j}{k-1} \mid \underline{t}^{(i-1)}\right).$$

The first expression on the right-hand side of the preceding equation is the likelihood; it is obtained by replacing the C in Equation (4.39) by $j/(k-1)$. The second expression is the posterior of C at $j/(k-1)$ given the data $\underline{t}^{(i-1)}$; for $i = 1$, the quantity $\mathcal{P}(C = j/(k-1) \mid \underline{t}^{(0)})$ is simply the prior $1/k$.

Once the posterior distribution of C has been computed, by repeating the procedure described for $j = 0, 1, \ldots, (k-1)$, the posterior distribution of θ_i, and the predictive distribution of T_{i+1} can be obtained by averaging out C in the Equations (4.37) and (4.39). The averaging will be done numerically, and with respect to the posterior distribution of C.

For an analysis of the interfailure time data given in column 2 of Table 4.7, C was discretized at 200 points, and the hyperparameters were chosen as $\omega_i = \nu_i = \sigma_i = 2$ and $u_0 = 500$. In Figure 4.11 we show a plot of the mean of the posterior distribution of C. It has been noted [see Chen and Singpurwalla (1994)] that the posterior distribution is quite sharp, and has a mean of about 0.425. This value of the mean is attained after about 15 iterations, and remains stable thereafter. With C being in the vicinity of 0.425, we cannot conclusively claim evidence either for or against growth in reliability. Column 6 of Table 4.7 gives the mean of the one-step-ahead predictive distribution of T_{i+1}, $i = 1, 2, \ldots$, 100. For assessing the predictive performance of the non-Gaussian Kalman filter model, we need to compare the entries in column 6 against those in column

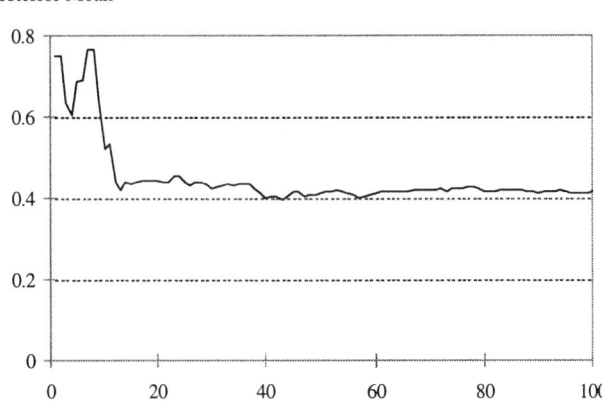

FIGURE 4.11. Mean of the Posterior Distribution of C for System 40 Data Using the Non-Gaussian Model.

2. A formal comparison of the predictive performance of this model versus the adaptive Kalman filter model of the previous section is described in Section 4.6.3; it shows a superiority of the non-Gaussian model over the adaptive model. Thus it appears that of the three dynamic models considered here, it is the non-Gaussian Kalman filter model of this section that provides the best predictivity. However, none of the models reveals strong evidence either for or against growth in reliability. This type of information is useful to a manager of the software development effort who is required to make decisions about when to stop testing and whether to make changes in the testing and debugging procedures. More on reliability growth for these data is discussed later, in Section 4.7.

4.6 Prequential Prediction, Bayes Factors, and Model Comparison

In the previous sections we discussed several models for describing software failure data. Each model provided a one-step-ahead predictive distribution for the time to next failure. The means of these predictive distributions can be compared with the actual data to obtain an assessment of the predictive ability of a proposed model. This point of view stems from the "Popperian" attitude of validating a model against data. The notion here is that any attempt at describing reality must be measured against empirical evidence, and be discarded if it proves inadequate [cf. Dawid (1992)]. Opposing this point of view is the (Bayesian) position which does not support the notion of the "absolute assessment" of a solo model [cf. Raftery (1992), who says, "You don't abandon a model unless you have a better one in hand"]. Also see Bernardo and Smith (1994), p. 419, who claim that " ... from a decision-theoretic point of view, the problem of accepting that a particular model is suitable, is ill defined unless an alternative is formally considered." However, the tradition of checking the adequacy of a given model without explicit consideration of an alternative is still attractive, partly because of the heritage of significance testing. A useful strategy is that given by Dawid (1992), who suggests testing for the goodness of fit via the statistic

$$Z_n = \frac{\sum_{i=1}^{n}(t_i - \mu_i)}{(\sum_{i=1}^{n} \sigma_i^2)^{\frac{1}{2}}},$$

where μ_i and σ_i^2 are the mean and the variance, respectively, of the predictive distribution of the observed t_is. Were the proposed model adequate, then under some mild conditions, the distribution of Z_n, as n gets large, is a Gaussian with mean 0 and variance 1. We do not pursue this tradition of testing for the goodness of fit of the models described before. Rather, we address the question of comparing the predictive performance of a proposed model versus one or more of its competitors. The purpose of this section is to describe methods by which models for tracking software failure data can be compared, and if appropriate, combined, so that better predictions are obtained.

4.6.1 Prequential Likelihoods and Prequential Prediction

As mentioned before, software failure data, like data from time series, arise sequentially. Thus, for example, if Y_i represents the time to failure of the ith version of the software, $i = 1, 2, \ldots$, and if y_i is a realization of Y_i, then we would observe y_1 first, y_2 next, y_3 subsequent to y_i, and so on. Given $y_1, y_2, \ldots, y_n$, which of the several software reliability models that are available should be used to predict Y_{n+1}? That is, which of the available models provides us with the "best" prediction of Y_{n+1}, given the data $y_1, \ldots, y_n$, where by best we mean closest to the actual observed values? There are several formal and informal approaches to model selection, an informal one being an examination of the mean square errors; see, for example, Table 4.5. In the following, we describe a formal approach.

Consider a model $\mathcal{M}$ that involves an unknown parameter θ on which a prior distribution $\mathcal{P}(\theta \mid \mathcal{H})$ has been assessed; assume that θ is continuous. Suppose that the data consist of n consecutive observations, $y_1, \ldots, y_n$, where y_i represents a failure time or a failure count, that is, the number of failures in a specified interval of time. Note that, in principle, the data need not be consecutively observed, although in the context of software failures this will

4.6 Prequential Prediction, Bayes Factors, and Model Comparison

naturally be so. At any stage of the analysis, say the ith, we may use $y_1, \ldots, y_{i-1}$, to predict Y_i via its predictive distribution [cf. Equation (4.7)], suitably annotated to reflect dependence on $\mathcal{M}$, as

$$f_{Y_i}(y \mid y_1, \ldots, y_{i-1}, \mathcal{M}, \mathcal{H}) = \int_\theta f_{Y_i}(y \mid \theta, \mathcal{M}) \, \mathcal{P}(\theta \mid y_1, \ldots, y_{i-1}, \mathcal{H}) d\theta;$$

note that when $i = 1$, $\mathcal{P}(\theta \mid y_1, \ldots, y_{i-1}, \mathcal{H})$, the posterior distribution of θ, is simply its prior $\mathcal{P}(\theta \mid \mathcal{H})$.

Once Y_i gets observed as y_i, the left-hand side of the preceding expression becomes the likelihood for $\mathcal{M}$ under y_i, so that the product, called the *prequential likelihood*,

$$\mathcal{L}_n(\mathcal{M}; y_1, \ldots, y_n) \stackrel{\text{def}}{=} \prod_{i=1}^{n} f_{Y_i}(y_i \mid y_1, \ldots, y_{i-1}, \mathcal{M}, \mathcal{H}), \quad (4.41)$$

is the likelihood for the model $\mathcal{M}$ under $y_1, \ldots, y_n$. If, for each observation, our model were able to predict the data well, then each term on the right-hand side of Equation (4.41) would be large and so would $\mathcal{L}_n(\mathcal{M}; y_1, \ldots, y_n)$. The taking of a product in the preceding expression is motivated by the joint predictive density at $(y_1^*, \ldots, y_n^*)$ of the observables

$$f(y_1^*, \ldots, y_n^* \mid \mathcal{M}, \mathcal{H}) = \prod_{i=1}^{n} f_{Y_i}(y_i^* \mid y_1^*, \ldots, y_{i-1}^*, \mathcal{M}, \mathcal{H}); \quad (4.42)$$

it is called the *prequential prediction* [see Dawid (1984)].

The prequential likelihood can be used as a basis for comparing the predictive performance of two models, say $\mathcal{M}_1$ and $\mathcal{M}_2$; see Roberts (1965). For this we need to compute the *prequential likelihood ratio*

$$\mathcal{R}_n(\mathcal{M}_1, \mathcal{M}_2; y_1, \ldots, y_n) \stackrel{\text{def}}{=} \frac{\mathcal{L}_n(\mathcal{M}_1; y_1, \ldots, y_n)}{\mathcal{L}_n(\mathcal{M}_2; y_1, \ldots, y_n)}. \quad (4.43)$$

If $\mathcal{R}_n(\mathcal{M}_1, \mathcal{M}_2; y_1, \ldots, y_n)$ is greater than one, then the evidence at hand, namely, $y_1, \ldots, y_n$ (and also $\mathcal{H}$), suggests that model $\mathcal{M}_1$ performs *relatively better* than model $\mathcal{M}_2$, and vice versa if the preceding ratio is less than one. The magnitude of the value of $\mathcal{R}_n(\mathcal{M}_1, \mathcal{M}_2; y_1, \ldots, y_n)$ indicates the degree to which $\mathcal{M}_1$ outperforms $\mathcal{M}_2$, and vice versa. How big should $\mathcal{R}_n(\bullet)$ be in order for us to judge the superiority of $\mathcal{M}_1$ over $\mathcal{M}_2$? Jeffreys (1961, Appendix B) provides some ground rules for doing this; these are given in Table 4.8.

It is useful to note that there is no assurance that $\mathcal{R}_n(\bullet)$ will continue to be greater than or less than one, as a function of n, for $n = 1, 2, \ldots$. Rather, $\mathcal{R}_n(\bullet)$ can fluctuate above and below one, the fluctuations reflecting the

Table 4.8. Strength of Evidence Provided by $\mathcal{R}_n(\mathcal{M}_1, \mathcal{M}_2; y_1, \ldots, y_n)$

$R_n(\mathcal{M}_1, \mathcal{M}_2; y_1, \ldots, y_n)$	Strength of evidence
< 0.01	Decisively against $\mathcal{M}_1$
0.01 to 0.1	Strongly against $\mathcal{M}_1$
0.1 to 0.32	Substantially against $\mathcal{M}_1$
0.32 to 3.2	Neither for nor against $\mathcal{M}_1$
3.2 to 10	Substantially for $\mathcal{M}_1$
10 to 100	Strongly for $\mathcal{M}_1$
> 100	Decisively for $\mathcal{M}_1$

changing nature of evidence, for or against $\mathcal{M}_1$, as a function of n. Increasing values of $\mathcal{R}_n(\bullet)$ suggest an accumulation of evidence in favor of $\mathcal{M}_1$ over $\mathcal{M}_2$, with $y_1, \ldots, y_n$, for $n = 1, 2, \ldots$.

Model comparison based on $\mathcal{R}_n(\bullet)$ alone, as discussed previously, can be criticized on two grounds: it does not have a justification within the calculus of probability, and the proposed approach offers a mechanism for comparing any two models at a time. How should one proceed when faced with the selection of a model among more than two models? This, after all, is the circumstance under which software engineers often operate. These concerns can be addressed via the notion of Bayes' factors and posterior weights. Also germane to this discussion is the notion of model averaging; these topics are discussed next.

4.6.2 Bayes' Factors and Model Averaging

We start our discussion by considering the case of two models $\mathcal{M}_1$ and $\mathcal{M}_2$. The comparing of $\mathcal{M}_1$ and $\mathcal{M}_2$ can be thought of as a test between two hypotheses as to which of the two models is the better descriptor of the data that will be generated. Suppose that before observing the data, we assign a weight π to $\mathcal{M}_1$ and a weight $(1 - \pi)$ to $\mathcal{M}_2$, for $0 < \pi < 1$. We may interpret π as our prior probability that $\mathcal{M}_1$ is the better of the two models as a descriptor of the data to be generated. The quantity $\pi/(1-\pi)$ is known as the *prior odds* for $\mathcal{M}_1$. Upon receipt of y_1, we can use Bayes' Law to change π to π_1, and then π_1 to π_2 upon receipt of y_2, and so on. In general, we can show that the *posterior odds* for $\mathcal{M}_1$, $\pi_i/(1-\pi_i)$, $i = 1, 2, \ldots$, is related to its prior odds via the expression

$$\frac{\pi_i}{1-\pi_i} = \mathcal{R}_i(\mathcal{M}_1, \mathcal{M}_2; y_1, \ldots, y_i) \frac{\pi}{1-\pi}, \tag{4.44}$$

where $\mathcal{R}_i(\mathcal{M}_1, \mathcal{M}_2; y_1, \ldots, y_i)$ is the prequential likelihood ratio.

The ratio of the posterior odds for $\mathcal{M}_1$ to its prior odds is called the *Bayes' factor* in favor of $\mathcal{M}_1$; clearly, the Bayes factor is simply the prequential likelihood ratio. When $\pi = 0.5$, the Bayes' factor in favor of $\mathcal{M}_1$ is simply the posterior odds in favor of $\mathcal{M}_1$ over $\mathcal{M}_2$, and that

$$\pi_i = \frac{\mathcal{R}_i(\mathcal{M}_1, \mathcal{M}_2; y_1, \ldots, y_i)}{1 + \mathcal{R}_i(\mathcal{M}_1, \mathcal{M}_2; y_1, \ldots, y_i)}. \tag{4.45}$$

In general, it is easy to verify, using Bayes' Law, that the *posterior weight* π_i is of the form

$$\pi_i = \frac{\pi \times \mathcal{L}_i(\mathcal{M}_1; y_1, \ldots, y_i)}{\pi \times \mathcal{L}_i(\mathcal{M}_1; y_1, \ldots, y_i) + (1-\pi) \times \mathcal{L}_i(\mathcal{M}_2; y_1, \ldots, y_i)}. \tag{4.46}$$

The posterior weight can be considered as a measure of the relative performance of the two models, on the basis of the first i observations. If π_i is greater than 0.5, then $\mathcal{M}_1$ is judged superior to $\mathcal{M}_2$; vice versa, otherwise. The connection between π_i and $\mathcal{R}_i(\bullet)$ is apparent from Equation (4.46). The former, which has a motivation within the calculus of probability, is a transformation of the latter, and is restricted to the interval (0,1).

The preceding idea extends easily to the case of k models, $\mathcal{M}_1, \ldots, \mathcal{M}_k$. Let $\pi^{(j)}$ be the prior weight assigned to model $\mathcal{M}_j$, with $0 < \pi^{(j)} < 1$, and $\sum_{j=1}^{k} \pi^{(j)} = 1$. Then, Equation (4.46) generalizes to give the posterior weight

$$\pi_i^{(j)} = \frac{\pi^{(j)} \times \mathcal{L}_i(\mathcal{M}_j; y_1, \ldots, y_i)}{\sum_{j=1}^{k} \pi^{(j)} \times \mathcal{L}_i(\mathcal{M}_j; y_1, \ldots, y_i)}. \tag{4.47}$$

When we are faced with the task of selecting *one* among the k proposed models, then we will choose that model which has the largest posterior weight. If, prior to observing the data, we have no preference for one model over another, then our prior weights will be $\pi^{(j)} = 1/k$, $j = 1, \ldots, k$.

The posterior weights not only provide us with a mechanism for model selection, but also come to play a role in the context of *model averaging*. To make the case for model averaging, suppose that in Table 4.8 we happen to arrive upon the scenario of the category "neither for nor against $\mathcal{M}_1$." That is, for all intents and purposes, the available evidence does not favor one model over the other vis-à-vis their predictivity. When such is the case predictions that are a weighted average of the predictions of the individual models make intuitive sense, and this is the idea behind model averaging. Another motivation for model averaging stems from the notion that "all models are useful, but some are more useful than the others," and so predictions that are a weighted average of the individual model predictions would be more encompassing than, and therefore superior to, the individual predictions. The weights assigned to the individual predictions are the posterior weights $\pi_i^{(j)}$ of Equation (4.47). Thus, given $y_1, \ldots, y_n$, we can use the law of total probability to argue that the predictive distribution of Y_{n+1} should be of the form

$$f_{Y_{n+1}}(y \mid y_1, \ldots, y_n, \mathcal{M}_1, \ldots, \mathcal{M}_k, \mathcal{H})$$
$$= \sum_{j=1}^{k} \pi_n^{(j)} \times f_{Y_{n+1}}(y \mid y_1, \ldots, y_n, \mathcal{M}_j, \mathcal{H}), \qquad (4.48)$$

rather than $f_{Y_{n+1}}(y \mid y_1, \ldots, y_n, \mathcal{M}_i, \mathcal{H})$, the predictive distribution associated with model $\mathcal{M}_i$ alone, when $\pi_n^{(j)}$ is the largest of the k posterior weights.

4.6.3 Model Complexity—Occam's Razor

Model complexity refers to how elaborate we have been in specifying a model. Since it is generally true that a model with more parameters will have a predictive advantage over a simpler model, particularly when the amount of data is small, one is tempted to select the most complex model. However, in statistical modeling, and in science as a whole, the principle of choosing a model that is as simple as possible has a high standing. This simplicity principle is known as "Occam's Razor," and may be stated as follows.

> *Model complexity must not be increased unless sufficiently justified in terms of improved observational prediction.*

In practice then, some sort of tradeoff between model complexity and predictive power is needed. It has been shown that the Bayesian approach to model selection that has just been described naturally incorporates Occam's razor [cf. Tversky, Lindley, and Brown (1979)]. How it does so is outside the scope of this book, but suffice to say that, in general, if we have two models—a complex $\mathcal{M}_1$ and a simpler $\mathcal{M}_2$—and both are assigned equal prior weight, then the posterior weight of $\mathcal{M}_1$ will be higher only if it shows a significant advantage in predictive ability over $\mathcal{M}_2$.

4.6.4 Application: Comparing the Exchangeable, Adaptive, and Non-Gaussian Models

The methodology of Sections 4.6.1 and 4.6.2 can be applied for comparing the predictive ability of the three dynamic models discussed in Section 4.5; a precedent here is the work of Raftery (1988). In what follows we perform a pairwise comparison of two models at a time, starting with the exchangeable model (henceforth, $\mathcal{M}_1$) and the adaptive model (henceforth $\mathcal{M}_2$) and then a comparison of $\mathcal{M}_1$ with the non-Gaussian model (henceforth $\mathcal{M}_3$). In all cases we assume that the prior odds is one; that is, a priori we have no preference for one model over the other.

For a comparison of $\mathcal{M}_1$ with $\mathcal{M}_2$, we first compute the prequential likelihood ratio $\mathcal{R}_{100}(\mathcal{M}_1, \mathcal{M}_2; y_1, \ldots, y_{100})$, using Equation (4.43), and the data in column 3 of Table 4.7. Note that to compute the prequential likelihood $\mathcal{L}_n(\mathcal{M}_1; y_1, \ldots, y_n)$, we must successively evaluate the predictive distribution, Equation (4.33), at the observed $y_1, y_2, \ldots, y_n$. Thus, for example, $\mathcal{L}_2(\mathcal{M}_1; y_1, y_2)$ is the Gaussian distribution $\mathcal{N}(m_1 y_1, y_1^2 s_n^2 + r_2)$ evaluated at y_2, and so on. Similarly, to compute the prequential likelihood $\mathcal{L}_n(\mathcal{M}_2; y_1, \ldots, y_n)$ we must use the predictive distribution of Equation (4.35). The computation of $\mathcal{R}_n(\mathcal{M}_1, \mathcal{M}_2; y_1, \ldots, y_n)$, although cumbersome, is relatively straightforward. For the data of Table 4.7, the prequential likelihood ratio turned out to be 490.50; this according to Jeffreys (see Table 4.8) would suggest a decisive evidence in favor of $\mathcal{M}_1$, the exchangeable model. Recall that with the prior odds of one, the prequential likelihood ratio (or the Bayes' factor in favor of $\mathcal{M}_1$) is also the posterior odds for $\mathcal{M}_1$.

It is interesting to monitor the behavior of $\mathcal{R}_n(\mathcal{M}_1, \mathcal{M}_2; y_1, \ldots, y_n)$ as a function of n. This indicates how evidence in favor of, or against, $\mathcal{M}_1$ evolves with the accumulation of data. Figure 4.12 shows a plot of $\mathcal{R}_n(\mathcal{M}_1, \mathcal{M}_2; y_1, \ldots, y_n)$ versus n, for the System 40 data. The predictive superiority of $\mathcal{M}_1$ over $\mathcal{M}_2$ is consistent, and clearly evident starting with about the fortieth failure time. In general, it need not be so that the plot of $\mathcal{R}_n(\bullet)$ versus n is always increasing (or decreasing) with n. It could, for example, decrease and then increase, suggesting that the initial evidence favors model $\mathcal{M}_2$ over $\mathcal{M}_1$, but then later on, the reverse is true; see Section 4.6.4 for an example.

When an exercise analogous to the preceding is performed to compare models $\mathcal{M}_3$ and $\mathcal{M}_1$ by computing

$$\mathcal{R}_n(\mathcal{M}_3, \mathcal{M}_1; t_1, \ldots, t_n, y_1, \ldots, y_n)$$

$$= \frac{\mathcal{L}_n(\mathcal{M}_3; t_1, \ldots, t_n)}{\mathcal{L}_n(\mathcal{M}_1; y_1, \ldots, y_n)}, \qquad (4.49)$$

152   4. Statistical Analysis of Software Failure Data

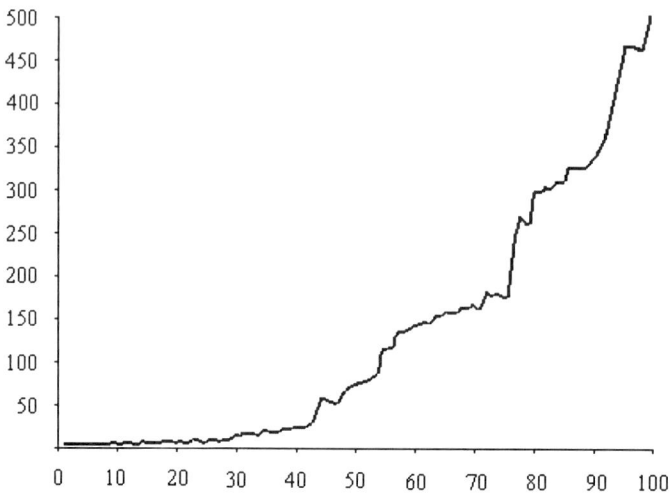

FIGURE 4.12. Prequential Likelihood Ratio $\mathcal{R}_n(\mathcal{M}_1, \mathcal{M}_2; \bullet)$ for System 40 Data.

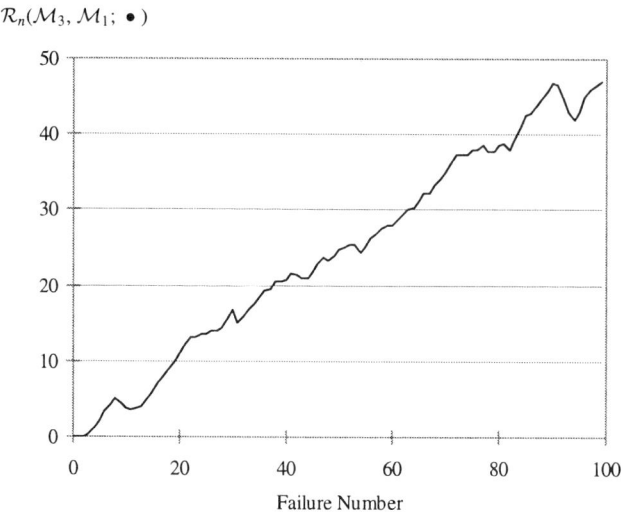

FIGURE 4.13. Logarithm of the Prequential Likelihood Ratio $\mathcal{R}_n(\mathcal{M}_3, \mathcal{M}_1; \bullet)$ for System 40 Data.

4.6 Prequential Prediction, Bayes Factors, and Model Comparison 153

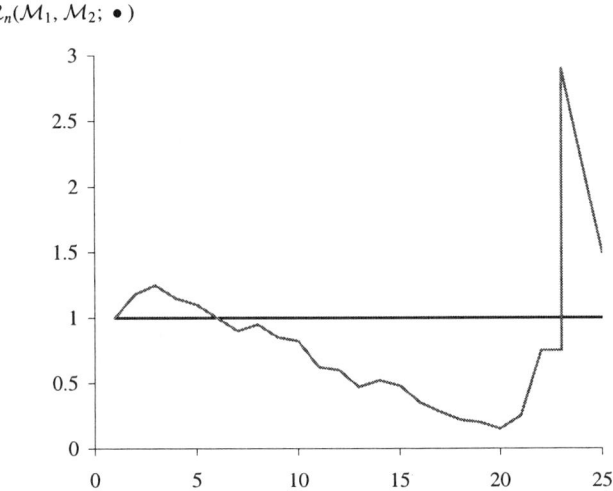

FIGURE 4.14. Prequential Likelihood Ratio $\mathcal{R}_n(\mathcal{M}_1, \mathcal{M}_2; \bullet)$ for the NTDS Data

it is found that the non-Gaussian model $\mathcal{M}_3$ clearly outperforms the exchangeable model $\mathcal{M}_1$. Figure 4.13 shows a plot of the logarithm of $\mathcal{R}_n(\mathcal{M}_3, \mathcal{M}_1; \bullet)$ versus n. Note that for computing $\mathcal{L}_n(\mathcal{M}_3; t_1, \ldots, t_n)$, the predictive distribution of Equation (4.39) is used. Observe that in computing (4.49) the numerator involves the actual observed data whereas the denominator involves the logarithms of the observed data. This asymmetry does not pose any problems because the prequential likelihoods are based on evaluations of predictive distributions.

4.6.5 *An Example of Reversals in the Prequential Likelihood Ratio.*

The plots of Figures 4.12 and 4.13 show a monotonic behavior of the prequential likelihood ratios suggesting the consistent superiority of one model over its competitor. To illustrate that this need not always be so, consider the NTDS data of Table 4.6, and Mazzuchi and Soyer's (1988) analysis of it using a hierarchical model (henceforth $\mathcal{M}_1$), and its special case (henceforth model $\mathcal{M}_2$); see Section 4.4.1. Recall that $\mathcal{M}_1$ is the more complex model; it involves five hyperparameters whereas $\mathcal{M}_2$ requires only three. $\mathcal{M}_1$ is specified via Equations (4.16) and (4.18), whereas $\mathcal{M}_2$ is its special case. The predictive distribution is given by Equation (4.22).

Figure 4.14 shows a plot of $\mathcal{R}_n(\mathcal{M}_1, \mathcal{M}_2; t_1, \ldots, t_n)$, the prequential likelihood ratio, as a function of the observed interfailure time t_n, $n = 1, \ldots, 26$. Observe that after initially favoring $\mathcal{M}_1$, $\mathcal{R}_n(\mathcal{M}_1, \mathcal{M}_2; \bullet)$ becomes less than one for most of the data, so that $\mathcal{M}_2$ is preferred. Both $\mathcal{M}_1$ and $\mathcal{M}_2$ provide good predictions, and so the selection procedure acts in according to Occam's law and chooses $\mathcal{M}_2$, the simpler of the two models. It is only when the surprising observation t_{24} (91 days) occurs that the more complex model $\mathcal{M}_1$ becomes favored again, presumably because of its greater flexibility. However, since t_{25} and t_{26} are more in line with the rest of the data, support for $\mathcal{M}_1$ over $\mathcal{M}_2$ begins to diminish and $\mathcal{R}_n(\mathcal{M}_1, \mathcal{M}_2; \bullet)$ begins to decrease. The example illustrates the role of $\mathcal{R}_n(\bullet)$ for comparing models in the presence of surprising evidence and its adherence to the principle of "Occam's Razor."

4.7 Inference for the Concatenated Failure Rate Model

In Section 3.6 we introduced a generic model for assessing software reliability growth that is potentially useful for applications other than software. As was pointed out in Section 3.6.2, the model is adaptive, has two parameters b and k, and possesses features that capture a software engineer's intuition and views about software failure and software quality. The model capitalizes upon some of the key features of existing software reliability models that have proved to be attractive. In particular, the model, as specified by a concatenation of failure rate functions given by Equation (3.28) exhibits the features reviewed in the following.

(a) For any fixed n, T_n has a decreasing failure rate.

(b) For any fixed n, $r_{T_n}(0)$ is the proportional intensity of failures up to time S_{n-1}.

(c) The failure rate takes an upward jump at S_n, if $(n-1) > k/b$, $r_{T_{n+1}}(t \mid S_n) < r_{T_n}(t \mid S_{n-1})$, if and only if T_{n+1} is greater than the average of times up to the $(n+1)$th failure.

(d) The parameter b tunes the initial failure rate, and the parameter k the rate at which the failure rate decreases.

(e) With $b < k/(k-1)$ the model reflects growth in reliability and also the feature that removal of early bugs contributes more to growth.

The aim of this section is to exploit these features for specifying prior distributions for b and k so that statistical inference based on n observed times between software failure $\underline{t} = (t_1, \ldots, t_n)$ can be conducted. To do so, we need to have a likelihood function, and for this we may use the one-step-ahead forecast density [Equation (3.30)]. Specifically, given $\underline{t}$, a likelihood for b and k is

$$\mathcal{L}_n(b, k; \underline{t}) = \prod_{i=1}^{n-1} \frac{ib}{s_i} \left(\frac{ib}{s_i} \frac{t_{i+1}}{k} + 1 \right)^{-(k+1)}, \quad (4.50)$$

where $s_i = \sum_{j=1}^{i} t_j$.

4.7.1 Specification of the Prior Distribution

To specify $\pi(b, k)$, a joint prior distribution for b and k, it is useful to recall the following results from Section 3.6.2.

$$E(T_{n+1} \mid S_n = s_n) = \frac{k}{(k-1)nb} s_n; \quad (A)$$

$$V(T_{n+1} \mid S_n = s_n) = \frac{k^3}{(k-1)^2(k-2)} \left(\frac{s_n}{nb} \right)^2, \text{ and} \quad (B)$$

$$E(T_{n+1} - w \mid T_{n+1} > w) = \frac{k}{k-1} w + \frac{k}{k-1} \left(\frac{s_n}{nb} \right). \quad (C)$$

From (A) we see that $b \geq 0$; similarly, from (B) we see that $k \geq 2$, and that large values of k do not influence the variance of T_{n+1}. From (A) and (B) we see that, for any fixed n, both the mean and the variance of T_{n+1} decrease in b. Thus, to generate conservative one-step-ahead forecasts, small values of b are to be preferred.

Prior Distribution for k

The prior distribution for k is largely dictated by (C). Because k must be greater than or equal to 2, (C) essentially says that the mean residual life (MRL) depends on k but is between w and $2w$ plus a constant. Values of k close to 2 make the MRL close to $2w$ and large values of k make it close to w. A compromise is to choose k such that the MRL $\approx 1.5w$ plus a constant. That is, to make $k/(k-1) \approx 1.5$, which suggests that $k \approx 3$. Thus a suitable prior for k is a gamma on $[2, \infty)$ with parameters λ (scale) and θ (shape) chosen such that $E(k \mid \lambda, \theta) \approx 3$; that is, $2 + (\theta/\lambda) \approx 3$, which suggests that $\theta = \lambda$. Hence a possible prior on k is a shifted (at 2) exponential density with scale λ; specifically,

$$\pi_K(k \mid \lambda) = \lambda e^{-\lambda(k-2)}, \quad k \geq 2, \lambda > 0. \quad (4.51)$$

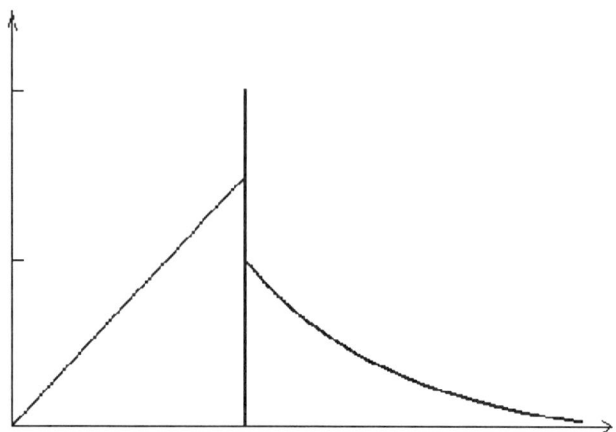

FIGURE 4.15. Illustration of the "Objective" Prior for b.

In what follows, interest centers around the quantity $u = k/(k-1)$. With the prior on k given by Equation (4.51), it is easy to verify (see Exercise 6), that the prior on u is

$$\pi(u \mid \lambda) = \tfrac{1}{(u-1)^2} \lambda e^{-\lambda((1/(u-1)) - 1)}, \quad 1 \leq u \leq 2. \tag{4.52}$$

Prior Distribution for b

For the prior on b, conditional on k, Al-Mutairi, Chen, and Singpurwalla (1998) argue that if the software is believed to experience a growth in reliability with a prior probability p_1, a decay with a prior probability p_2, and neither growth nor decay with probability $(1 - p_1 - p_2)$, then an *omnibus* prior is a composite distribution with components that have beta and shifted gamma densities; specifically,

$$\pi(b \mid k; \bullet) = \begin{cases} p_1 \dfrac{\Gamma(\alpha+\beta)}{\Gamma(\alpha)\Gamma(\beta)} \dfrac{b^{\alpha-1}(u-b)^{\beta-1}}{u^{\alpha+\beta-1}}, & 0 < b < u, \\[6pt] p_1 \dfrac{\Gamma(\alpha+\beta)}{\Gamma(\alpha)\Gamma(\beta)} \dfrac{b^{\alpha-1}(u-b)^{\beta-1}}{u^{\alpha+\beta-1}}, & 0 < b < u, \\[6pt] 1 - p_1 - p_2, & b = u, \\[6pt] p_2 \dfrac{(\alpha^*)^{\beta^*}(b-u)^{\beta^*-1} e^{-\alpha^*(b-u)}}{\Gamma(\beta^*)}, & b > u. \end{cases} \tag{4.53}$$

4.7 Inference for the Concatenated Failure Rate Model

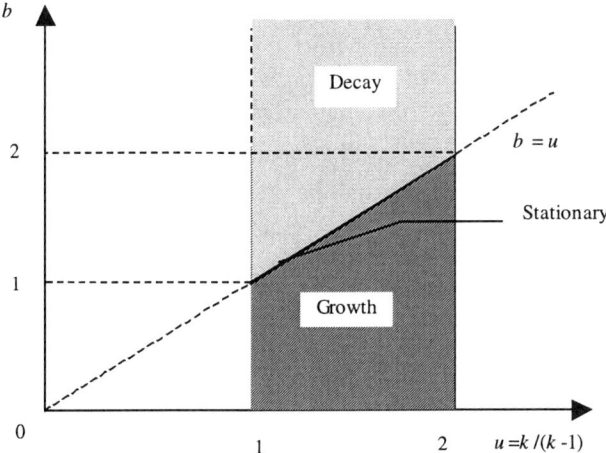

FIGURE 4.16. Regions Where the Joint Prior is Defined.

To make this choice as "objective," or as neutral, as is possible, α should be chosen to equal 2, and $\beta = \alpha^* = \beta^* = 1$. Pictorially, this prior takes the form shown in Figure 4.15.

Joint Prior for b and k

Once the preceding has been done, $\pi(b, k \mid \bullet)$, a joint prior on b and k, and also the unconditional prior for b can be easily induced. Figure 4.16 shows the regions over which the joint prior is defined; it delineates the regions of reliability growth, decay, and stationarity (i.e., neither growth nor decay).

4.7.2 Calculating Posteriors by Markov Chain Monte Carlo

The joint prior on b and k, together with the likelihood function Equation (4.50), enables us to obtain the joint posterior distribution of b and k via the relationship

$$\pi(b, k \mid \underline{t}; \bullet) \propto \mathcal{L}_n(b, k; \underline{t}) \, \pi(b, k \mid \bullet).$$

Once the preceding is done, the marginal posterior distributions $\pi(k \mid \underline{t}, \bullet)$ and $\pi(b \mid \underline{t}, \bullet)$ can be obtained; these are useful for testing hypotheses about reliability growth. The complicated nature of the likelihood and the prior makes it difficult to obtain closed-form expressions for the posteriors. The same is also true with numerical approximations. When such is the case simulation via the Gibbs sampling algorithm has proven to be a useful technique. Gibbs sampling is a Markov Chain Monte Carlo (MCMC) method, an excellent description of

158   4. Statistical Analysis of Software Failure Data

which is given by Casella and George (1992). We have given an overview of this technique in Appendix A. For our particular application, a computer program that implements the Gibbs sampling algorithm has been developed by Lynn (1996). It obtains the posterior distributions of k and b, and also the predictive distribution of T_{n+1}, the time to next failure. The program is also able to compute Bayes factors for testing hypotheses about reliability growth, decay, or stationarity.

To implement the Gibbs sampling algorithm (see Appendix A), we first need to obtain the following conditional distributions.

$$\pi(k \mid b, \underline{t}, \bullet) \quad \propto \mathcal{L}_n(k; b, \underline{t}) \, \pi(k \mid b, \bullet)$$

$$\propto \mathcal{L}_n(k; b, \underline{t}) \, \pi(b \mid k, \bullet) \, \pi(k \mid \bullet), \text{ and}$$

$$\pi(b \mid k, \underline{t}, \bullet) \quad \propto \mathcal{L}_n(b; k, \underline{t}) \, \pi(b \mid k, \bullet), \text{ where} \quad (4.54)$$

$\mathcal{L}_n(k; b, \underline{t})$ is the likelihood of k for fixed values of b and $\underline{t}$; similarly $\mathcal{L}_n(b; k, \underline{t})$. Note that the second expression of Equation (4.54) follows from the fact that $\pi(k \mid b, \bullet) \propto \pi(b \mid k, \bullet) \, \pi(k \mid \bullet)$. We now generate the *Gibbs sequence* $(k'_0, b'_0), (k'_1, b'_1), \ldots, (k'_m, b'_m)$ for some $m > 0$, as follows.

(i)   Choose an initial value of k, say k'_0,

(ii) Generate a b'_0 via $\pi(b \mid k'_0, \underline{t}, \bullet)$,

(iii) Generate a k'_1 via $\pi(k \mid b'_0, \underline{t}, \bullet)$,

(iv) Generate a b'_1 via $\pi(b \mid k'_1, \underline{t}, \bullet)$,

(v)   Generate a k'_2 via $\pi(k \mid b'_1, \underline{t}, \bullet)$ and so on.

For large values of m, k'_m is a sample observation from the posterior distribution $\pi(k \mid \underline{t}, \bullet)$, and b'_m an observation from $\pi(b \mid \underline{t}, \bullet)$. If we repeat the foregoing procedure N times, choosing N different starting values $k^1_0, k^2_0, \ldots, k^N_0$, and obtaining, at the end of each cycle, the sample points $(k^j_m, b^j_m), j = 1, \ldots, N$, then we have generated a sample of size N from the required posterior distributions $\pi(k \mid \underline{t}, \bullet)$ and $\pi(b \mid \underline{t}, \bullet)$. Thus an estimate of $\pi(k \mid \underline{t}, \bullet), k \in [2, \infty)$ is

$$\widehat{\pi}(k \mid \underline{t}, \bullet) = \tfrac{1}{N} \sum_{j=1}^{N} \pi(k \mid b^j_m, \underline{t}, \bullet);$$

similarly,
$$\widehat{\pi}(b \mid \underline{t}, \bullet) = \tfrac{1}{N} \sum_{j=1}^{N} \pi(b \mid k^j_m, \underline{t}, \bullet),$$

4.7 Inference for the Concatenated Failure Rate Model

for b within its defined range is an estimate of $\pi(b \mid \underline{t}, \bullet)$.
Finally, since

$$E(T_{n+1} \mid s_n, b, k) = \frac{k}{(k-1) \, nb} \, s_n \,,$$

$$E(T_{n+1} \mid \underline{t}) = \int_{(k, b)} E(T_{n+1} \mid s_n, b, k, \underline{t}) \, \pi(b, k \mid \underline{t}, \bullet) \, db \, dk.$$

Thus a Gibbs sequence based estimate of $E(T_{n+1} \mid \underline{t})$ is

$$\widehat{E}(T_{n+1} \mid \underline{t}) = \frac{1}{N} \sum_{j=1}^{N} \frac{k_m^j}{(k_m^j - 1) \, nb_m^j} \, s_n \,, \tag{4.55}$$

and from Equation (3.30), a Gibbs sequence based estimate of the posterior predictive density of T_{n+1} at $\tau \geq 0$ is

$$\widehat{f}_{T_{n+1}}(\tau \mid \underline{t}) = \frac{1}{N} \sum_{j=1}^{N} \frac{n \, b_m^j}{s_n} \left(\frac{n \, b_m^j}{s_n} \frac{\tau}{(k_m^j)} + 1 \right)^{-(k_m^j + 1)}. \tag{4.56}$$

Thus the Gibbs sampling algorithm provides an easy to implement procedure for obtaining Monte Carlo based estimates of the required posterior and predictive distributions.

4.7.3 Testing Hypotheses About Reliability Growth or Decay

The marginal posterior distributions $\pi(b \mid \underline{t}, \bullet)$ and $\pi(k \mid \underline{t}, \bullet)$ provide evidence about growth, decay, or stationarity of reliability. A way to capture this is via "Bayes factors." Suppose that H_1 denotes the hypothesis that there is a growth in reliability, H_2 the hypothesis that there is decay, and H_3 the hypothesis that the debugging process results in the stationarity of reliability. Then, we can verify (see Exercise 7) that the posterior probability of H_i, $i = 1, 2, 3$, is

$$\mathcal{P}(H_i \mid \underline{t}, \bullet) = \left(\sum_{j=1}^{3} \frac{p_j}{p_i} B_{ji} \right)^{-1}, \tag{4.57}$$

where p_i is our prior probability of H_i, and B_{ji} is the Bayes factor in favor of H_j against H_i. Furthermore, $B_{ii} = 1$, $B_{ji} = (B_{ij})^{-1}$ and $B_{ik} = B_{ij}/B_{kj}$, for all i and j. The calculation of B_{ij} is facilitated by the fact that B_{ij} is the "weighted" (by the priors) likelihood ratio of H_i against H_j [cf. Lee (1989), p. 126]. For the problem at hand, the estimation of the needed Bayes factors is accomplished via the Gibbs sequence $(b_m^j, k_m^j), j = 1, \ldots, N$. Specifically, B_{12} is estimated by

$$\widehat{B}_{12} = \frac{p_1(\underline{t})/p_2(\underline{t})}{p_1/p_2},$$

where $p_1(\underline{t})$ is the proportion of times (out of N) that b_m^j is strictly less than $k_m^j/(1-k_m^j)$, and $p_2(\underline{t})$ is the proportion of times that b_m^j is strictly greater than $k_m^j/(1-k_m^j)$. Note that $p_i(\underline{t})$ is an estimator of $\mathcal{P}(H_i \mid \underline{t})$, $i = 1, 2$. Similarly, B_{13} and B_{23} are estimated via

$$\widehat{B}_{13} = \frac{p_1(\underline{t})/(1-p_1(\underline{t})-p_2(\underline{t}))}{p_1/(1-p_1-p_2)},$$

$$\text{and } \widehat{B}_{23} = \frac{p_2(\underline{t})/(1-p_1(\underline{t})-p_2(\underline{t}))}{p_2/(1-p_1-p_2)}. \tag{4.58}$$

4.7.4 Application to System 40 Data

The concatenated failure rate model of this section and its methodology for inference has been applied to the System 40 Data of Section 4.5; see column 2, Table 4.7. The hyperparameters of the prior distributions, Equations (4.51) and (4.53), were chosen as: $\lambda = 1$, $\alpha = 2$, $\beta = \alpha^* = \beta^* = 1$, and $p_1 = p_2 = 0.25$. Details about the rationale behind this choice are given by Al-Mutairi, Chen, and Singpurwalla (1998). In column 7 of Table 4.7 we give the means of the one-step-ahead predictive distributions of T_{n+1}, for $n = 2, \ldots, 100$; see Equation (4.55). Figure 4.17 shows a superposition of a plot of these predictive means on a plot of the observed data; the plots are on a logarithmic scale. These plots provide an informal assessment of the ability of the proposed model to track the data.

We note from Figure 4.17 that the adaptivity of our model reflects its ability to track the data, and the tracking is particularly good during the latter stages of testing. The predictive mean times to the next failure show an initial growth in reliability followed by a period of general decay (or perhaps stationarity), which is then followed by a slow upward trend. The growth in reliability appears to stabilize towards the very end, but the large fluctuations in the data signal the need for caution in making this claim.

A comparison of the entries in columns 6 and 7 of Table 4.7 via the prequential likelihood ratio method of Section 4.6.1 demonstrates the predictive superiority of the concatenated failure rate model over the non-Gaussian Kalman filter model; the details are in Lynn (1996). Recall (see Section 4.6.4) that since the non-Gaussian model outperforms its competitors in the class of dynamic models, we claim that among the models considered here, the concatenated failure rate model provides the best predictivity for the System 40 data.

4.7 Inference for the Concatenated Failure Rate Model   161

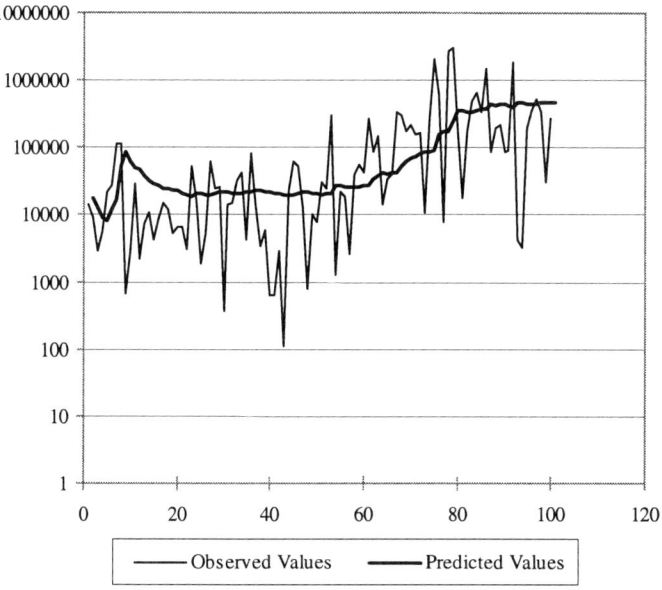

FIGURE 4.17: A Plot of the System 40 Data and Its One-Step-Ahead Predicted Means (on a Logarithmic Scale).

A final issue that remains to be addressed is that pertaining to reliability growth. Our previous analysis of the System 40 Data did not lead to conclusive results about reliability growth or decay. Will the model of this section provide better insights? To explore this matter, let $\underline{t}^{(j)} = (t_1, \ldots, t_j), j = 1, 2, \ldots$. Using the approach of Section 4.7.3, we obtain an estimate of $\mathcal{P}(H_i \mid \underline{t}^{(j)})$, for $i = 1, 2, 3$, where $\mathcal{P}(H_i \mid \underline{t}^{(j)})$ is the posterior probability of H_i given the first j interfailure times. Figure 4.18 is interesting; it shows *band plots* of the estimated posterior probabilities $\mathcal{P}(H_i \mid \underline{t}^{(j)})$, as a function of j, for $j = 1, 2, \ldots, 101$. The bottom band of Figure 4.18 shows $\mathcal{P}(H_1 \mid t_{(j)}, \bullet), j = 1, \ldots, 101$, and the central band $\mathcal{P}(H_1 \mid t_{(j)}, \bullet) + \mathcal{P}(H_3 \mid t_{(j)}, \bullet)$. From these bands we may infer how the three posterior probabilities relate to each other over the various stages of debugging and testing. That is, how the evidence for, say growth, fluctuates and evolves over time. Roughly speaking, one conclusion is that the initial 70% or so of the effort has not resulted in a growth in reliability, and the 30% effort that has led to growth has occurred during the latter stages. Also, the absence of reliability growth can be attributed more to the consequence of stationarity than to decay. Observe that the top band of Figure 4.18 indicates very small values for $\mathcal{P}(H_2 \mid t_{(j)}, \bullet), j = 1, \ldots, 101$. Could it be that the initial stages of

162   4. Statistical Analysis of Software Failure Data

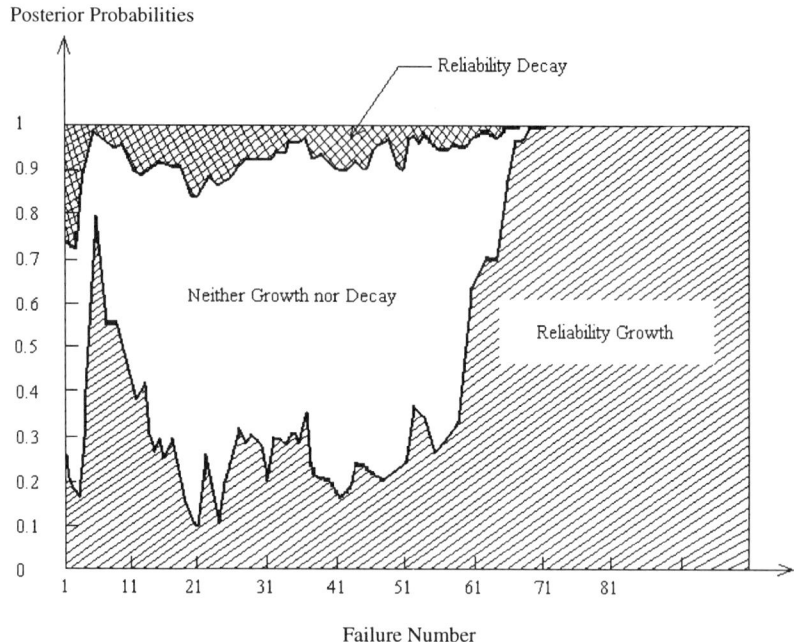

FIGURE 4.18. Band Plots of Estimated Posterior Probabilities for System 40 Data.

debugging have provided a learning environment for the process with the fortunate circumstance that there was not a preponderance of reliability decay? One could draw several other such conclusions, but the fact remains that plots of the posterior probability bands for testing hypotheses—like Figure 4.18—are valuable inferential devices.

Another attractive feature of the concatenated failure rate model, as compared to the non-Gaussian Kalman filter model, is smoothness of the predictive means; see Figure 4.19. An appreciation of reliability growth and/or decay can be visually better assessed via the predictive means of the former. Thus, to conclude, the improved predictivity of the concatenated failure rate model (for the System 40 Data), the availability of band plots of Figure 4.18, plus the fact that the model has more structure to it compared to the "black-box models" of Section 4.5, suggest that the model be given serious consideration for application after validation against other sets of software failure data.

4.7 Inference for the Concatenated Failure Rate Model   163

Posterior Mean of Concatenated Model

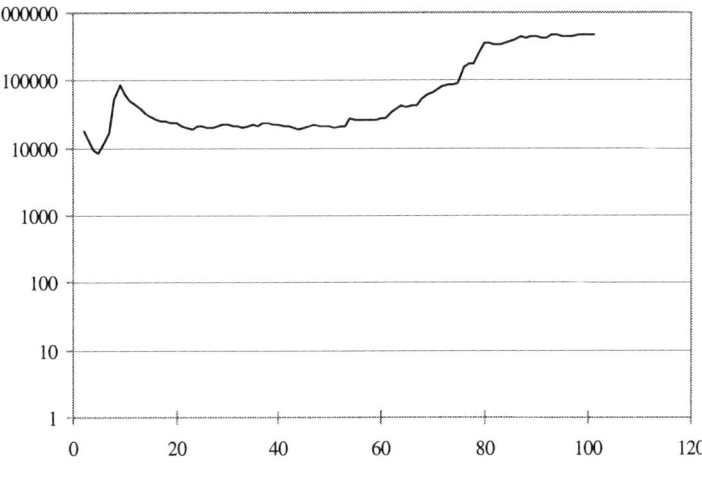

Failure Number

Posterior Means of Non-Gaussian Model

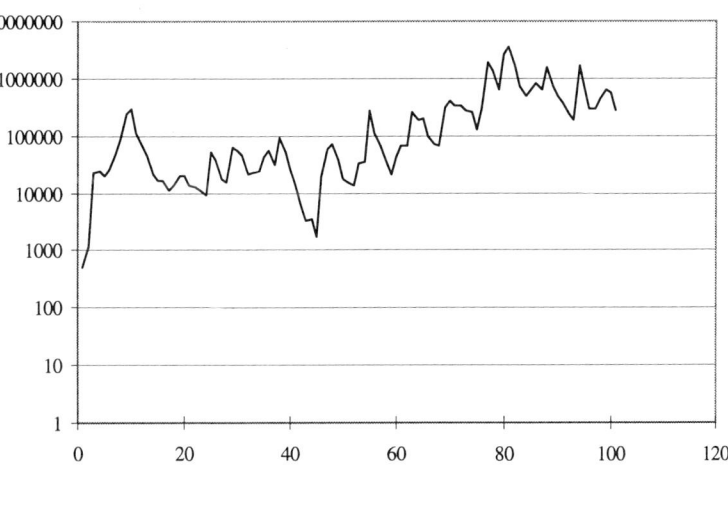

Failure Number

FIGURE 4.19. Predictive Means of the Concatenated and the Non-Gaussian Models for System 40 Data (on Logarithmic Scale).

4.8 Chapter Summary

The focus of this chapter (by far the biggest) is on the incorporation of observed failure history when assessing uncertainties. The driving theme here was Bayes' Law and its components, the likelihood function and the prior distribution. Thus the role of the data is to revise (or to update) prior uncertainties in the light of the information that the data provide. For purposes of completeness and also connections with current practice, the method of maximum likelihood and its associated confidence interval estimation was reviewed and illustrated.

Prediction and Bayesian inference using the models by Jelinski and Moranda, and by Goel and Okumoto were then described by us using some real data and proper prior distributions. In the sequel, we alluded to the computational difficulties that the Bayesian approach generally tends to create, and suggested approaches for overcoming these. Since proper prior distributions having well-known mathematical forms need not reflect true beliefs, an approach for eliciting priors based on the opinions of experts was described. The approach boils down to treating the expert's inputs as data, and for reflecting our assessment of the expertise of the expert through the likelihood. The case of multiple experts entails possible correlations between the expert announcements, and the treatment of such correlations via the likelihood function was mentioned. The preceding approach was applied to some software failure data that were analyzed by the nonhomogeneous Poisson process model of Musa and Okumoto. The expert input was based on several empirically developed ground rules suggested by software engineers with coding and testing experience.

Prediction and Bayesian inference using three dynamic models, the random coefficient autoregressive model, the adaptive Kalman filter model, and non-Gaussian Kalman filter model were next described, and their predictive performance on a single set of data, namely, the System 40 Data was compared. It was found that the non-Gaussian model gave predictions having the smallest mean squares error, but that implementing this model entailed simulation by the Markov Chain Monte Carlo (MCMC) method; this method is described by us in Appendix A.

An important feature of this chapter is our discussion of the topics of model comparison, model selection, and model averaging, using Bayes' factors and prequential likelihood ratios. Bayes' factors are weights that are assigned to a class of plausible models. These weights are developed a posteriori; that is, after incorporating the data. Bayes' factors can be used for the tasks of model selection and model averaging. We also give Jeffreys' ground rules for judging the superiority of one model over another. Prequential likelihood ratios compare one model over another based on their predictivity.

The chapter concluded with inference and hypotheses testing using the newly proposed adaptive concatenated failure rate model. The noteworthy features of the material here are the development of suitable proper prior distributions, and the MCMC computation of Bayes' factors for testing hypotheses about reliability growth, decay, or neither. One-step-ahead predictions provided by this model using the System 40 Data turned out to be superior to those provided by the non-Gaussian Kalman filter model; this comparison was made using prequential likelihood ratios. A plot of the profile of the Bayes' factors, as a function of testing, showed that it is only during the latter stages of testing that true growth in reliability occurred. During the bulk of the testing effort, there was neither a growth nor a decay in reliability. A profile plot of the Bayes' factors facilitates the preceding type of conclusions.

Exercises for Chapter 4

1. Verify statements (i), (ii), and (iii) about the posterior distributions given in Sections 4.2.3 and 4.2.4.

2. Verify the results given by Equations (4.10), (4.14), (4.19), (4.20), and (4.22).

3. For the "de-eutrophication" model of Moranda (1975) [see Equation (3.6)] show that if the prior distribution on D is a gamma with a scale (shape) parameter $b(a)$, and the prior distribution on k is a beta with parameters β_1 and β_2, independent of the distribution of D, then, given the interfailure times $\underline{t}^{(n)} = (t_1, \ldots, t_n)$:

   (i) the joint posterior distribution of D and k is proportional to
   $$D^{n+b-1} k^{\beta_1 + ((n(n-1))/2) - 1} (1-k)^{\beta_2 - 1} \exp\left\{ -\left(a + \sum_{i=1}^{n} k^{i-1} t_i\right) D \right\};$$

   (ii) the posterior distribution of k is proportional to
   $$\frac{k^{\beta_1 + ((n(n-1))/2) - 1} (1-k)^{\beta_2 - 1}}{(a + \sum_{i=1}^{n} k^{i-1} t_i)^{n+b}}; \quad \text{and}$$

   (iii) the predictive density of T_{n+1}, the next time to failure, at the point t, conditional on k, is proportional to
   $$\left\{ a + \sum_{i=1}^{n} k^{i-1} t_i + k^n t \right\}^{-(n+b+1)}.$$

   (iv) What is the predictive density of T_{n+1} conditional on $\underline{t}^{(n)}$ alone?

   (v) Numerically evaluate the predictive density that you have obtained in part (iv) using the NTDS data, for $n = 1, 2, \ldots, 26$, and show its plot for T_{27}. Compare the means of your predictive densities with the entries in column 3 of Table 4.6.

   (vi) Does the de-eutrophication model provide better predictivity of the NTDS data than the hierarchical model of Section 4.4?

4. Verify Equations (4.29) through (4.33) of Section 4.5.1.

5. Consider the interfailure time data given in column 3 of Table 4.7. Obtain the means of the predictive distributions of Y_i, $i = 3, 4, \ldots, 100$, using the adaptive Kalman filter model of Section 4.5.2, for $\alpha = -1, -0.5, +0.5$, and $+1$. Which of the preceding values of α provides the best fit for the observed data? You may use the mean square error as a criterion for assessing the goodness of fit.

6. Verify Equations (4.37) through (4.39) of Section 4.5.3, and Equation (4.52) of Section 4.7.1.

7. Verify Equations (4.57) and (4.58) on Bayes' factors of Section 4.7.3.

8. Using the data of Tables 4.1 and 4.2, perform an analysis of the Jelinski–Moranda model via the Gibbs sampling algorithm outlined in Section A.3.1 of Appendix A. Assume the same values for the hyperparameters as those of Section 4.2.3, and compare your answers with those in the preceding section.

9. For the NTDS Data of Table 4.2, replicate the analysis shown in Table 4.6 and Figure 4.4 using the Gibbs sampling algorithm outlined in Section A.3.2 of Appendix A. Assume the same values for the hyperparameters as those in Section 4.4.1, and compare your answers with those given in the preceding section.

5
SOFTWARE PRODUCTIVITY AND PROCESS MANAGEMENT

5.1 Background: Producing Quality Software

In this chapter we address two topics that are of interest to managers of software development teams. The first pertains to producing software within reasonable cost and time constraints. The second pertains to ensuring that the software produced is of acceptable quality. Statistical techniques have a role to play here, and the aim of this chapter is to highlight this role.

When developing software, an issue faced by a manager is the prediction of development time and effort. A common approach is to first estimate the size of the program, say the number of lines of code, and then use some guidelines on productivity rates to arrive at the time and effort needed to complete the project. Our focus here is on the use of historical data to estimate productivity rates. Good estimates of the number of lines of code are essential, and a plausible approach is to elicit specialized knowledge, and then use the techniques of Section 4.3.2 to assess the required quantity.

Typical approaches for assessing productivity rates are based on intuition or, at best, rough averages of historical data on the design, implementation, and unit tests of small programs. The introduction of a structured statistical framework for collecting and analyzing such data should be an improvement over current practice; this is what is described in Section 5.2. It is important to bear in mind that the time required to design, code, and unit test a program is only a modest, although important, part of the total project activity. The remainder of a programmer's time is used for other activities such as meetings, negotiations, documentation, and the like. Although these activities are necessary, a professional's time spent on them cannot be accurately measured, nor can it be controlled. Thus what is discussed in Section 5.2 does not cover the full spectrum of activities that go into determining productivity rates. However, until meaningful data on these "other activities," can be collected, the best that we can hope to do is to carefully analyze the available data on designing, coding, and testing by an individual programmer or a group of programmers.

In Section 5.3 we address the issue of process management for software quality. This issue is of concern not only to managers of software development houses, but also to those charged with the task of choosing a vendor of software. It is not uncommon for the developers of large systems to charter a software development organization, a software house, to produce the software needed to run the system. Vendor evaluation models have been used since the earliest days of competitive marketing. Comparing like products by price should not be the sole criterion for purchase. To make sound purchasing decisions myriad factors should be considered and evaluated. To facilitate this, several models have been proposed and used by industry, many of which result in classifying a vendor (or a supplier) by one of several classes that constitute a hierarchy. The *Capability Maturity Model* (CMM) is one such hierarchical classification scheme. It was developed by the Software Engineering Institute (SEI) of Carnegie Mellon University, and is specifically designed for software development. The material of Section 5.3 is geared towards the CMM. There we describe a normative approach for classifying a supplier by one of several categories, with the classification being probabilistic. That is, we are able to assign a positive probability that a supplier belongs to each of the several classes. This is in contrast to the traditional use of the CMM wherein classification is made with certainty. Despite criticisms, the CMM has been widely used by government and industry. Thus what is described in Section 5.3 should be viewed as an enhancement of the CMM, an enhancement that we hope helps overcome one of its weaknesses.

5.2 A Growth-Curve Model for Estimating Software Productivity

Human performance indices, such as software productivity, can be assessed by what are known as *growth-curve models* [cf. Rao (1987)]. Such models have been used for describing human learning experiences and have provided suitable fits to data on performance-based activities. A characteristic feature of learning experience data is that the successive measurements tend to be *autocorrelated* (i.e., related to each other). When developing software code, there appears to be the presence of an underlying learning process that has the tendency to improve the successive development activities. Consequently, the development times for individual programmers, or teams of programmers, tend to be correlated. It is because of this dependence that the software productivity process is predictable. If the data were not correlated, a level of stability is likely to have been reached, and the best prediction is the arithmetic average of the data.

5.2 A Growth-Curve Model for Estimating Software Productivity

As a general rule, moderate to long sequences of human performance data exhibit a degree of learning so that there is an underlying trend. Superimposed on this trend are often unanticipated and sharp deviations from normal expectations, followed by a reversal to the normal. Such fluctuations occur when individuals (or groups of individuals) alternate between performing unfamiliar tasks and familiar extensions of previously accomplished tasks. The methodology proposed here is geared towards drawing conclusions about productivity rates using data that consist of trends and fluctuations.

5.2.1 The Statistical Model

Consider a database of n programs, each program containing several lines of code. The number of lines of code per program need not be the same. We assume that each program in this database is developed by a single programmer, or by the same team of programmers, under circumstances that are for all intents and purposes identical. By "identical circumstances," we mean a similar working environment, such as policies, procedures, management, and the like. Of course each program in the database is distinct, in the sense that it is required to perform its own specific task. The observed data consist of the minutes per line of code needed to develop each program, and by the term "develop" we mean design, implement, and unit test. It is important to bear in mind that since the database pertains to an individual programmer, or to a team of programmers, all the derived measures of software productivity, and also the projected productivity figures, are valid for that individual or that team.

Let $X(1)$ denote the minutes taken per line of code to develop the first program that is written, and let $Y(1) = \log_e X(1)$. Similarly, $X(2)$ denotes the minutes per line of code for the second program that is developed, and so on. Thus $X(t)$, $t = 1, \ldots, n$, constitutes a time series that is indexed by the program number, $1, 2, \ldots, n$, instead of the usual time. If the assumption that there is an underlying learning phenomenon in writing the programs were to be true, then $X(t)$ would tend to be smaller than $X(t-1)$, for all values of t; it is otherwise if there is a degradation (i.e., the opposite of learning). In what follows, we assume that $X(t)$ has a lognormal distribution; this implies that $Y(t) = \log_e X(t)$ has a normal distribution. The assumption of lognormality can be justified on grounds of a subjective choice; however, it can also be supported on the basis of empirical evidence. It has often been claimed that time required to complete tasks such as maintenance and repairs tends to have a lognormal distribution. The development of software code is not unlike these tasks.

Suppose that the $X(t)$s are scaled so that $X(t) \geq 1$, for all t. Then, since $X(t)$ bears a relationship to $X(t-1)$, we propose that

$$X(t) = (X(t-1))^{\theta(t)}, \tag{5.1}$$

where $\theta(t) < 1$ suggests that there is a growth in productivity (or learning) in going from program $t-1$ to program t. Similarly, $\theta(t) > 1$ suggests a degradation of productivity. With $\theta(t) < 1$, the model of Equation (5.1) is a *growth-curve model*.

The power function relationship of (5.1) was postulated based on assumptions about the software development process mentioned previously. To introduce a measure of uncertainty into this formulation, we suppose that the relationship between $X(t)$ and $X(t-1)$ incorporates a multiplicative random quantity as follows,

$$X(t) = (X(t-1))^{\theta(t)} \epsilon(t) . \qquad (5.2)$$

Here $\epsilon(t)$ is an assumed lognormal error term with parameters 0 and $\sigma^2(u)$. The parameter $\sigma^2(u)$ is positive; its magnitude reflects the extent of our uncertainty about the relationship (5.1).

If we let $\eta(t) = \ln(\epsilon(t))$, then $Y(t)$ can be written as a *first-order nonhomogeneous autoregressive process* of the form

$$Y(t) = \theta(t) Y(t-1) + \eta(t) . \qquad (5.3)$$

Since the $X(t)$s are assumed lognormal, it follows that the $Y(t)$s are normally distributed and that $\eta(t)$ has a normal distribution with a mean 0 and variance $\sigma^2(u)$. Here, for example, a large value of $\eta(t)$ would represent a high degree of uncertainty about the appropriateness of the relationship $Y(t) = \theta(t)Y(t-1)$. The relationship (5.3) is an *autoregressive process of order 1* [see, for example, Box and Jenkins (1970)] except that the coefficient $\theta(t)$ is allowed to change from stage to stage. This is the reason for using the qualifier *nonhomogeneous* when describing (5.3). The model of Equation (5.3) is identical to the random coefficient autoregressive process model of Equation (3.21), except that the latter pertains to interfailure times, whereas the former pertains to the minutes per line of software code.

Even though the model given previously is simple in construction, it can incorporate a wide variety of user opinions about the growth or decay in productivity. This is achieved by assuming that the parameters $\theta(t)$ and $\sigma^2(u)$ themselves have distributions whose hyperparameters are specified. To implement this, we first focus on the *model uncertainty* parameter $\sigma^2(u)$, and then the *growth/decay* parameter $\theta(t)$.

Prior on the Model Uncertainty Parameter $\sigma^2(u)$

The parameter $\sigma^2(u)$ can either be specified by a user or can be estimated from the data; here we assume that it is estimated. For this we must specify a distribution that describes our uncertainty about $\sigma^2(u)$. The traditional approach is to assume that ϕ, the reciprocal of $\sigma^2(u)$, has a *gamma distribution* with a scale hyperparameter $(\delta_1/2)$ and a shape hyperparameter $(\gamma_1/2)$ [see West, Harrison, and Migon (1985)]. The advantage of letting the data determine the value of $\sigma^2(u)$ is flexibility in model choice and, as a consequence, greater

5.2 A Growth-Curve Model for Estimating Software Productivity

accuracy with respect to a given set of data. The disadvantage is that, when the working process has productivity growth or decay characteristics, and when the limited available data do not clearly reflect this trend, predictions of future data would be distorted. This is always a matter of concern: the degree to which predictions are controlled by data rather than by prior knowledge of process fundamentals.

Prior on the Growth/Decay Parameter $\theta(t)$

Since $\theta(t)$ can take values in the range ($-\infty$, $+\infty$), with $\theta(t) < 1$ implying gains in productivity, it is reasonable to assume that the $\theta(t)$s have a normal distribution with a mean of λ, and a variance of $(\sigma^2(v))/\phi$, where ϕ is $1/\sigma^2(u)$. The hyperparameter $\sigma^2(v)$ is specified by us and is based on our opinions about the underlying learning process. Thus, for example, with $\lambda < 1$, small values of $\sigma^2(v)$ would suggest productivity growth, whereas with $\lambda > 1$, they would describe a decay in productivity. Large values of $\sigma^2(v)$ also suggest that productivity fluctuates widely. Although values of λ less than or greater than 1 would again describe productivity growth or decay, the degree of fluctuation would generally mask this trend. Large values of $\sigma^2(v)$ would be appropriate, for example, with drastic changes in the nature of the programming tasks, with significant changes in the programming methods used, or with major changes in personnel staffing or capability.

At this point, we should note that, with the productivity of individual programmers, a slowly evolving pattern of productivity growth or decay should be expected with an occasional disruption due to unplanned interventions. Thus it seems reasonable that the $\theta(t)$s would be related, possibly in some mild form. There are different ways of describing a mild form of dependence between the $\theta(t)$s. One way is to assume that the hyperparameter λ is itself normally distributed with a mean of m_1 and a variance of s_1/ϕ. A value of $m_1 < 1$ would suggest the user's prior belief in an overall gain in productivity, and an $m_1 > 1$ would suggest the opposite. The value s_1 reflects the strength of the user's belief about the value of m_1. The value $m_1 = 1$ suggests the user's prior belief in neither a gain nor a loss in productivity. In many applications this would be a convenient and neutral choice.

Note that when the dependence of the $\theta(t)$s is described by a two-stage scheme, the $\theta(t)$s are exchangeable. The previously described model is identical to that of Section 3.4.1 wherein the θ_is of equation (3.21) were judged exchangeable. The main difference is that in the present case, $\sigma^2(u)$ is assumed unknown and estimated from the data, whereas in Section 3.4.1, σ_1^2, the analogue of $\sigma^2(u)$, was assumed known.

Some Guidelines on Specifying Hyperparameters

Hyperparameters capture a user's prior subjective opinions about the software productivity process. These opinions are held *before* any data are observed. For the hyperparameters of the prior distributions of $\sigma^2(u)$ and $\theta(t)$ mentioned previously, Humphrey and Singpurwalla (1998) provide some guidelines. These are summarized in the following.

The parameter γ_1 is used for calculating γ_t, which in turn is used for determining the prediction intervals. Furthermore, since γ_t increases linearly with the accumulation of data, γ_1 is only significant for determining projection intervals of the very early programs. For all intents and purposes, therefore, the value of γ_1 is not of much concern. However, in practice one sets δ_1 at 1 and tries out a range of values of γ_1 to see which one gives better predictions.

Regarding m_1, unless it is known a priori that the process has a steady trend, m_1 should be selected as 1. On the other hand, if it is known that the (logarithms of) successive terms will generally have a ratio of, say β, then the assumed value for m_1 should be β.

For s_1, an initial value of 0 would force all subsequent values of s_t to be 0 and thus restrict m_t to the initial value m_1. This would only be appropriate when a constant rate of productivity change was known with certainty. Conversely, a relatively large value of s_1 (i.e., 1.0) would imply relatively little confidence in the value of m_1 and would result in large initial fluctuations in the value of m_t until sufficient data had been gathered to cause it to stabilize. A compromise is to choose $s_1 = 0.35$ or $s_1 = 0.5$.

Similarly, the value selected for $\sigma^2(v)$ reflects the experimenter's views on the degree of fluctuation of productivity. Small values presume relative stability whereas large values (near 1.0) reflect wide variations. A compromise is to choose $\sigma^2(v) = 0.35$ or 0.5.

5.2.2 Inference and Prediction Under the Growth-Curve Model

Let $y(t)$ denote the observed value of $Y(t)$, $t = 1, 2, \ldots$, and suppose that $\underset{\sim}{y}(t) = (y(1), \ldots, y(t))$ have been observed. Let m_t denote the mean of the posterior distribution of λ given $\underset{\sim}{y}(t)$; that is, m_t is the updated value of m_1 in the light of $\underset{\sim}{y}(t)$. Since m_t conveys information about an overall growth or decay in productivity, a plot of m_t versus t, for $t = 2, 3, \ldots$, would suggest a steady growth in productivity if the values of m_t were to lie below one; otherwise, there is evidence of a decay in productivity.

The determination of m_t is relatively straightforward; it is left as an exercise for the reader. Specifically, with m_1, s_1, and $\sigma^2(v)$ specified, it can be shown [see Singpurwalla and Soyer (1992)], that for $t = 2, 3, \ldots$,

$$m_t = m_{t-1} + \frac{s_{t-1} \times y(t-1) \ (y(t) - y(t-1) \times m_{t-1})}{P(t)},$$

where

$$P(t) = (y(t-1))^2 [s_{t-1} + \sigma^2(v)] + 1,$$

5.2 A Growth-Curve Model for Estimating Software Productivity

and

$$S_t = S_{t-1} - \frac{(s_{t-1})^2 \times (y(t-1))^2}{P(t)}. \tag{5.4}$$

Assessments about productivity growth or decay, from program to program, are provided by the parameter $\theta(t)$, $t = 1, 2, \ldots$. Let $\theta^+(t)$ denote the mean of the posterior distribution of $\theta(t)$, given $\underset{\sim}{y}(t)$. Then it can be seen that

$$\theta^+(t) = \pi_t m_{t-1} + (1 - \pi_t) \frac{y(t)}{y(t-1)},$$

where

$$\pi(t) = [1 + (y(t-1))^2 (\sigma^2(v) + s_{t-1})]^{-1}. \tag{5.5}$$

Similarly, if $Y^+(t+1)$ denotes the predicted value of $Y(t+1)$, given $\underset{\sim}{y}(t) = (y(1), \ldots, y(t))$, then

$$Y^+(t+1) = m_t \times y(t), \text{ for } t = 1, 2, \ldots. \tag{5.6}$$

Note that $Y^+(t+1)$ is a projection of productivity. It is needed to estimate the time and effort required to complete future tasks involving design, implementation, and unit testing. Associated with such estimates are measures of uncertainty. The upper (lower) projection limit for $Y^+(t+1)$, with a coverage probability of approximately 68%, is given by the formula $(m_t \times y(t) + \sqrt{\omega_t}) \times [m_t \times y(t) - \sqrt{\omega_t}]$, where

$$\omega_t = \frac{\delta_t [1 + (y(t))^2 (s_t + \sigma^2(v))]}{\gamma_t - 2};$$

γ_t and δ_t are the updated values of γ_1 and δ_1, respectively, in the light of $\underset{\sim}{y}(t)$. They go to determine the degrees of freedom parameter of the *Student's t-distribution* that is used to obtain ω_t. They are given as

$$\gamma_t = \gamma_{t-1} + 1,$$

and

$$\delta_t = \delta_{t-1} + \frac{[y(t) + m_{t-1} \times y(t-1)]^2}{[1 + (y(t-1))^2 (s_{t-1} + \sigma^2(v))]}; \tag{5.7}$$

the details are in Singpurwalla and Soyer (1992).

5.2.3 Application: Estimating Individual Software Productivity

One problem with software productivity studies is the difficulty of gathering sufficient data to support a credible statistical analysis. It is desirable that programmers gather basic data on their personal performance on every program they produce. The data described in the following have been gathered in response to this theme. Regarding the productivity of large-scale tasks or team activities, there are problems with gathering data. First is that team projects take several months to complete, and so this would entail much time and effort vis-à-vis the data collection. Second, the stability of such data would depend on the composition of the team. Thus, for example, if team members were to leave or new members were to be added during the project, the prior data would not be likely to be comparable to the new data.

In Table 5.1, column 2, we show the minutes taken by a highly experienced programmer to develop a line of code, for 20 programs of varying sizes, taken from a Pascal textbook. Development includes designing, implementation, and unit testing. The 20 programs in question were not similar, but the environment under which they were developed was, for all intents and purposes, identical; that is, it was carefully controlled. Thus conclusions and projections of productivity based on these data would be valid only for this programmer, working under the tightly controlled environment mentioned previously.

Humphrey and Singpurwalla (1991) have analyzed these data using the methods of classical time series analysis. Their productivity projections, based on an *exponential smoothing* formula, were reasonable when compared with actual data. However, their approach relied only on past data to make future projections and did not take into consideration a knowledge of learning theories and application environments. Furthermore, their approach did not provide any insights about growth (or decay) in productivity. Specifically, was this programmer still experiencing a learning phenomenon or did he or she reach a point of saturation whereby learning was de facto minimal? Can the techniques described here provide an answer to the preceding question? In what follows we explore this and related issues.

In column 3 of Table 5.1 we show the values of $Y(t) = \log_e X(t)$, $t = 1, \ldots, 20$, and in Figure 5.1 we show a plot of $X(t)$ versus t. This plot shows that $X(t)$ fluctuates quite a bit, alternating between an up and a down, but otherwise fails to reveal any underlying trend. A plot of $Y(t)$ versus t is shown in Figure 5.2.

To apply the methodology of this section, we follow the guidelines for choosing hyperparameters mentioned in Section 5.2.1, and make the following choices: $\sigma^2(v) = 0.35$, $\delta_1 = 1$, $\gamma_1 = 5$, $m_1 = 1$, and $s_1 = 0.35$. These choices reflect a strong commitment to the proposed model, and a strong a priori opinion that there is neither growth nor decay of productivity. The latter position is appropriate because even though the programmer is an experienced one, the textbook exercises tend to increase in difficulty. Columns 4, 5, and 6 of Table 5.1 show the values of m_t, $\theta(t)$, and $Y^+(t)$; these are obtained via Equations (5.4) to (5.6), respectively. Column 7 of Table 5.1 compares the one-step-ahead

5.2 A Growth-Curve Model for Estimating Software Productivity

Table 5.1. Minutes per Line of Code, their Projected Values, and Estimates of Growth Parameters for a Specific Programmer

Program Number t	$X(t)$ Observed Min/LOC	$Y(t): \ln X(t)$	m_t	Estimate of the Growth Parameter $\theta^+(t)$	Predicted Values $Y^+(t)$	Absolute Difference Between Actual and Predicted Value $|Y(t) - Y^+(t)|$	68% Prediction Intervals Upper	68% Prediction Intervals Lower
1	3.619	1.28620	1					
2	2.461	0.90077	0.9195986	0.8391972	1.286197	0.385427	3.3637000	0.7913000
3	7.571	2.02438	1.1045470	1.3573140	0.828347	1.196033	1.8593549	0.2026611
4	7.634	2.03269	1.0773630	1.0342000	2.236019	0.203329	3.8813680	0.5906700
5	6.156	1.81754	1.0382370	0.9530569	2.189946	0.372406	3.5364390	0.8435430
6	4.214	1.43848	0.9981901	0.9873286	1.887038	0.448558	2.9790120	0.7950640
7	6	1.79176	1.0260910	1.1182870	1.435879	0.355881	2.3100929	0.5616651
8	4.095	1.40982	0.9963292	0.8654826	1.838507	0.428687	2.7502930	0.9267210
9	7.169	1.96977	1.0315900	1.1815760	1.404640	0.565130	2.1603556	0.6489244
10	4.312	1.46152	0.9997581	0.8512981	2.031991	0.570471	2.9103802	1.1536018
11	5.726	1.74505	1.0144200	1.0912430	1.461163	0.283887	2.1802163	0.7241090
12	9.721	2.27431	1.0385220	1.1751240	1.770215	0.504095	2.5231596	1.0172702
13	4.535	1.51198	1.0032630	0.7852389	2.361920	0.849940	3.2239487	1.4998913
14	6.036	1.79776	1.0146160	1.0821310	1.516913	0.820847	2.1937491	0.8400769
15	4.512	1.50692	1.0026170	0.9688724	1.510861	0.107561	2.1447327	0.8769893
16	4.068	1.40330	0.9987838	0.9688724	1.510861	0.107561	2.1447327	0.8769893
17	11.451	2.43891	1.0336100	1.3207760	1.401591	1.036529	1.9949138	0.8082682
18	5.120	1.63315	1.0072750	0.7793983	2.520064	0.886614	3.3811449	1.6589831
19	3.320	1.19996	0.9938651	0.8687636	1.645035	0.445075	2.3200237	0.9700463
20	2.820	1.03674	0.9895756	0.9474869	1.192602	0.155862	1.7738205	0.6113835

178   5. Software Productivity and Process Management

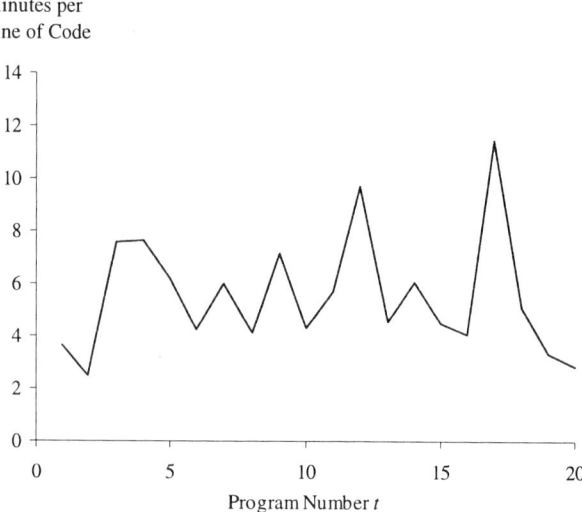

FIGURE 5.1. A Plot of $X(t)$, Minutes/Line of Code versus Program Number t.

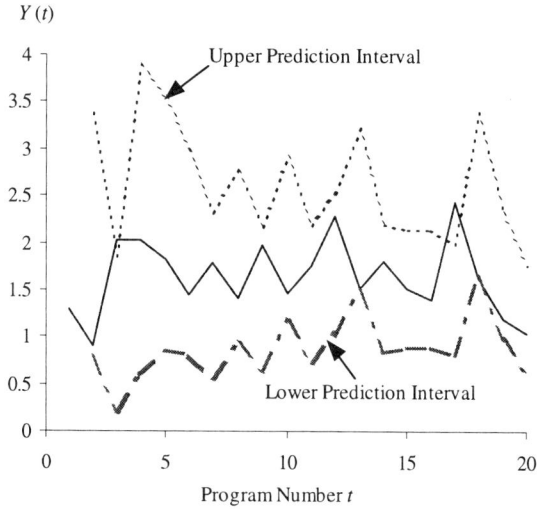

FIGURE 5.2. A Plot of $Y(t)$ and the Upper and Lower Prediction Intervals.

5.2 A Growth-Curve Model for Estimating Software Productivity 179

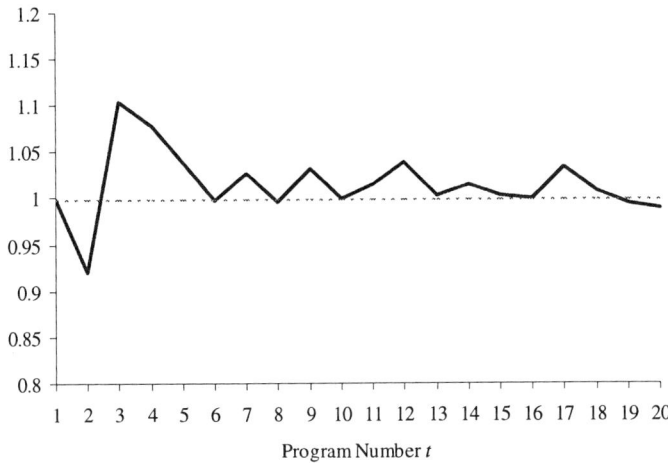

FIGURE 5.3. A Plot of m_t, an Indicator of Overall Productivity, Versus t, the Program Number.

predictions $Y^+(t)$ versus $Y(t)$, $t = 2, 3, \ldots, 20$, via their absolute differences, and columns 8 and 9 give the 68% prediction limits for the one-step-ahead projections. A plot of these prediction limits is shown in Figure 5.2; the plot indicates the extent to which the prediction limits cover the observed $Y(t)$. Since $Y^+(t + 1) = m_t \times y(t)$ [see Equation (5.6)] the predicted values tend to be relatively close to the most recent observed value; this is borne out by a comparison of the entries in columns 3 and 6 of Table 5.1.

In Figure 5.3, we show a plot of m_t versus t, for $t = 1, \ldots, 20$. Since the values of m_t tend to remain above 1, for most values of t, there does not appear to be present any evidence of productivity growth.

Figure 5.4 shows a plot of $\theta^+(t)$ versus t, for $t = 2, \ldots, 20$. The uneven nature of this plot suggests there is no steady pattern of growth or decay from program to program. However, the number of times the plotted values exceed one, and the magnitudes of these variations, indicate a slight decay in productivity. This is also suggested by the plot of Figure 5.2. In this example, this slight negative learning trend might be caused by the fact that the 20 programs from the Pascal textbook have a problem (and thus program) sequence of progressively increasing difficulty.

Thus to conclude, we have presented a Bayesian approach for assessing and evaluating productivity data, as with software development. Even though our discussion was focused on a particular measure of productivity, the approach is general and can be used to study the behavior of any data that evolve over time and are suspected to have growth or decay characteristics. The main virtue of our

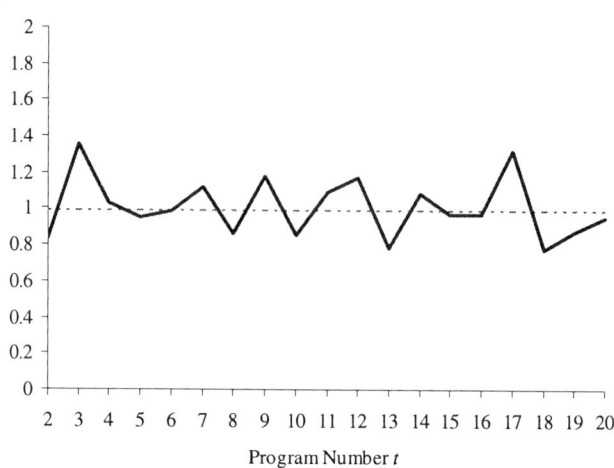

FIGURE 5.4. A Plot of $\theta^+(t)$, an Indicator of Program to Program Productivity Change, Versus t, the Program Number.

approach is its ability to assess underlying trends in the presence of wild fluctuations, and its ability to assess stage-by-stage growth or decay. With highly trended and relatively stable data, this approach should lead to superior predictions over the standard time series approaches, both because of its underlying structure, and because it is adaptive; that is, the model parameters are constantly updated in the light of new data.

It is easy to see that this approach can be extended so that $X(t)$ can be made to depend on more than one of its previous values, namely, $X(t-1)$, $X(t-2)$, $X(t-3)$, ..., and so on. The advantage of such a scheme would be to make the model more *robust* with respect to large oscillations in the data. With the current mode, the predictions are strongly influenced by the last observation, with the consequence that if it is an outlier, its impact will dominate the prediction. A model with a longer memory of the previous observations would modulate the effect of outlying observations. However, the resulting model would be quite complex and would require even more input parameters. Whether such an extension would provide more useful interpretations and predictions is a question that depends on the application.

5.3 The Capability Maturity Model for Process Management

Humphrey (1989, Chapter 1) describes the *software development process* as a set of actions that efficiently transform a user's needs into an effective software solution. The development focuses on schedules, standards, and practices rather

5.3 The Capability Maturity Model for Process Management

than on technologies and abilities of people. Over the last few years software process management has been touted as the key to developing reliable software. Documented studies by Paulk et al. (1993) have found that successful software process improvement efforts result in a return on investment in the range of 5:1 to 8:1.

The SEI's Capability Maturity Model (CMM) is a tool for evaluating an organization's software development process; see Humphrey and Sweet (1987). It focuses on the establishment of a systematic process for software development. The model identifies key software processes and skills that in the aggregate comprise a process management approach to software. The CMM has been used by administrators of software houses for improving practices and processes, and by program managers in government and industry for selecting contractors. It is therefore a tool that can provide inputs to a decision-making scheme (see Section 6.2) that a program manager may wish to use.

5.3.1 The Conceptual Framework

Based on a framework envisioned by Crosby (1979), the CMM classifies an organization into one of five "maturity levels," where level 1 is the lowest level of a hierarchy, and level 5 the highest. The placement is based on responses to a series of questions, called the *maturity questionnaire*, and follow-up visits to the organization for clarifying and validating responses. Each maturity level is defined by several attributes, called *key process areas* (KPA); see Figure 5.5. The separate sections of the maturity questionnaire focus on each KPA. The CMM requires that for an organization to be classified at a certain maturity level, say i, *all* the KPAs associated with level i must be satisfied. The judgment as to whether a key process area is satisfied has been generally based on the proportion of affirmative responses to the questions pertaining to the key process area. In Figure 5.5, M_1 through M_5 denote the five maturity levels whereas K_{ij}, $i = 1, \ldots, 5$ and $j = 1, 2, \ldots, n_i$, denote the jth key process area associated with the ith maturity level. Finally, R_{ijk} denotes the response to the kth question pertaining to K_{ij}, with $R_{ijk} = 1(0)$ denoting the fact that the response is in the affirmative (negative).

The maturity levels M_i, and key process areas K_{ij} are to be viewed as unobservable constructs, like parameters in probability models.

In what follows, we present an approach, given in Singpurwalla (1999), for probabilistically classifying a software house into the five maturity levels. That is, we are able to specify the probabilities with which an organization belongs to the five levels. This is in contrast to those approaches that classify with certainty. Our approach is based on responses to the questionnaire as well as expert judgment about the organization that an assessor may have. This is because Crosby (1979) conceptualizes a manufacturing organization maturing through the five nonquantifiable stages which he labels "uncertainty," "awakening," "enlightenment," "wisdom," and "certainty." Indicators of these stages are

182   5. Software Productivity and Process Management

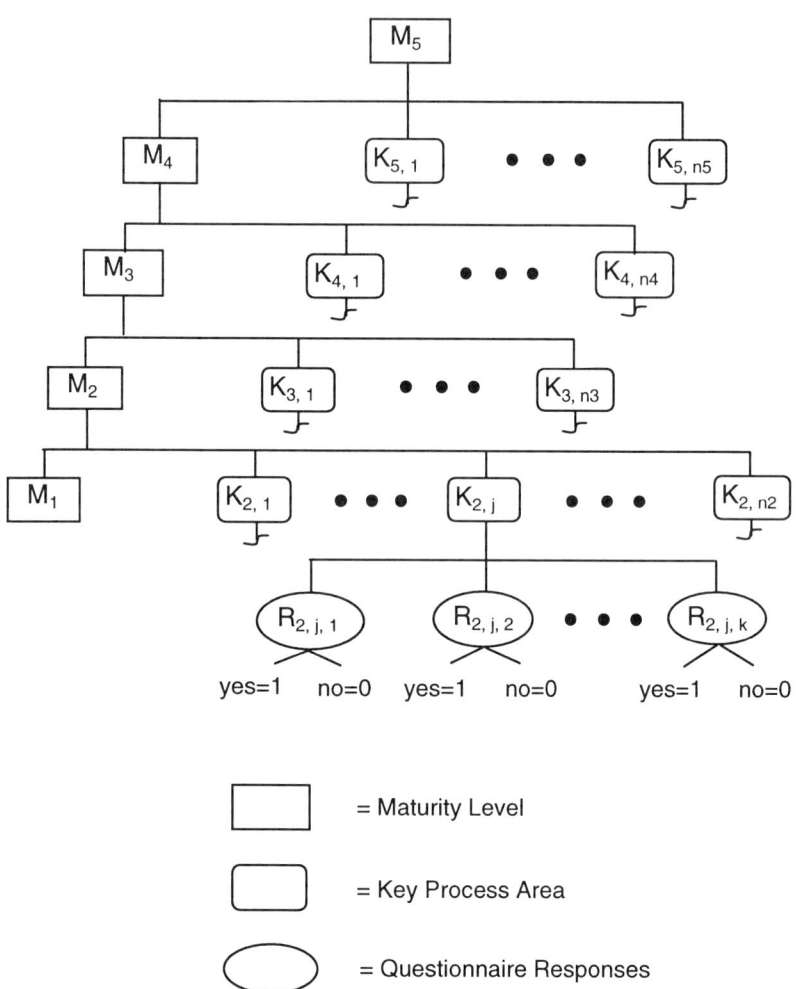

FIGURE 5.5. The Hierarchical Structure of the Capability Maturity Model.

attributes such as management understanding and attitude, quality organization status, problem handling, cost of quality as a percentage of sales, quality improvement actions, and the like.

5.3 The Capability Maturity Model for Process Management

5.3.2 The Probabilistic Approach for Hierarchical Classification

We begin by introducing some notation that helps us to distinguish between two events that are of interest. Let $M_i = 1(0)$ denote the event that a software house has (has not) attained a maturity level i or higher, and let $L_i = 1(0)$ denote the event that the highest maturity level attained (not attained) by the software house is i. By default, all software houses belong to level 1, and so $\mathcal{P}(M_1 = 1) = 1$; $\mathcal{P}(E)$ denotes the probability of event E. Also, since the five maturity levels form a hierarchy, $\mathcal{P}(M_{i-1} = 1) \geq \mathcal{P}(M_i = 1)$, $i = 2, \ldots, 5$. To help the reader appreciate the difference between the events $M_i = 1$ and $L_i = 1$, we consider the following simple illustration.

Level i	$P(M_i = 1)$	$P(L_i = 1)$
1	1.00	0.40
2	0.60	0.35
3	0.25	0.20
4	0.05	0.04
5	0.01	0.01
Sum	—	1

If $\mathcal{R}$ denotes the responses to the questionnaire data, then our goal is to assess $\mathcal{P}(L_i = 1 \mid \mathcal{R})$, $i = 1, \ldots, 5$. To do the preceding, we suppress $\mathcal{R}$, and observe that

$$\mathcal{P}(L_i = 1) = \begin{cases} \mathcal{P}(M_i=1|M_{i+1}=0)=\mathcal{P}(M_i=1) - \mathcal{P}(M_{i+1}=1), & i = 1, \ldots, 4 \\ \mathcal{P}(M_5=1). \end{cases}$$

Thus to assess $\mathcal{P}(L_i = 1 \mid \mathcal{R})$ we need to assess $\mathcal{P}(M_i = 1 \mid \mathcal{R})$, $i = 1, \ldots, 5$, and the bulk of what follows is devoted to an assessment of this latter quantity. But first we need to introduce some additional notation.

Let n_i denote the number of key process areas associated with maturity level i; by convention $n_1 = 0$. Let $K_{ij} = 1(0)$ denote the event that the jth key process area associated with level i is (is not) satisfied; $i = 1, \ldots, 5$ and $j = 1, \ldots, n_i$. Finally, let the vector $\underline{R}_{ij}$ denote the collection of responses to the questions associated with K_{ij}; thus $\mathcal{R} = (\underline{R}_{11}, \ldots, \underline{R}_{1n_1}, \ldots, \underline{R}_{51}, \ldots, \underline{R}_{5n_5})$.

To implement our approach several probabilities and likelihoods have to be specified; this is best done by experts who are knowledgeable and experienced

184   5. Software Productivity and Process Management

about the software development process. The probabilities are $\mathcal{P}(M_i \mid M_{i-1})$, $\mathcal{P}(K_{ij})$, and $\mathcal{P}(K_{ij} \mid M_i)$; the likelihoods are $\mathcal{L}(K_{ij}; \underline{R}_i)$.

Model Specification and Assumptions

To assess $\mathcal{P}(M_i = 1 \mid \mathcal{R})$, $i = 2, \ldots, 5$, we extend the conversation to all events that go to define the level M_i, and apply the law of total probability. Thus

$$\mathcal{P}(M_i \mid \mathcal{R}) = \sum_{(M_{i-1}, K_{i1}, \ldots, K_{in_i})} \mathcal{P}(M_i \mid M_{i-1}, K_{i1}, \ldots, K_{in_i}, \mathcal{R}) \times$$
$$\mathcal{P}(M_{i-1}, K_{i1}, \ldots, K_{in_i} \mid \mathcal{R}), \quad (5.8)$$

where (symbolically) the summation is over all possible permutations of the binary conditioning variables.

To evaluate the preceding expression, we need to make a series of assumptions. These assumptions are very reasonable and form the crux of our approach. In what follows, the notation $(E_1 \perp E_2) \mid (E_3)$ denotes the fact that given event E_3, the events E_1 and E_2 are independent.

The following Assumption A1 is prompted by the first term under the summation sign of the preceding expression, whereas the other assumptions are motivated by an application of the multiplication rule of probability to the second expression under the summation sign.

- A1. The attainment (or not) of level i is independent of the responses, given the status regarding level $(i - 1)$, and the status of all key process areas associated with level i. That is, for $i = 2, \ldots, 5$,

$$(M_i \perp \mathcal{R}) \mid (M_{i-1}, K_{i1}, \ldots, K_{in_i}).$$

- A2. The binary variables associated with the key process areas, within any level, are independent of each other. Specifically, for $i = 2, \ldots, 5$,

$$(K_{ih} \perp K_{i\ell}), \quad \text{for all } h \neq \ell.$$

- A3. Given the responses to the questionnaire, the attainment (or not) of level $(i - 1)$ is independent of whether the key process areas for level i are satisfied. That is, for $i = 2, \ldots, 5$,

$$(M_{i-1} \perp K_{i1}, \ldots, K_{in_i}) \mid \mathcal{R}.$$

- A4. Given an organization's disposition with respect to level i, the satisfaction (or not) of the K_{ij}s, $j = 1, \ldots, n_i$ is independent of the organization's disposition with respect to level $(i - 1)$. That is, for $i = 2, \ldots, 5$ and $j = 1, \ldots n_i$,

5.3 The Capability Maturity Model for Process Management

$$(K_{ij} \perp M_{i-1}) \mid M_i .$$

- A5. Only those questions (and their responses) that pertain to K_{ij} are relevant for determining its satisfaction (or not). That is, for $i = 2, \ldots, 5$,

$$(K_{ij} \perp \mathcal{R}_{i\ell}), \quad \text{for } j \neq \ell .$$

The Recursive Relationship

The hierarchical structure of the CMM enables us to develop a recursive relationship wherein the assessment of $\mathcal{P}(M_i \mid \mathcal{R})$ is facilitated by an assessment of $\mathcal{P}(M_{i-1} \mid \mathcal{R})$, $i = 2, \ldots, 5$ and the fact that $\mathcal{P}(M_i = 1) = 1$. The assumptions given before make this recursion possible; the recursive scheme eases the computational burden. Because of $A1$, $A2$, and $A3$, Equation (5.8) becomes

$$\mathcal{P}(M_i \mid \mathcal{R}) = \sum_{(M_{i-1}, K_{i1}, \ldots, K_{in_i})} \mathcal{P}(M_i \mid M_{i-1}, K_{i1}, \ldots, K_{in_i}) \times \prod_{j=1}^{n_i} \mathcal{P}(K_{ij} \mid \mathcal{R}) \, \mathcal{P}(M_{i-1} \mid \mathcal{R}).$$

To assess $\mathcal{P}(M_i \mid M_{i-1}, K_{i1}, \ldots, K_{in_i})$ we use Bayes' Law; consequently

$$\mathcal{P}(M_i \mid M_{i-1}, K_{i1}, \ldots, K_{in_i}) \propto \mathcal{P}(K_{i1}, \ldots, K_{in_i} \mid M_i, M_{i-1}) \, \mathcal{P}(M_i \mid M_{i-1}) ,$$

where $\mathcal{P}(K_{i1}, \ldots, K_{in_i} \mid M_i, M_{i-1})$ can, by virtue of $A2$ and the multiplication law, be written as

$$\mathcal{P}(K_{i1}, \ldots, K_{in_i} \mid M_i, M_{i-1}) = \prod_{j=1}^{n_i} \mathcal{P}(K_{ij} \mid M_i, M_{i-1}).$$

Invoking $A4$, and resubstituting in $\mathcal{P}(M_i \mid M_{i-1}, K_{i1}, \ldots, K_{in_i})$, gives

$$\mathcal{P}(M_i, M_{i-1}, K_{i1}, \ldots, K_{in_i}) \propto \prod_{j=1}^{n_i} \mathcal{P}(K_{ij} \mid M_i) \bullet \mathcal{P}(M_i \mid M_{i-1}) .$$

The final step pertains to the assessment of $\mathcal{P}(K_{ij} \mid \mathcal{R})$. This too is done via Bayes' Law, as

$$\mathcal{P}(K_{ij} \mid \mathcal{R}) \propto \mathcal{L}(K_{ij}; \mathcal{R}) \, \mathcal{P}(K_{ij}) .$$

186   5. Software Productivity and Process Management

But by virtue of A5, the likelihood $\mathcal{L}(K_{ij}; \mathcal{R})$ is $\mathcal{L}(K_{ij}; \underline{R}_{ij})$, so that

$$\mathcal{P}(K_{ij} \mid \mathcal{R}) \propto \mathcal{L}(K_{ij}; \underline{R}_{ij}) \mathcal{P}(K_{ij}) .$$

To evaluate $\mathcal{L}(K_{ij}; \underline{R}_{ij})$ we need to make some assumptions about the probabilistic structure of the collection of responses $\underline{R}_{ij} = (R_{ij1}, \ldots, R_{ijn_{ij}})$, assuming that the total number of questions pertaining to K_{ij}, the (ij)th key process area, is n_{ij}. A natural assumption would be conditional independence given K_{ij}, so that

$$\mathcal{P}(\underline{R}_{ij} \mid K_{ij}) = \prod_{k=1}^{n_{ij}} \mathcal{P}(R_{ijk} \mid K_{ij}) .$$

When such is the case

$$\mathcal{L}(K_{ij}; \underline{R}_{ij}) = \prod_{k=1}^{n_{ij}} \mathcal{L}(K_{ij}; R_{ijk}) .$$

Instead of conditional independence mentioned previously we may consider other possibilities, such as Markov dependence of the sequence $(R_{ij1}, \ldots, R_{ijn_{ij}})$ or exchangeability of this sequence. These, however, are not pursued here. Thus to summarize, the recursive probabilistic classification scheme reduces to the expression

$$\mathcal{P}(M_i \mid \mathcal{R}) \propto \sum_{M_{i-1}, \underline{K}_i} \prod_{j=1}^{n_i} \mathcal{P}(K_{ij} \mid M_i) \mathcal{P}(M_i \mid M_{i-1}) \times \prod_{j=1}^{n_i} \prod_{k=1}^{n_{ij}} \mathcal{L}(K_{ij}; R_{ijk}) \frac{\mathcal{P}(K_{ij})}{\mathcal{P}(M_{i-1} \mid \mathcal{R})}, \quad (5.9)$$

where $\underline{K}_i = (K_{i1}, \ldots, K_{in_i})$. The constant of proportionality is to be numerically evaluated; its role is to ensure that each of the $\mathcal{P}(M_i \mid \mathcal{R})$s, $i = 2, \ldots, 5$, is less than one. Recall that $\mathcal{P}(M_1 \mid \mathcal{R}) = \mathcal{P}(M_1) = 1$, so that the recursive scheme begins by first evaluating $\mathcal{P}(M_2 \mid \mathcal{R})$, and the indices pertaining to the summation sign take binary values.

5.3.3   Application: Classifying a Software Developer

We have been fortunate in having access to a software developer's responses to the 1987 version of the CMM questionnaire. The nature of the questions is shown in Section B.1 of Appendix B. The 1987 version of the CMM

Table 5.2. Probabilistic Classification of a Software Developer

L_i the Highest Maturity Level Attained	Probability of Attaining L_i
1	0.3771
2	0.4986
3	0.0979
4	0.0281
5	< 0.001

did not subdivide the questionnaire with respect to the KPAs; the 1994 version of the CMM does. The subdivisions shown in Appendix B were performed, at our request, by a software development analyst familiar with the various versions of the CMM. This exercise resulted in three KPAs each for maturity levels 2 and 3, two KPAs for maturity level 4, and one KPA for maturity level 5. There are 5 questions per KPA, making the total number of questions equal 45; the others were judged (by our analyst) to be no longer relevant, and were therefore discarded. The software developer's binary (Y = yes, N = no) responses to each of the 45 questions are given in Section B.2.

Recall that to implement our approach we first need to assess our probabilities $\mathcal{P}(M_i \mid M_{i-1})$, $i = 2, \ldots, 5$, and $\mathcal{P}(K_{ij})$, $i = 1, \ldots, 5, j = 1, \ldots, n_i$, and $\mathcal{P}(K_{ij} \mid M_i)$. These probabilities were elicited via experts from government, industry, and academia. They are given in Section B.3 of Appendix B. Next, we also need to assess the likelihoods $\mathcal{L}(K_{ij}; \underline{R}_{ij})$. These too were assessed by the aforementioned experts, and are also given in Section B.3. We emphasize that the specified values of the priors and the likelihoods are judgmental; they are therefore subject to discussion and change. For simplicity, the likelihoods chosen here happen to be the same across all key process areas; this could be a matter of debate.

An application of the entries of Appendix B to the recursive classification scheme of Section 5.3.2 results in the probabilistic classification shown in Table 5.2.

From Table 5.2 we see that our classification assigns the largest probability at level 2, and that the probability at level 1 is not much smaller. The probability that the highest level attained by this developer is 3 or greater is small, namely, 0.1270 (= 0.0979 + 0.0281 + 0.001).

Thus to conclude, we have described the probabilistic approach for classification based on a hierarchical structure. The problem of classification in a hierarchy is a generic one, and arises in the contexts of quality control (supplier

188   5. Software Productivity and Process Management

rating and defect classification), personnel management, educational placement, and perhaps even medical diagnosis. What distinguishes our probabilistic scheme from the prevailing (deterministic) ones, is that here there is a probability that an item belongs to a particular class, and these probabilities are spread out among all classes. Such probabilities reflect the inherent uncertainties behind the underlying information that is used to make the classification. The probabilistic approach mentioned here has the attractive feature of adaptivity. That is, the classification can be updated in the light of new information via Bayes' Law. The current classification serves as a prior, which with the likelihoods and new information provides an updated (i.e., posterior) classification. Finally, the proposed approach can be enhanced by a consideration of multinomial (instead of the binomial) responses, and the likelihood function can be used as a device for giving weights to those questions that are deemed more important than the others.

5.4   Chapter Summary

The purpose of this chapter is to show how statistical techniques can be used to manage the software development process, be it for productivity assessment or for source selection.

When software productivity data can be indexed, as in a time series, then growth-curve models can be used to track the data for trends, and for making projections. There is a vast amount of literature on growth-curve models and consequently the choice of models is large. However, for purposes of illustration, we selected a simple power rule model, and motivated its relevance for monitoring software productivity. The chosen model when suitably transformed is a random coefficient autoregressive process which, we recall, is also one of the dynamic linear models used to describe software interfailure times. A Bayesian approach for inference and predictions using this model was described by us, and this was illustrated via an application to real data on the times taken to develop a line of code. The bulk of our discussion pertained to guidelines for choosing priors for the model parameters.

The second part of this chapter focused on the Capability Maturity Model (CMM) that is widely used in government and industry to select and to rate software development houses. As currently practiced, the CMM requires binary responses to a series of questions about the software development process, and uses these to classify an organization into one of five categories. The categories form a hierarchy, and the classification is done with certainty. The problem described previously is quite generic; our discussion of course centered around software development. The approach described by us enables one to classify probabilistically, and this is the key feature of the material presented here. Classification with certainty is a limitation. The probabilistic classification is based on a repeated application of the law of total probability and a series of assumptions about independence. The hierarchical structure of the problem simplifies the computation. Bayes' Law comes into play during several phases of the development and the priors and likelihoods are based on the available experience of those who have performed such ratings over a broad spectrum of industries. We illustrated the workings of our approach via a consideration of data from a real scenario.

Exercises for Chapter 5

1. Verify Equations (5.4) through (5.6) of Section 5.2.2.

2. Analyze the data of Table 5.1 assuming:

   (i) a strong commitment to the proposed model of Section 5.2.1, but a strong prior opinion that there is a growth in productivity;

   (ii) a weak commitment to the proposed model of Section 5.2.1, but a strong prior opinion that there is neither growth nor decay of productivity.

   Discuss your choice of prior parameters and contrast your results with those of Section 5.2.3. Based on your comparisons, would you judge the approach of Sections 5.2.1 and 5.2.2 robust to prior assumptions?

3. In Section 5.3.2 we assumed that for the (ij)th key process area, the n_{ij} responses $R_{ij1}, \ldots, R_{ijn_{ij}}$, are conditionally independent. Show how Equation (5.9) for $\mathcal{P}(M_i \mid \mathcal{R})$ would change if instead of the conditional independence mentioned previously, we assumed exchangeability of the $R_{ij1}, \ldots, R_{ijn_{ij}}$. To address this question you will need to propose a simple model for the exchangeability of Bernoulli random variables.

4. Address Exercise 3 if instead of exchangeability you are required to assume that the Bernoulli variables are dependent, the dependence described by a first-order Markov chain. Propose a simple Markov chain model, and assume any values for the underlying parameters.

5. Repeat the analysis of Section 5.3.3 by assuming the dependence structure of Exercises 3 and 4, and the appropriate data from Appendix B. What effect does the dependence have on the classification given in Table 5.2?

6
THE OPTIMAL TESTING AND RELEASE OF SOFTWARE

6.1 Background: Decision Making and the Calculus of Probability

In Chapters 3 and 4 we have described how probability models can be used to quantify uncertainties about the software failure process, and to make predictions about failure times. In Chapter 5 we have seen how probability models can be used to predict productivity, and how the calculus of probability can be used for the classification of software development houses. However, the quantification of uncertainty, inferences from failure data, and the placement of software houses are not necessarily the final goals of an engineering endeavor. Rather, they are intermediate steps for taking actions or making decisions, and these are characteristic of an engineer's activities. The consequences of such decisions depend on the outcomes of uncertain quantities. The making of decisions under uncertainty is the aim of *statistical decision theory*, be it based on the Bayesian or the frequentist paradigm. Thus statistical decision theory has a natural place in software engineering, and indeed in the general area of design, engineering, and manufacturing [cf. Singpurwalla (1992), (1993), and (1998c)]. The purpose of this chapter is to give an overview of the key elements of decision theory, and to describe how this theory can be used to address a basic concern that software engineers face.

Decision theory is intimately associated with probability. Although probability theory is a coherent method of quantifying uncertainties, decision theory tries to build an analogous approach to the problem of making decisions. Most commonly, we make decisions in an atmosphere of uncertainty, not

192   6. The Optimal Testing and Release of Software

knowing the consequences of the decision; in software engineering, decisions under uncertainty might include the selection of an organization to write software, the uncertainty being the capability of the tendering companies, or choosing when to release software, the uncertainty being the reliability of the code. In this chapter we focus on aspects of the latter; the former needs development and is a topic for future research; also see Exercise 1.

The testing phase of software development is central to the production of a reliable system, and an important question for the software engineer to address, at the start of this stage, is how much time should be devoted to this process. The optimal testing time is a function of many variables: size of the software, level of reliability desired, personnel available, market conditions, and penalties of in-process failure.

One purpose of the testing phase is to satisfy the development team that the software is operating satisfactorily. This will involve subjecting the software to a variety of inputs in order to see if it is producing the required output; during this process, errors are observed, located, and eliminated. Due to the very large number of possible inputs into the software, exhaustive testing is almost never feasible, and so only a certain number of the possible inputs are tried. Even so, debugging the software to be highly reliable could take a long time. Balancing the desire for high reliability are criteria that favor a short testing time, such as the cost of testing and debugging, and the risk of product obsolescence. The testing time that is chosen should be a compromise between these two sets of conflicting criteria.

Finding the optimal testing time is a decision problem; we must make a decision as to the time we feel that it is best to test. Since we must make our decision before testing begins (and so before we observe any data on software failure), this decision—at least initially—must be made using our prior knowledge of the performance of the software, and the costs and consequences of the testing procedure. Also, decision theory is ideally suited to sequential testing, where a decision is made after the first testing stage on whether to continue further testing. This second decision should be made in light of what has happened in the first stage, and decision theory provides a method of incorporating this new information.

In Sections 6.2 through 6.4 we give an overview of the key elements of decision theory, from a Bayesian point of view. This is followed by an application of this theory to the problems of software testing mentioned previously.

6.2   Decision Making Under Uncertainty

A piece of software has been developed and is ready to enter the testing phase. Before testing begins, a management decision must be made as to the length of time it should be tested. Too short a testing time will result in unreliable software being released to the user, with the attendant costs of postrelease fixing of bugs and loss of consumer confidence. Conversely, too long a testing time adds to the cost of the project and risks product obsolescence. In between these two extremes is a time that most effectively balances these competing costs; this will be the optimal time to test.

Optimal testing is therefore a decision problem and, like most decision problems, can be divided into three components.

Actions

There is a set of available actions or decisions that we can take. This set may be discrete or continuous; the decision problem is to choose the "best" action from this group. In software testing, the actions are all the times that it is possible to test for, together with the action to release the software to the user, without further testing.

States of Nature

These are the parts of the problem that are unknown or outside the control of the decision maker. The state of nature will reveal itself only after a particular action is taken, and will affect the outcome of that action. As the states of nature are uncertain, it will be necessary to assign probabilities to them. Here, the states of nature are the unknown quantities related to the performance of the software and the testing phase, such as the number of bugs discovered by the testing team and the number of bugs that remain after its release.

Consequences

Associated with every action and state of nature combination, there is an outcome or consequence. This is the final result of a particular action and a particular state of nature. Often, the consequences will be monetary (profit or loss), but not necessarily. As with uncertainty, it will be necessary to somehow quantify the consequences, especially when they are qualitative, such as customer satisfaction. As with probability, this is done by subjectively assigning a number to each consequence called its *utility*. With software testing a utility is based on the costs of testing plus the consequences of an in-service failure or success.

The general approach to a decision problem is to enumerate the possible actions and states of nature, assign probabilities and utilities where needed, and use these assignments to solve the problem by producing the "best," or optimal, action. The assignment of probabilities has been discussed at length in previous chapters. That leaves two more issues to address: assigning utilities, and the definition of what constitutes the best action.

6.3 Utility and Choosing the Optimal Decision

We have alluded to the need to numerically express the consequences of a decision problem. This is done by assigning a *utility* to every possible outcome. A utility describes the worth of the given consequence to the decision maker and, together with probability, must be assessed before an optimal decision is found. Thus, like probability, a utility must be interpreted as a *subjective* quantity, depending upon the individual (or group of individuals) making the decision. This is only natural since, in a particular decision, it is perfectly legitimate for two people to have different priorities, and so they assess the worth of the consequences differently. The concept of utility dates back to the times of Nicholas Bernoulli and the St. Petersburg paradox. Its role in decision theory was laid out by many of the people who promoted subjective probability, such as de Finetti (1974), Ramsey (1964), and Savage (1972). Hill (1993) contains a narrative on the development of utility theory.

6.3.1 Maximization of Expected Utility

Consider a decision problem where there are m possible actions $a_1, a_2, \ldots, a_m$ and n possible states of nature $s_1, s_2, \ldots, s_n$. If action a_i is chosen, then we denote the probability that s_j occurs by p_{ij}. Suppose that after choosing action a_i, the state of nature s_j occurs, and consequence c_{ij} results. We denote the utility of that consequence by $\mathcal{U}(c_{ij})$, for some utility function $\mathcal{U}$ that maps c_{ij} into $[0, 1]$. It is important to note that $\mathcal{U}(c_{ij})$ should map into a bounded interval, say $[a, b]$; the bounded interval can then be transformed to the interval $[0, 1]$.

The optimal action is that which yields the highest utility to the decision maker. Since it is not known which state of nature will occur when an action is taken, this utility is unknown. However, we can calculate the *expected utility* of a particular action, say a_i, using the probabilities p_{ij} as

$$E(\mathcal{U}(a_i)) = \sum_{j=1}^{n} p_{ij}\mathcal{U}(c_{ij}), \qquad (6.1)$$

for $i = 1, \ldots, m$. We choose that action for which the expected utility is a maximum. Thinking of utility as a monetary gain, this translates to saying that we pick the action that, in our opinion, has the greatest expected profit (or perhaps smallest expected loss). This is called the *principle of maximization of expected utility*, and is the decision criterion for choosing a decision under uncertainty. The principle generalizes in the usual way when there are a continuum of possible actions and possible states of nature; the p_{ij}s are replaced

Table 6.1 A Simple Decision Table

Action	States of Nature		$E(U(a_i))$
	S_1	S_2	
a_1	$100,000	-$1,000,000	45,000
a_2	$1	-$10	0.45

by densities, and one obtains expected utilities by integration. Differential calculus can be used to find the maximum of the expected utility.

There are strong mathematical arguments that back this principle as a decision rule. These arguments rely on the concept of *coherence*; roughly speaking, any other decision rule lacks the property of coherence and leaves the decision maker vulnerable to making decisions where loss is inevitable (a so-called Dutch book). Lindley (1982b) discusses coherent decision making in detail, and shows that the principle of maximization of expected utility is a consequence of the laws of probability, once utility is viewed as obeying the calculus of probability.

6.3.2 The Utility of Money

Since we are often concerned with decisions whose outcome is measured financially, it is important to remark that utility is not necessarily linear in monetary value; that is, the utility of x is not necessarily x, or some linear function of x. To see why, consider the simple decision problem described in Table 6.1. There are two actions, a_1 and a_2, and two possible states of nature, s_1 and s_2. The entries in the table enumerate the consequences of each possible decision and state of nature. For example, if the decision maker chooses action a_2 and the state of nature turns out to be s_1, then the decision maker stands to gain $1. Suppose that we assess the probability of s_1 occurring to be 0.95 and of s_2 to be 0.05, regardless of which action we take. If we assume that the utility of $x is x then, by employing the principle of maximizing expected utility, we see that action 1 is the preferred one, it having an expected utility of 45,000, compared to action 2 whose expected utility is only 0.45.

However, we would argue that one would not choose a_1 because one could not afford to lose $1,000,000, even though the chances of winning $100,000 are high. Instead, we would "play safe" and decide on a_2. Such behavior is known as *risk aversion* and can be modeled by assuming a utility function that is concave; typical examples would be $\mathcal{U}(\$x) = 1 - \exp(-x/r)$, for a constant r, or $\mathcal{U}(\$x) = \log(x)$, for positive x. Risk aversion is also the reason why one buys insurance at a premium to the expected monetary loss of the item to be insured.

Other forms of behavior towards risk are possible. The opposite of risk aversion is *risk proneness*, where the decision maker actively seeks out risky situations. Buying a ticket in a lottery is an example of risk-prone behavior, since the expected financial gain from a lottery is less than zero. Risk proneness is modeled by a convex utility function. Finally, there is *risk neutral* behavior, where one's utility of money is in fact linear. It is important to note that in the preceding example, the states of nature are not influenced by the action taken. In many examples, for instance, the testing of software, the action taken, such as debugging during testing, will have an influence on the state of nature. The attributes of risk aversion and risk proneness are apparent in the software industry wherein some organizations tend to release software that is knowingly not thoroughly tested.

6.4 Decision Trees

A decision table, such as Table 6.1, is a natural way of representing a decision problem. Although the table is a useful device for laying out the various ingredients of the problem, it does not show its evolution over time, from the action taken to the state of nature occurring, to the final outcome. Being able to show this progression is a valuable aid to visualizing the decision process, particularly in more complex decision problems. This is especially so when there may be a sequence of actions and states of nature before a final outcome is reached.

A *decision tree* is one way of graphically portraying a decision problem so that this temporal progression is captured. A decision tree is like a directed graph, composed of nodes and branches as in Figure 6.1. This figure shows the decision tree associated with the decision problem of Table 6.1.

Whenever a decision is to be made, there is a *decision node* in the tree, denoted by a square box. Branches that sprout from this box represent the various possible actions that can be taken. The revealing of a state of nature pursuant to an action is represented by a *random node*, denoted by a circle. Branches sprouting from a random node represent the various possible states of nature that might occur, each with its attached probability. The terminus of a tree denotes the utility associated with the consequence resulting from the path of actions and the states of nature in that branch.

A probability must be assessed for each branch that emanates from a random node. If a decision tree contains more than one random node [see, e.g., Figure (6.2)], then the probabilities assessed at a node should be conditional on $\mathcal{H}$, and the path of all actions and all outcomes from the nodes that precede it. Since the outcomes from all the nodes, the node of interest and its predecessors, are unknown at the time a decision is made, we have to average the outcomes with respect to their probabilities. This can be a formidable task; it is called a *preposterior analysis* [cf. Lindley (1972), p. 21]; see Section 6.6.1.

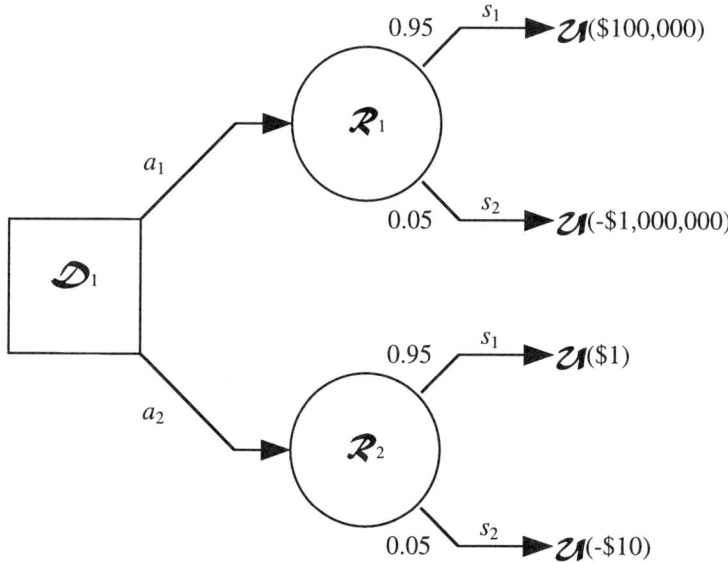

FIGURE 6.1. A Simple Decision Tree.

6.4.1 Solving Decision Trees

Once a decision tree has been constructed, the optimal decision (i.e., the action that maximizes expected utility) can be obtained by a sequence of steps that are akin to "pruning" the tree back to its root. Starting from the terminating branches and working backwards, branches and the nodes of the tree are eliminated in the following way.

- Each random node and its branches are replaced with the node's expected utility, which is the sum of the product of the probability and the utility of each branch emanating from the node.

- At each decision node, the action that has the maximum expected utility is identified and all the other branches of that decision node are removed.

This pruning continues until we have reached the leftmost node. We are left with the optimal expected utility, and a path through the decision tree that shows the optimal decision. This method of solving decision problems is of particular use when the decision problem is multistage, that is, with a sequence of decision and random nodes following each other. In this more complex type of tree, we may have a sequence of actions to take, depending on the outcome of intermediary random nodes, so the solution is not an optimal action but a *policy* of what to do at each eventuality.

Thus, for example, in the decision tree of Figure 6.1, at the random node labeled $\mathcal{R}_1$, we compute the expected utility $\mathcal{U}(\mathcal{R}_1)$ as $0.95 \times (100,000) + 0.05 \times (-1,000,000) = 45,000$; similarly, at $\mathcal{R}_2$ we obtain $\mathcal{U}(\mathcal{R}_2) = 0.95 \times (1) + 0.05 \times (-10) = 0.45$. At the decision node labeled $\mathcal{D}_1$, we choose the larger of $\mathcal{U}(\mathcal{R}_1)$ and $\mathcal{U}(\mathcal{R}_2)$ which is 45,000, so that $\mathcal{U}(\mathcal{D}_1)$, the expected utility at $\mathcal{D}_1$, is 45,000. This corresponds to action a_1, which is our optimal decision.

6.5   Software Testing Plans

Returning to the problem at hand, we first describe a variety of different ways in which one may organize the testing phase of software. The simplest is *one-stage testing*, where a decision is made that the software is to be tested for a period of time T and then released, regardless of the results of testing. The only decision to be made is the size of T. Most of the work on testing procedures has addressed this form of the problem, which can be represented by the decision tree of Figure 6.2. Here, the multiple arrows issuing from a node indicate that more than one decision is available, or that more than one state of nature is possible. In this tree, the first decision node $\mathcal{D}_1$ refers to the decision on T, how long to test, followed by the unknown consequences of testing, where $N(T)$, an unknown number of bugs, are discovered and corrected. Then comes $\mathcal{D}_2$, the decision to release the software, followed by the unknown results of release, with $N - N(T)$ bugs discovered by users. Subsequent to the preceding, a final utility $\mathcal{U}[T, N(T), (N - N(T))]$ for the testing and release of the software is realized. We are assuming that the terminal utility depends only on $(N - N(T))$ and not the times at which these bugs are encountered.

An elaboration of the preceding is *two-stage testing*, where a second stage of testing is conducted after the first, but only if needed. The decision is to choose (in advance of testing) the length of the two testing periods, say T_1 and T_2, and a criterion for deciding whether to release after the first test, or to proceed with the second; this criterion is influenced by N^*, the number of bugs observed in the first test period. The choices T_1, T_2, and N^* are made *prior* to the first test. Figure 6.3 is an example decision tree for a two-stage test, where the second stage is only conducted if more than N^* bugs are observed in the first stage. One can of course extend this type of plan to three or more stages.

A common feature of both the one- and the two-stage tests is that all decisions are made before any testing occurs, on the basis of prior information alone. However, the decision to proceed with the second stage is based on the information gained in the first stage. The results of testing are not incorporated into the decision criteria. This is in contrast to *sequential* testing, wherein the number of stages to test is random, and the decision on whether to test, and if so

6.5 Software Testing Plans 199

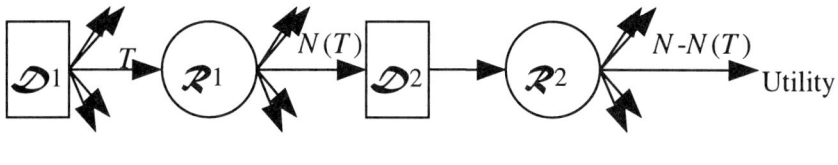

	Bugs		Bugs
Decide on	Observed		Observed
Test Time T	During Test	Release	After Release

$\mathcal{U}[T, N(T), (N - N(T))]$

FIGURE 6.2. One-Stage Testing.

for how long, is made at the completion of the previous stage. This allows the incorporation of information that becomes available during the previous testing stages into the decision process. A decision tree with the first two stages of a sequential test is given in Figure 6.4. Note that the decision tree continues at the top right to a third testing stage; indeed, the tree is actually infinite in extent, as we do not know beforehand how many stages of testing may be required.

The sequential testing plan seems to be the most satisfactory, because it allows for any number of stages and can adapt to experience gained from earlier testing periods. However, the infinite nature of the sequential testing tree presents a problem, since the solution of the tree requires us to peel back the branches of the tree from the terminal nodes, an operation that is not possible for a tree whose terminus is not known. Another way of thinking about this is to say that at any stage, the decision to test and for how long must take into account the possibility of an unknown number of further stages. Although theoretically unsatisfactory, in practice one can impose some upper bound on the number of testing stages allowed so that there is some stage after which release must take place. But, for even a moderately small upper bound on the number of testing stages, the resulting decision tree for the sequential plan can be large and difficult to solve, because of the successive expectations and maximizations that are required.

One proposal, that avoids the preceding computational difficulties, is to consider each testing stage on its own, that is, a testing plan that is simply a sequence of single-stage tests. At each stage, the decision must be made to release or to test, and information obtained from previous tests is used. In contrast to single-stage testing, described before, we call this *one-stage lookahead* testing, to reflect the fact that decisions are made without an accounting of all the possible future stages of testing. Such plans are a lot easier to implement, and still have the advantage of incorporating information learned through the testing process. One-stage lookahead plans cannot be represented as the full

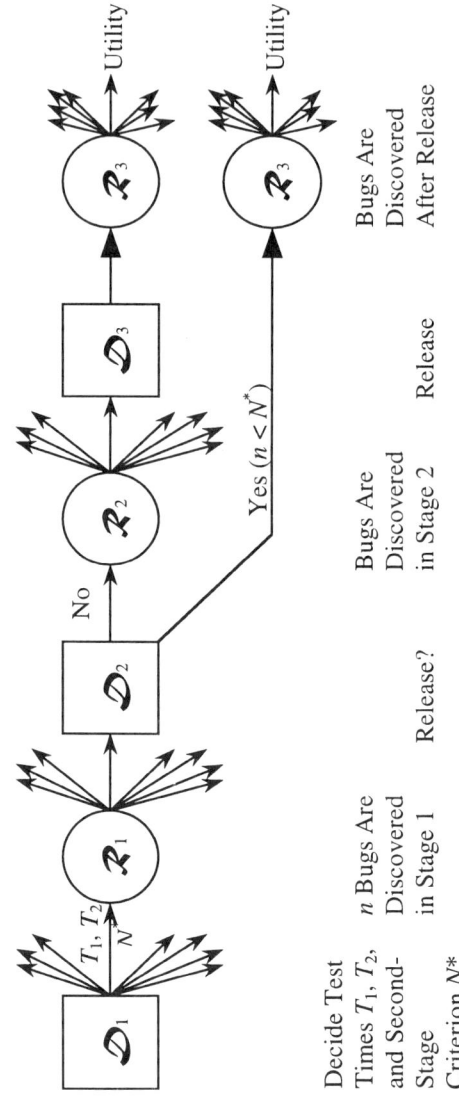

FIGURE 6.3. Two-Stage Testing.

6.5 Software Testing Plans 201

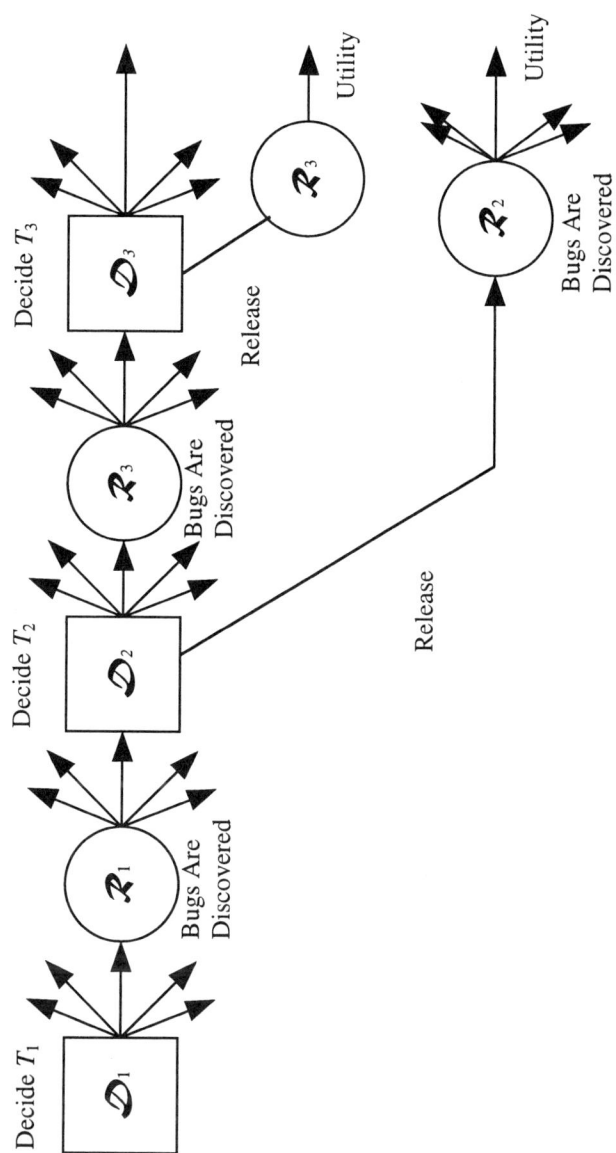

FIGURE 6.4. Sequential Testing.

blown decision tree of Figure 6.4, because of the lack of dependence on the future; they are an approximation of the sequential testing plan. However, under certain testing scenarios and conditions, it can be proved that one-stage lookahead tests are optimal in the sense that it is sufficient to look to the next stage and no further, in deciding whether to release; see, for example, Theorem 6.1.

Other types of one-stage lookahead plans are also possible. In the *fixed-time lookahead* plan, a sequence of times is specified at which testing is to stop and, when each time is reached, a decision is made whether to release or test until the next time in the sequence. These times may correspond to each day or week of testing. In *one-bug lookahead* plans, testing is conducted until a bug is found and fixed, at which point it is decided whether to release or continue testing until another bug is found.

As discussed before, there are two parts to a decision problem that must be assessed: the utility model (to quantify the economic aspects of the situation), and the probability model (to quantify the random aspects of the situation). The optimal action is to take the decision that maximizes expected utility. The rest of this chapter gives some examples of optimal testing under different probability models and utilities, and the different testing plans that have been described. In this way, we endeavor to give a fairly broad picture of the types of optimal testing that might be considered. We should mention that other approaches to the problem, that often lack the decision-theoretic character described here, have been proposed in the literature. Some suitable references are Dalal and Mallows (1988) (1990), Okumoto and Goel (1980), Ross (1985a), Forman and Singpurwalla (1977), Yamada, Narihisa, and Osaki (1984), and Randolph and Sahinoglu (1995). Clearly, this is an important issue where more research is needed.

It is also useful to note that there exists related work, not necessarily directed towards software testing, that is germane to the problems discussed here. For example, the work of Benkherouf and Bather (1988) on oil exploration, that of Ferguson and Hardwick (1989) on stopping rules for proofreading, the work of Efron and Thisted (1976) on estimating the number of unseen species, and the work of Andreatta and Kaufman (1986) on estimation in finite populations.

6.6 Examples of Optimal Testing Plans

6.6.1 One-Stage Testing Using the Jelinski–Moranda Model

As one of the most commonly discussed models, we now investigate the one-stage test using the Jelinski and Moranda model of Section 3.2.2. Recall that the ith failure time is exponentially distributed and has mean $(\Lambda(N - i + 1))^{-1}$, where N is the total number of bugs in the code and Λ is a constant:

$$P(T_i \geq t \mid N, \Lambda) = e^{-\Lambda(N-i+1)t}. \tag{6.2}$$

When statistical inference for this model was discussed in Section 4.2.3, we considered as a prior for N, a Poisson distribution with mean θ, and as a prior model for Λ, a gamma distribution with parameters α and β.

The decision tree for one-stage testing under this model was given in Figure 6.2, but with the model in place we can be precise about what is observed at the random nodes. Following a decision to test for a time T, suppose that $N(T)$ bugs are observed, with interfailure times t_i, $i = 1, \ldots, N(T)$, and a final period of length $T - \sum_{i=1}^{N(T)} t_i$ in which no bugs occur. Thus the random quantities at node $\mathcal{R}_1$ are $[t_1, \ldots, t_{N(T)}, T - \sum_{i=1}^{N(T)} t_i, N(T)]$. There is a chance that $N(T) = 0$, in which case the data consist only of $N(T) = 0$.

After testing for time T, release occurs. The second random node $\mathcal{R}_2$ generates the number of bugs, say $(N - N(T))$, that are discovered after release. The distribution used to describe this number must be conditional on all the random events and decisions that have occurred prior to reaching $\mathcal{R}_2$; in other words, the distribution over all the possible outcomes at the node is the posterior distribution for the number of remaining bugs $N - N(T)$, given the data from the testing phase. Using arguments similar to those used in Section 4.2.3, we can show (see Exercise 3) that the posterior distribution of N is

$$\mathcal{P}(N = n \mid t_1, \ldots, t_{N(t)}, T - \sum_i t_i, N(T))$$

$$= \frac{We^{-\theta}\theta^n}{(n-N(T))!} (\alpha + tT + (n - N(T))T)^{-(\beta+N(T))}, \tag{6.3}$$

where W is a normalizing constant, and

$$t = \begin{cases} 0, & \text{if } N(T)=0, \\ \frac{1}{T}\sum_{j=1}^{N(T)} (N(T)-j+1)t_j, & \text{otherwise.} \end{cases} \tag{6.4}$$

We note that only T, $N(T)$, and t are needed from the branches of the tree to specify the posterior distribution. Thus, from now on, we denote the data from the testing stage as $(t, N(T))$, where t is as previously defined.

Specification of a Utility Function

The costs and benefits of testing software fall into two groups: the cost of testing and removing bugs during the testing phase, and the cost of leaving bugs

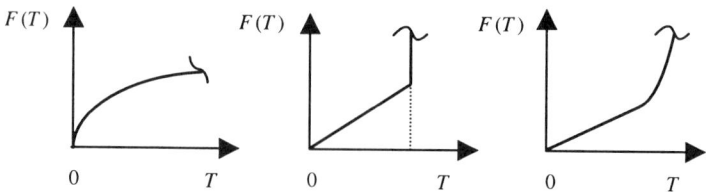

FIGURE 6.5. Candidate Shapes for Testing Cost $F(T)$ Versus Testing Time T.

in the code that are discovered after release. As regards the testing phase, we assume that there is a fixed cost C associated with removing each bug encountered, so the cost of fixing bugs is $C \cdot N(T)$. The other costs of testing for a time T are quantified by an increasing function of T, say $F(T)$. This function is supposed to account for other costs that are a function of time, such as payment to staff, possible penalties for late release, and the "opportunity cost" of not releasing the software. Even though this function is difficult to specify, there are several possibilities: linear, with $F(T) = f \times T$; a power law, with $F(T) = f \times T^g$; linear up to a threshold, say t_{max}, and infinite beyond (so that the testing period has a deadline); or linear initially and then exponential to reflect the increasing costs of product obsolescence. These forms are illustrated in Figure 6.5; they have been suggested by Dalal and Mallows (1990).

After release, we assume that the costs of encountering and eliminating bugs are on an exponential scale; that is, if $\bar{N}(T)$ is the number of bugs found after testing to T, then the disutility of the $\bar{N}(T)$ bugs is $D_1 - D_2 \exp(-D_3 \bar{N}(T))$, where D_1, D_2, and D_3 are nonnegative constants. Thus if a large number of bugs are discovered after release, then the cost is D_1, and if none are discovered the cost is $D_1 - D_2$. It is meaningful to suppose that D_2 equals D_1, and that D_1 is large.

Finally, we may also include a discounting factor e^{-hT} to account for the fact that these costs are incurred at a future date. Thus $\mathcal{U}[T, (t, N(T)), \bar{N}(T)]$, the utility of testing for a time T, during which time data $(t, N(T))$ are observed, and then releasing after which $\bar{N}(T)$ bugs are discovered, is

$$e^{-hT}[D_2 e^{-D_3 \bar{N}(T)} - D_1 - CN(T) - F(T)], \qquad (6.5)$$

where $C_1, D_1, D_2, D_3, h > 0$.

The preceding specification is for a very simple model of utilities; it can be criticized on several grounds. For one, the model supposes that each bug is equally costly to fix. For another, it assumes that the disutility of bugs discovered

6.6 Examples of Optimal Testing Plans

after release is independent of the time at which they are encountered. We would expect that bugs discovered during the latter phases of a software's lifecycle would have a smaller impact on customer goodwill than those discovered early on. However, the proposed model has the advantage of tractability, and allows us to illustrate the workings of the general principles. The model can be extended to represent other more realistic scenarios. It is useful to note that the notion of time does not necessarily mean clock time—the CPU time may be a better standard. But no matter what time scale is used, the entities $F(T)$, $N(T)$, and $\bar{N}(T)$ convey the same message.

Armed with a probability model and a utility function, we can now solve the single-stage problem. Recall that this involves peeling back the branches of the tree, taking the expectation over random nodes, and taking the maximum utility over decision nodes. The process starts with taking the expectation over $\bar{N}(T)$, the number of remaining bugs, using the posterior distribution, and continues with taking the expectation over the distribution of $N(T)$ and t. This gives an expected utility for testing to any time T, denoted by $\mathcal{U}(T)$, and $\mathcal{U}(T)$ turns out to be [see Singpurwalla (1991) for details]

$$\sum_{k=1}^{\infty} \left(\int_t \left\{ \sum_{j=0}^{\infty} \mathcal{U}(T, (t, N(T) = k), \bar{N}(T) = j) \frac{W_1 e^{-\theta} \theta^{k+j}}{j!} (\alpha + tT + jT)^{-(\beta+k)} \right\} \right.$$

$$\left. \times \int_0^{\infty} \frac{e^{-\alpha\lambda}(\alpha\lambda)^{\beta-1}\alpha\,(\lambda T)^k e^{-\lambda Tt}}{\Gamma(\beta)\,(1-e^{-\lambda T})^k\,(k-1)!} \sum_{i=0}^{k_0} (-1)^i \binom{k}{i} (t-i)^{k-1}\,d\lambda\,dt \right)$$

$$\times \left(\int_0^{\infty} \frac{\exp(-\theta(1-e^{-\lambda T}))}{k!} (\theta(1-e^{-\lambda T}))^k \frac{e^{-\alpha\lambda}(\alpha\lambda)^{\beta-1}}{\Gamma(\beta)} \alpha d\lambda \right)$$

$$+ \left(\left\{ \sum_{j=0}^{\infty} \mathcal{U}(T, (t=0, N(T)=0), \bar{N}(T) = j) \frac{W_2 e^{-\theta} \theta^j}{j!} (\alpha + jT)^{-\beta} \right\} \right.$$

$$\left. \times \left\{ \int_0^{\infty} \exp(-\theta(1-e^{-\lambda T})) \frac{e^{-\alpha\lambda}(\alpha\lambda)^{\beta-1}}{\Gamma(\beta)} \alpha d\lambda \right\} \right), \quad (6.6)$$

where

$$W_1 = \left(\sum_{j=0}^{\infty} \frac{e^{-\theta}\theta^{k+j}}{j!(\alpha+tT+jT)^{\beta+k}} \right)^{-1}, \qquad (6.7)$$

$$W_2 = \left(\sum_{j=0}^{\infty} \frac{e^{-\theta}\theta^{j}}{j!(\alpha+jT)^{\beta}} \right)^{-1}, \qquad (6.8)$$

k_0 is the integer part of t, and the range of integration for t is $k_0 < t < k_0 + 1$, for each $k_0 = 0, 1, \ldots, k-1$.

Although Equation 6.6 looks formidable, its numerical evaluation is not too difficult, and maximization with respect to T is made easier by the observation that $\mathcal{U}(T)$ is generally unimodal. The complicated nature of Equation (6.6) is brought about by the required preposterior analysis which requires us to average out the quantity t; see Equation (4.4). An extension of this model to the two-stage testing plan is discussed in Singpurwalla (1989b).

6.6.2   One- and Two-Stage Testing Using the Model by Goel and Okumoto

Optimal testing using the model of Goel and Okumoto turns out to be considerably more tractable than that involving the Jelinski–Moranda model. This is because here we monitor only the number of failures, and not the times of failure. Indeed, under a simple utility function one can even obtain a closed form solution to the one-stage test. Solutions for the two-stage and also the one-stage lookahead tests require only a moderate amount of numerical computation.

Recall (see Section 3.3.1) that in the Goel–Okumoto model, bugs are encountered as a Poisson process with a mean value function $a(1 - e^{-bt})$. Following the strategy of Section 4.2.4, we place independent gamma priors on the parameters a and b; that is, given λ, τ, α, and μ, we let

$$\pi(a, b) = \left(\frac{\lambda^\tau}{\Gamma(\tau)} a^{\tau-1} e^{-\lambda a} \right) \times \left(\frac{\alpha^\mu}{\Gamma(\mu)} b^{\mu-1} e^{-\alpha b} \right). \qquad (6.9)$$

In Section 4.2.4, expressions for the posterior distribution of a and b were also given.

A simple utility function, very similar in form to that used in the previous example, is also adopted here. Testing for time T, where $N(T)$ bugs are discovered and fixed, followed by release where $\overline{N}(T)$ bugs are discovered, the utility function is

$$\mathcal{U}[T, N(T), \overline{N}(T)] = P - C \times N(T) - D \times \overline{N}(T) - f \times T^g, \quad (6.10)$$

where P is the profit associated with releasing a perfect code without any testing (i.e., with $T = 0$), D is the cost of fixing a bug after release (typically, $D > C$), and $f \times T^g$ represents, as before, the cost of lost opportunity when testing to time T. This form of the utility function is similar to that used by Dalal and Mallows (1988) in their famous paper on the optimal time to release software.

For the case of a one-stage test, it can be shown (see Exercise 4) that the expected utility has the simple form:

$$\mathcal{U}(T) = E[\mathcal{U}(T, N(T), \overline{N}(T))]$$

$$= P - f \times T^g - \frac{\tau}{\lambda}[C + (D - C)\left(\frac{\mu}{\mu+T}\right)^\alpha]. \quad (6.11)$$

It is now easy to show that $\mathcal{U}(T)$ has a unique maximum at T^*, and that T^* satisfies the equation

$$[\mu + T^*]^{\alpha+1}[T^*]^{g-1} = \frac{\alpha \tau \mu^\alpha (D-C)}{\lambda f g}. \quad (6.12)$$

For $g = 1$ (i.e., if the cost of testing and of lost opportunity is linear in T), then we can obtain an explicit formula for T^*, namely,

$$T^* = \mu \left[\left(\frac{\tau \alpha (D-C)}{\lambda \mu f}\right)^{1/(\alpha+1)} - 1\right]; \quad (6.13)$$

the details are in McDaid and Wilson (1999).

Observe that T^* is a function of the parameters of the distribution a with only the difference $(D - C)$, the costs of fixing a bug in the field and in the test environment. Also, T^* is a function of the parameters of the distribution only through its prior mean τ/λ, and that T^* increases as τ/λ increases.

In the case of a two-stage test, the software is initially tested for a time T_1, at the end of which $N(T_1)$ bugs are encountered. Then a decision is made to release the software if $N(T_1)$ is less than N^*, our predetermined decision criterion. Once the software is released it may experience $\overline{N}(T_1)$ failures in the field. This sequence of events results in a utility $\mathcal{U}_1[T_1, N(T_1), \overline{N}(T_1)]$. If $N(T_1)$ is equal to or greater than N^*, the software is tested until time T_2, where $T_2 > T_1$, and then released. Let $N(T_2)$ be the cumulative number of bugs encountered when testing until T_2, and $\overline{N}(T_2)$ the number of bugs experienced by the software in the field. This latter sequence of events results in a utility $\mathcal{U}_2[T_1, N(T_1), T_2, N(T_2), \overline{N}(T_2)]$.

The decision at node $\mathcal{D}_1$ is to choose T_1, T_2, and N^* such that the expected utility

$$\sum_{N(T_1)=0}^{N^*-1} \sum_{\overline{N}(T_1)=0}^{\infty} \mathcal{U}_1[T_1, N(T_1), \overline{N}(T_1)] \times \mathcal{P}(N(T_1), \overline{N}(T_1)) +$$

6. The Optimal Testing and Release of Software

$$\sum_{N(T_1)=N^*}^{\infty} \sum_{N(T_2)=N(T_1)}^{\infty} \sum_{\bar{N}(T_2)=0}^{\infty} \mathcal{U}_2[T_1, N(T_1), T_2, N(T_2), \bar{N}(T_2)] \times$$

$$\mathcal{P}(N(T_1), \bar{N}(T_2), \bar{N}(T_2)),$$

is maximized.

In choosing the utility functions $\mathcal{U}_1$ and $\mathcal{U}_2$ we adopt forms that are identical to that chosen for the one-stage case. Specifically,

$$\mathcal{U}_1[T_1, N(T_1), \bar{N}(T_1)] = P - C \times N(T_1) - D \times \bar{N}(T_1) - f \times T_1^g,$$

and

$$\mathcal{U}_2[T_1, N(T_1), T_2, N(T_2), \bar{N}(T_2)]$$
$$= P - C \times N(T_2) - D \times \bar{N}(T_2) - f \times T_2^g.$$

Thus the decision at node $\mathcal{D}_1$ boils down to finding those values of T_1, T_2, and N^* that maximize

$$P - f \times T_2^g - \frac{C\tau}{\lambda} - \frac{(D-C)\tau}{\lambda} \left(\frac{\mu}{\mu+T_2}\right)^{\alpha}$$

$$+ f \times T_1^g \frac{\mu^{\alpha}\lambda^{\tau}}{\Gamma(\alpha)\Gamma(\tau)} \times \sum_{n=0}^{N^*-1} \frac{\Gamma(n+\tau)}{n!} \int_0^{\infty} \frac{(1-e^{-bT_1})^n b^{\alpha-1} e^{-\mu b}}{(1+\lambda-e^{-bT_1})^{n+\tau}} db$$

$$- \frac{(D-C)\mu^{\alpha}\lambda^{\tau}}{\Gamma(\alpha)\Gamma(\tau)} \sum_{n=0}^{N^*-1} \frac{\Gamma(n+\tau+1)}{n!} \int_0^{\infty} \frac{(1-e^{-bT_1})^n b^{\alpha-1} e^{-\mu b}(e^{-bT_1}-e^{-bT_2})}{(1+\lambda-e^{-bT_1})^{n+\tau+1}} db$$

over $T_1 > 0$, $T_2 > T_1$, and $N^* = 0, 1, 2, \ldots$. Let us denote these optimal values by T_1^*, T_2^*, and $\widehat{N}^*$, respectively.

The preceding maximization can be done numerically. Observe that the foregoing expression will degenerate to a one-stage equivalent if T_2^* turns out to be zero, or if $\widehat{N}^*$ turns out to be infinite.

Eliciting Parameters of the Prior Distribution and the Utility Function

It is evident from Equations (6.12) and (6.13) that decision making in a single-stage testing environment depends solely on the prior parameters and the constants in the utility function. Thus it is crucial that these quantities be judiciously elicited and selected.

A simple approach for eliciting the prior parameters is as follows. To start with, a location (mean) m_1, and spread (standard deviation) s_1, are assessed for the total expected number of bugs that will be discovered over the lifetime of the software. These quantities are then equated to $E(a) = \tau/\lambda$, and $Var(a) = \tau/\lambda^2$, respectively. This implies that

$$\tau = \frac{m_1^2}{s_1^2}, \quad \text{and} \quad \lambda = \frac{m_1}{s_1^2}. \tag{6.14}$$

To help specify m_1 and s_1, we recall that several empirical formulae for the number of bugs per line of code, for different languages, have been proposed; see Section 4.3.4 on the elicitation of priors for the logarithmic-Poisson model of Musa and Okumoto (1984).

To specify α and μ, a time T' is selected, and a location m_2 and spread s_2 are elicited for $\Lambda(T')$, the expected number of bugs that are discovered by T'. A good choice for T' may be the software development team's initial estimate of the testing time, prior to any calculations. We equate m_2 to $E[\Lambda(T')]$, where

$$E(\Lambda(T')) = E(a(1 - e^{-bT'}))$$

$$= E(a)(1 - E(e^{-bT'}))$$

$$= m_1 \left(1 - \left(\frac{\mu}{\mu+T'}\right)^\alpha\right), \tag{6.15}$$

and s_2^2 to $Var(\Lambda(T'))$ which, after calculations similar to those of Equation 6.15, becomes

$$Var(\Lambda(T')) = (s_1^2 + m_1^2)\left[1 - 2\left(\frac{\mu}{\mu+T'}\right)^\alpha + \left(\frac{\mu}{\mu+2T'}\right)^\alpha\right]$$

$$- m_1^2 \left[1 - \left(\frac{\mu}{\mu+T'}\right)^\alpha\right]^2. \tag{6.16}$$

Equations 6.15 and 6.16 are then solved numerically to obtain the parameters α and μ. As with the elicitation for the Musa and Okumoto model,

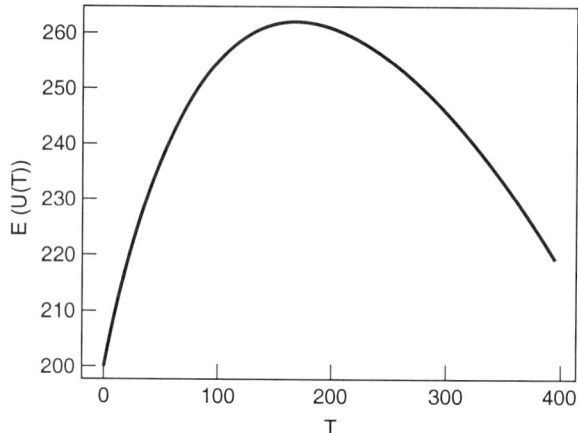

FIGURE 6.6. Expected Utility for the Single-Stage Test.

the elicited values m_1, s_1, m_2, and s_2 may be elicited from an expert and can be modulated by the analyst.

As regards the parameters of the utility function, one might start with g, the exponent of the opportunity cost function. A linear or quadratic cost would suffice in most situations, restricting this to a choice of $g = 1$ or 2. Arbitrarily designating the in-testing fix cost as $C = 1$, one can then think about how many times more costly it would be to fix a bug after release; this is D. The profit P associated with releasing a perfect code can also be assessed as a multiple of C. Finally, to elicit a value of f, one might look at X, the opportunity cost of testing to a certain time in the future, say T^0; equating X to $f \times T^g$ implies that $f = X/((T^0)^g)$. As with the prior parameter elicitation, we may take T^0 to be the time until which it is thought testing might continue.

Illustrative Example

Suppose that a software testing team is deciding on the length of time to test a piece of software. They decide that the opportunity cost function is linear, so $g = 1$. Specifying $C = 1$, they decide that the cost of fixing bugs after release is 10 times C, so $D = 10$, and that the profit of the perfect software is $500C$, so $P = 500$. Finally, for the value of f, they look at X, the opportunity cost of testing to a time $T^0 = 20$, and decides that it is $10C$; thus f is 0.5.

As regards the parameters of the prior probability, the team decides, perhaps with the aid of the formulae given earlier, that they expect 30 bugs in the code, but that the uncertainty in this figure is quite large, so they set $m_1 = 30$ and $s_1 = 5.5$. Using the Equations (6.14) yields $\tau = 30$ and $\lambda = 1$. Looking at the number of bugs to be discovered by time $T^0 = 20$, they decide that it will be between 0 and 5, finally settling on $m_2 = 2.8$ and $s_2 = 1.3$, from which solutions of Equations (6.15) and (6.16) yield $\alpha = 5$ and $\mu = 1000$. In Figure 6.6 we plot, for these values of the prior parameters and the constants, $E[\mathcal{U}(T, N(T), \overline{N}(T))]$ as functions of T. The value of T, say T^* at which the preceding expected utility is a maximum is about 180. We emphasize that this value of T^* is based on prior information alone.

For the utility and the prior parameters that we have used in the illustrative example discussed before, the optimal two-stage strategy turns out to be $T_1^* = 116.5$, $T_2^* = 188.9$, and $\widehat{N}^* = 8$. It is instructive to note that with one-stage testing, the optimal strategy resulted in $T^* = 180$. Therefore, a consideration of a two-stage strategy has resulted in the reduction of the first-stage testing time from 180 to 116.5. This is reasonable because the anticipation of a possible second-stage testing provides us with a cushion of a smaller first-stage testing.

6.6.3 One-Stage Lookahead Testing Using the Model by Goel and Okumoto

The mathematical simplicity of this model allows us to design a one-stage lookahead test quite easily. By way of notation, define T_j^* to be the optimal time as measured from 0, at which testing is to stop for the jth stage of testing. With this notation, testing stops and the software released at the first stage j for which $T_j^* \leq T_{j-1}^*$. Note that it is conceivable to obtain $T_j^* < T_{j-1}^*$, since this feature implies that we may have already tested more than what is necessary.

The first stage of testing in the one-stage lookahead scheme would be to use the prior distributions on a and b, so that T_1^* is the solution to the single-stage test, based on Equation (6.12). After the first stage, suppose that stage j, $j \geq 2$, has been completed and a totality of $n_j = N(T_j^*)$ bugs discovered by time T_j^*. Suppose that the interarrival times $t_1, \ldots, t_{n_j}$ have been observed. Then the posterior distribution of a and b is given by

$$\pi(a, b \mid n_j, t_1, \ldots, t_{n_j}) = \frac{\pi(a, b)\, P(n_j, t_1, \ldots, t_{n_j} \mid a, b)}{\int\int \pi(a, b)\, P(n_j, t_1, \ldots, t_{n_j} \mid a, b)\, da\, db},$$

$$= a^{n_j + \tau - 1} e^{-(1+\lambda)a} b^{n_j + \alpha - 1}\, e^{-(\mu + s_j)b} \times$$

$$e^{\left\{ ae^{-T_j^* b} \Gamma(\tau + n_j) \int_0^\infty g_j(b)\, db \right\}}, \qquad (6.17)$$

where

$$g_j(b) = \frac{b^{n_j+\alpha-1} e^{-(\mu+s_j)b}}{(\lambda+1-e^{-T_j^* b})^{\tau+n_j}}, \qquad (6.18)$$

$s_j = \sum_{i=1}^{n_j} t_i$, is the time at which the last bug is encountered. For convenience, we have suppressed the times taken to rectify the bugs. Using this posterior distribution, the (posterior) expectations of $N(T)$ and $\overline{N}(T)$ are calculated to be

$$E(N(T) \mid T_j^*, n_j, s_j) = (n_j + \tau) \frac{\int_0^\infty \frac{1-e^{-Tb}}{(\lambda+1-e^{-T_j^* b})} g_j(b) db}{\int_0^\infty g_j(b) db}, \qquad (6.19)$$

and

$$E(\overline{N}(T) \mid T_j^*, n_j, s_j) = (n_j + \tau) \frac{\int_0^\infty \frac{e^{-Tb}}{(\lambda+1-e^{-T_j^* b})} g_j(b) db}{\int_0^\infty g_j(b) db}. \qquad (6.20)$$

We can now obtain the expected utility of testing to time T, at stage $j+1$, as

$$\begin{aligned}\mathcal{U}_j(T) &= E[\mathcal{U}(T, N(T), \overline{N}(T)) \mid T_j^*, n_j, s_j] \\ &= P - f \times T^g - C \times E[N(T) \mid T_j^*, n_j, s_j] - \\ & \quad D \times E[\overline{N}(T) \mid T_j^*, n_j, s_j]. \end{aligned} \qquad (6.21)$$

Let T_{j+1}^* be that value of T which maximizes $\mathcal{U}_j(T)$ given previously. Then if $T_{j+1}^* \leq T_j^*$ testing stops; otherwise testing continues for an additional $T_{j+1}^* - T_j^*$ units of time, and the process moves to stage $j+2$.

The evaluation of Equation (6.21) is not as straightforward as the equivalent expression for the single-stage test, but it is still easy to numerically compute, requiring nothing more complex than the one-dimension integrations of Equations (6.19) and (6.20).

6.6.4 Fixed-Time Lookahead Testing for the Goel–Okumoto Model

In allowing the testing team to adapt testing to the software's performance, the one-stage lookahead plan can present problems for management. This is because at time 0, when the testing plan is designed, it is not known how long the testing will take or even the stage at which the testing will terminate. One solution to these problems is to propose a predefined sequence of times, $T_1, T_2, \ldots, T_J$, at which testing stops, and a decision is made to either release or

6.6 Examples of Optimal Testing Plans

continue testing to the next time. This is a one-stage lookahead policy where each decision is binary; at T_j, one either releases or tests further to the prespecified time T_{j+1}. At some stage J, release is mandatory; in practice, J can be made large. The prespecification of the mandatory release time T_J is not necessarily contrary to practice. We often hear managers of software development houses say "we will release the software when we said we will do so."

This plan has some advantages. The nature of the decision makes the plan easy to implement. Usually, the predefined times will be a regular arithmetic sequence, perhaps corresponding to each day or week of testing. With decisions occurring at known times, planning for the testing phase is made easier. Theoretical results on the optimality of the plan can be proved. However, it does have some disadvantages. The success of the testing schedule depends on what times are specified; too short an interval between times causes the testing to stop too early, and too large an interval risks over-testing.

Let N_j represent the number of bugs discovered in the jth stage, that is, $N_j = N(T_j) - N(T_{j-1})$, and let $\mathcal{D}_j$ denote all the information that is available up to T_j; this consists of the prior hyperparameters, the number of failures in each interval $N_1, \ldots, N_j$, and the last failure time s_j prior to T_j. From the previous sections, we know that these are sufficient statistics under this model. We emphasize that the times $T_1, T_2, \ldots, T_J$ are fixed at the start and are therefore not optimal in any sense.

The utility associated with the release of the program at the end of a stage j is of the same form as in earlier plans; namely,

$$\mathcal{U}_j[T_j, (N_1, N_2, \ldots, N_j), \bar{N}(T_j)] = P - C \times \sum_{k=1}^{j} N_k - D \times \bar{N}(T_j) - F(T_j). \tag{6.22}$$

The constants are exactly as before; the only change is that now we allow the opportunity loss function to be of a more general form, say $F(T)$, rather than the previous $f \times T^g$. The reason for this departure becomes clear later, when the optimality of the plan is mentioned; see Section 6.6.6.

At the completion of a stage j, the decision to be made is to release or to continue testing to T_{j+1}. The utility of releasing now is given by the expectation of Equation (6.22), where the expectation is with respect to the only unknown quantity in Equation (6.22), namely, $\bar{N}(T_j)$. We write this as

$$E(\mathcal{U}_j \mid \mathcal{D}_j) = \sum_{\bar{N}(T_j)=0}^{\infty} \mathcal{U}_j[T_j, (N_1, N_2, \ldots, N_j), \bar{N}(T_j)] \times \mathcal{P}(\bar{N}(T_j) \mid \mathcal{D}_j). \tag{6.23}$$

The utility of testing to the next stage is

$$\mathcal{U}_{j+1}[T_{j+1}, (N_1, N_2, \ldots, N_{j+1}), \overline{N}(T_{j+1})]; \quad (6.24)$$

here, N_{j+1} and $\overline{N}(T_{j+1})$ are unknown and so we must take the expectation with respect to these two quantities. We write this expected utility as:

$$E(\mathcal{U}_{j+1} \mid \mathcal{D}_j) = \sum_{N_{j+1}, \overline{N}(T_{j+1})} \mathcal{U}_{j+1}[T_{j+1}, (N_1, \ldots, N_{j+1}), \overline{N}(T_{j+1})] \times$$

$$\mathcal{P}(N_{j+1}, \overline{N}(T_{j+1}) \mid \mathcal{D}_j). \quad (6.25)$$

For both expectations, the distributions used are the posterior distributions conditional on $\mathcal{D}_j$. Thus the decision criterion at stage j is to test to T_{j+1} if $E(\mathcal{U}_j \mid \mathcal{D}_j) \leq E(\mathcal{U}_{j+1} \mid \mathcal{D}_j)$; otherwise release. This reduces to a decision to test if and only if

$$(D - C)[E(\overline{N}(T_j) \mid \mathcal{D}_j) - E(\overline{N}(T_{j+1}) \mid \mathcal{D}_j)] \geq F(T_{j+1}) - F(T_j). \quad (6.26)$$

Thus it is necessary, upon completion of a stage j, to evaluate the two expectations $E(\overline{N}(T_j) \mid \mathcal{D}_j)$ and $E(\overline{N}(T_{j+1}) \mid \mathcal{D}_j)$ for which the appropriate formula is that given by Equation (6.20).

6.6.5 One-Bug Lookahead Testing Plans

An alternative to the fixed-time one-stage lookahead plan is the one-bug lookahead plan. This is the plan considered by Ozekici and Catkan (1993), and by Morali and Soyer (1999). Here instead of testing until a prespecified time (as was done in the fixed-time one-stage lookahead plan), we test until a bug is encountered. When the bug is discovered and fixed, a decision has to be made whether to release the software or to test it until we observe the next bug. For this scheme to make sense it is assumed that testing until the next bug does not entail testing forever. Since the uncertainty here is about the time to occurrence of the next bug, a natural probability model to use is one that belongs to the Type I category. Morali and Soyer consider the non-Gaussian Kalman filter model of Section 3.4.2, and we focus upon this for purposes of discussion. Recall [see Equations (3.23) and (3.24)] that the system equation for this model was specified in terms of the parameter θ_i, $i = 1, 2, \ldots$, where θ_i is the scale parameter of the gamma distribution of T_i; T_i is the ith interfailure time.

Recall that the utility function is made up of two cost components: the cost of testing, and the cost of an in-service failure. The former is a function of the amount of test time, and the latter can be based on the time to encounter an in-service failure, or on the failure rate of the software subsequent to its release. Morali and Soyer do the latter by considering θ_i as a proxy for the failure rate of T_i. Instead of basing the utility on the failure rate, we could also base it on the reliability of the software, as was done in Singpurwalla (1991). The reliability function being bounded between zero and one is a natural candidate for utilities; recall that utilities are probabilities and obey the rules of probability.

To move along, suppose that the ith bug has been encountered and corrected, and that the non-Gaussian Kalman filter model of Section 3.4.2 is used. Let $\mathcal{U}_T(T_{i+1})$ denote the utility of testing to the next bug, supposing that it is encountered at time T_{i+1}. Assuming, as before, a linear cost of testing, suppose that $\mathcal{U}_T(T_{i+1}) = -f \times T_{i+1}$, where f is a constant. Similarly, let $\mathcal{U}_R(\theta_{i+1})$ denote the utility of releasing the software subsequent to correcting the ith bug. Then a simple model for $\mathcal{U}_R(\theta_{i+1})$ would be $\mathcal{U}_R(\theta_{i+1}) = -D \times \theta_{i+1}$, where D is a constant. The subscripts T and R associated with $\mathcal{U}$ represent "test" and "release," respectively. Clearly, we would stop testing after correcting the ith bug, and release the software, if the expected value of $\mathcal{U}_R(\theta_{i+1})$ were greater than the expected value of $\mathcal{U}_T(T_{i+1})$. These expectations can be calculated (in actuality, computed) using the approach outlined in Section 4.5.3.

6.6.6 Optimality of One-Stage Lookahead Plans

The one-stage lookahead plans described in Sections 6.6.3 to 6.6.5 are restrictive because they do not consider future stages of testing. For this reason it is difficult, in general, to say in what sense they might be optimal. The usual strategy is to say that, were it decided to stop after the ith stage on the basis of looking at testing to the $(i + 1)$th stage, then the utility of subsequently testing to stages $(i + 2), (i + 3), \ldots$, would be no better. Thus, it is sufficient just to look one stage ahead when deciding on the optimal stopping time.

Such sufficiency results have been proved in a variety of settings. A general result, when $N(T)$ is a Markov chain is well known; see, for example, Ross (1970). With regard to the plans described here, McDaid and Wilson (1999) have shown, under certain conditions on the utility function and the probability model, that the fixed-time lookahead plan is optimal in the sense given by Theorem 6.1.

Theorem 6.1 [McDaid and Wilson (1999)]. *Consider the fixed-time lookahead testing plan with the utility function given by Equation (6.22) and with the maximum number of testing stages J. Let $F(T_i)$ be discrete convex in i, and let $E(\overline{N}(T_i) \mid \mathcal{D}_j)$ be discrete concave in i, for $i = j, \ldots, J$. Then, after the completion of j stages of testing, the following stopping rule is optimal.*

If

$$E(\mathcal{U}_{j+1} \mid \mathcal{D}_j) \geq E(\mathcal{U}_j \mid \mathcal{D}_j),$$

216 6. The Optimal Testing and Release of Software

then test to T_{j+1}; otherwise, release.

Morali and Soyer (1999) have proved an equivalent result for their one-bug lookahead plan.

6.7 Application: Testing the NTDS Data

Consider the NTDS data of Table 4.2. How would the testing plans described in Section 6.6 fare under these data, supposing that the model of Goel and Okumoto were to be entertained?

To start with, suppose that the necessary parameter values and the values of the constants are those that were specified in the illustrative example of Section 6.6.2. Then, under the one-stage test, the optimal test time would be $T^* = 180$, and the expected utility about 260; see Figure 6.6. Note that according to the entries in columns one and three of Table 4.2, testing until 180 units of time will reveal $n_1 = 23$ failures, with the last failure occurring at $s_1 = 156$.

For the case of a two-stage test, since 21 bugs were discovered by 116 units of test time a second stage of testing is necessary. This will result in the discovery of 26 (or possibly 27) bugs in total. By contrast a one-stage procedure calls for a testing time of 180 and results in the discovery of 23 (or possibly 24) bugs.

We now consider what would happen if the one-stage lookahead plan of Section 6.6.3 were applied to the NTDS data. Using $n_1 = 23$ and $s_1 = 156$ in Equation (6.21), we compute $T_2^* = 218.4$; the expected utility is about 288. Thus testing for the second stage involves $(218.4 - 180) = 38.4$ additional units of time; recall that under the one-stage test the optimal test time was 180. But according to the entries of Table 4.2, no further bugs were discovered during this additional 38.4 time units of testing, and so $n_2 = 23$ and $s_2 = 156$. Repeating the calculation involving Equation (6.21) we see that $T_3^* = 201.4$. Since $T_3^* < T_2^*$, our decision would be to stop testing. To conclude, if the model by Goel and Okumoto and the previously specified constants were to be invoked with a single-stage lookahead procedure on the NTDS data, then testing would have stopped after the second stage, at 218.4 units of time, and would have yielded an expected utility of about 289. This is larger than 260, the expected utility of the single-stage test.

Suppose now that instead of the single-stage lookahead procedure we considered the fixed-time lookahead procedure of Section 6.6.4, with $T_i = 50i$, for $i = 1, 2, \ldots$. The initial decision is to test until 50 units of time or to release. The expected utilities under these two actions are 232 and 200, respectively. Thus the decision is to test for 50 units of time. This results (see Table 4.2) in $N_1 = 7$ failures (bugs) with $s_1 = 50$. Using these values of N_1 and s_1 to obtain the required posterior distributions, we note that testing to $T_2 = 100$ gives an expected utility of 402 against a utility of release at $T_1 = 50$ of 335. So the decision is to test to T_2. The procedure continues until $T_4 = 200$, whereupon the expected utility of testing to T_5 is 276, against the expected utility of release at

T_4 of 289; thus testing stops and the software is released after 200 units of testing. Therefore, both the one-stage lookahead and the fixed-time one-stage lookahead testing plans lead to roughly the same decision; namely, release after about 200 units of testing time.

With the one-bug lookahead plan the first decision is to test until the first bug is discovered or to release immediately. The expected utilities are 208 and 200, respectively, so it is decided to test until the first bug; this occurs after 9 units of time. Next, using $N(9) = 1$ and $s_1 = 9$ in the appropriate posterior distribution, a decision is made again, with the expected utility of further testing of 217 against the expected utility of release of 212. Testing continues in this manner until the 24th bug occurs at time 247, at which the expected utility of testing to the 25th bug is 270 against 277 for release. Thus under the one-bug lookahead plan testing would stop at 247 time units with 24 of the 34 (recorded) bugs discovered. Contrast this with stopping at 200 time units with 23 of the 34 bugs discovered. The expected utility in the former case is 277; in the latter case it is 289. Thus, from a retrospective viewpoint, the one-stage lookahead plan would have been the optimal plan to use. It would have resulted in the decision to release the software after 200 units of test time and would have yielded an expected utility of 289.

6.8 Chapter Summary

This chapter pertains to a fundamental problem faced by software developers and managers, namely, how long to test a piece of software prior to its release? This is a problem of decision making under uncertainty and involves a tradeoff between costs and risks. Thus to facilitate such a discussion, the chapter begins with an overview of normative decision theory and utility theory. The theory specifies the maximization of expected utility (MEU) as a criterion for making optimal decisions. A key consideration that drives us to the MEU principle is that utility is a probability, and thus obeys the calculus of probability. The MEU principle then follows via the law of total probability.

The remainder of the chapter pertains to an application of the MEU principle for different types of software testing plans and under different models for failure. The different types of testing plans that are mentioned are: single-stage, two-stage, sequential, one-stage lookahead, fixed-time lookahead, and one-bug lookahead plans. The cost of testing pertains to the pre-fixing of discovered bugs and the loss of consumer confidence resulting from the release of unreliable code. On the other hand long test times add to the cost of the development effort, and contribute to the risk of obsolescence. The optimal test time balances these two competing criteria.

Sequential testing plans give rise to infinite decision trees. This problem is overcome in practice by imposing (arbitrarily) an upper bound on the number of testing stages that are allowed. The computational difficulties associated with sequential plans can also be overcome by the one-stage lookahead plans. That is, we make a decision without accounting for the possible future (i.e., two, three, four, etc.) stages of testing. However, such strategies are not optimal.

We gave a detailed example of an optimal single-stage testing plan using the model by Jelinski and Moranda, and described the setup when the testing was to be done for two stages. We made several assumptions about costs and utilities; these were based on the reported experiences of industry. The optimal single-stage problem, although simplistic in nature, poses difficult computational issues. These can only be addressed numerically. We also discussed optimal testing for the model by Goel and Okumoto under a single-stage test, and under a one-stage lookahead and a fixed-time one-stage look-ahead testing schemes.

The optimal testing of software is an important issue which calls for more research, especially research that will lead to approaches that are easy to use. We anticipate that more is going to be written on this topic. Perhaps the search for an omnibus simple-minded model for describing software failures is strongly justified by the need for developing an easy to use, but realistic, optimal testing strategy.

Exercises for Chapter 6

1. A decision maker wishes to select one of two software development organizations for producing software needed to run a large system. Let p_{i1} be the probability that the highest level achieved by organization 1 is i, for $i = 1, \ldots, 5$. Similarly, let p_{i2} denote the corresponding probabilities for organization 2. Let C_j, $j = 1, 2$, be the cost of developing the software quoted by organization j. Let U_i be the utility to the decision maker of software developed by an organization whose highest maturity level is i. Assume that, in general, $U_1 \leq U_2 \leq U_3 \leq U_4 \leq U_5$.

   Draw a decision tree to outline the steps that the decision maker takes to select one of the two software houses.

2. In Exercise 1 suppose that the p_{i1}s are as given in Table 5.2, and the p_{i2}s are as follows.

   $p_{12} = 0.4986,\qquad p_{22} = 0.3771,$
   $p_{32} = 0.0281,\qquad p_{42} = 0.0979,\qquad$ and $p_{52} < 0.001$.

   Assume that $U_1 = 0.2$, $U_2 = 0.4$, $U_3 = 0.6$, $U_4 = 0.8$, and $U_5 = 1$.

   (i) For what values of C_1 and C_2 will the decision maker choose organization 1 over organization 2?

   (ii) When will the decision maker flip an unbiased coin and choose organization 1 if the coin lands heads?

3. Verify Equation (6.3) of Section 6.6.1.

4. Consider the illustrative example of Section 6.6.2. Suppose that s_1 and s_2 are changed to 7 and 1, respectively. How does this change affect the optimal testing time of 180?

   What if m_1 and m_2 are changed to 15 and 5, respectively, with $s_1 = 5.5$ and $s_2 = 1.3$?

5. Verify Equations (6.11) through (6.13) of Section 6.6.2.

7
OTHER DEVELOPMENTS: OPEN PROBLEMS

7.0 Preamble

The scope of applicability of probabilistic ideas to address problems in software engineering is constantly expanding. Consequently, what has been covered is just a sample of the ultimate possibilities. Indeed, even now, there are several topics that are currently being researched that have not been highlighted by us. Some of these are: the use of stochastic process models (such as birth and death processes) for describing the evolution and maintenance of software, software certification and insurability, the incorporation of an operational profile for reliability assessment, embedding the CMM of Chapter 5 into a decision-theoretic framework, statistical aspects of software testing and using experimental designs for the testing of software, reliability assessment when testing reveals no failures, the integration of module and system testing, and so on. The aim of this chapter is to provide a bird's-eye view of some of these topics, and to put forth some open problems that they pose. It is hoped that this will inspire other researchers to pursue each topic in more detail than what we report. With this in mind, we have selected three topics for further discussion: dynamic modeling and the operational profile, statistical aspects of software testing and experimental designs for developing software testing strategies, and the integration of module and system performance.

222   7. Other Developments: Open Problems

7.1 Dynamic Modeling and the Operational Profile

Intuitively, a *dynamic model* is one wherein the future development of a process is explained, among other things, in terms of its past history. We have already encountered dynamic models in our discussion of self-exciting point processes and the software reliability models generated by concatenating failure rates. The model of Section 3.6 is a particularly instructive example. Dynamic models have played a key role in the biostatistical literature vis-à-vis their applications in survival analysis; see, for example, the survey by Andersen and Borgan (1985). Their importance derives from the famous Doob–Meyer decomposition which is fundamental to the development of martingale theory. Thus to get an appreciation of the general structure of dynamic models it is helpful to start with a brief overview of the martingale property of stochastic processes and its associated terminology. To keep our exposition simple, we focus attention on a discrete time stochastic process, and conclude with a passing reference to the continuous time version.

7.1.1   *Martingales, Predictable Processes, and Compensators: An Overview.*

Let X_t, $t = 0, 1, 2, \ldots$, be a discrete time stochastic process; for convenience we suppose that X_t is the tth interfailure time of software undergoing a test–debug cycle. Since $(X_t - X_{t-1})$ denotes the change in the $\{X_t; t = 0, 1, 2, \ldots,\}$ process at time t, our "best" prediction of this change, were we to know the past history of the process $X_1, \ldots, X_{t-1}$, could be of the form

$$E(X_t - X_{t-1} \mid X_1, \ldots, X_{t-1}) \stackrel{\text{def}}{=} V_t ,$$

where E denotes an expectation.

Let $U_t = \sum_{i=1}^{t} V_i$; then U_t is simply the cumulative sum of our expected changes up to time t. Indeed, U_t is our "best" prediction of X_t based on $X_1, X_2, \ldots, X_{t-1}$. Since U_t is merely a prediction of X_t, we define the *error of prediction* via a random variable M_t, where

$$M_t = X_t - U_t, \quad t = 1, 2, \ldots . \tag{7.1}$$

The random variable M_t is an interesting quantity. It has the easily verified property that

$$E(M_t \mid X_1, \ldots, X_{t-1}) = M_{t-1} . \tag{7.2}$$

Since knowing $X_1, \ldots, X_{t-1}$ boils down to knowing M_{t-1}, Equation (7.2) says that the expected error in predicting X_t using $X_1, \ldots, X_{t-1}$ is the actual observed error in predicting X_{t-1} (using $X_1, \ldots, X_{t-2}$). This seemingly innocuous property is of fundamental importance in probability theory. To appreciate why, let us rewrite (7.2) as

$$E(M_t - M_{t-1} \mid X_1, \ldots, X_{t-1}) = 0, \qquad (7.3)$$

and focus on the stochastic process M_t, $t = 1, 2, \ldots$. Equation (7.3) says that the increments of the process M_t have expectation zero; that is, the process M_t has *orthogonal increments*. Contrast this to the process of Section 2.3.1 which has independent increments. Processes having the orthogonal increments property are called *martingales*; their defining characteristics are either Equation (7.2) or, equivalently, Equation (7.3). If we rewrite Equation (7.1) as

$$X_t = U_t + M_t, \quad t = 1, 2, \ldots, \qquad (7.4)$$

then we can see that the stochastic process X_t can be decomposed into two parts: a process U_t that sums up our best predictions of the changes in the X_t process, and a process M_t that sums up the errors of the predictions. The decomposition of Equation (7.4) is called a *Doob decomposition*, and the quantity $(M_t - M_{t-1})$ is called a *martingale difference*, or an *innovation*. This latter terminology reflects the fact that it is $(M_t - M_{t-1})$ that is the uncertain (or the new) part in the development of the process. Since U_t depends on $X_1, \ldots, X_{t-1}$, it is *known* at time t; consequently, the process U_t is known as a *predictable process*. Furthermore, since $M_t = X_t - U_t$, U_t is called a *compensator* of X_t.

Because U_t is made up of a sum of the V_is, the predictability of U_t implies the predictability of the V_is as well. Thus the process V_t is also a predictable process.

We are now ready to introduce the concept of a *dynamic statistical model* as any statistical parameterization of the predictable process V_t. A simple example is the linear model

$$V_t = \alpha_t R_t, \qquad (7.5)$$

where R_t is composed of predictable and/or observable stochastic processes, and α_t is some unknown parameter. As an example of the preceding, suppose that $R_t = X_{t-1}$; then, from Equations (7.4) and (7.5), we see that

$$X_t - X_{t-1} = \alpha_t X_{t-1} + (M_t - M_{t-1}), \quad \text{or that}$$

$$X_t = (1 + \alpha_t) X_{t-1} + (M_t - M_{t-1}). \qquad (7.6)$$

Since $(M_t - M_{t-1})$ is an innovation, Equation (7.6) is an autoregressive process of order one, with a varying coefficient $(1 + \alpha_t)$. Thus we see that our dynamic model encompasses the class of autoregressive processes; such processes were considered by us as models for tracking software reliability (see Section 3.4.1).

Note that the process V_t, and hence the process R_t, can be any predictable process and conceivably can be any complicated function of the past. In particular, R_t may also include covariates (see Section 2.4.3), as long as the covariates are a predictable process! It is this feature that will allow us to incorporate the operational profile as a covariate of the software failure process; more is said about this later in Section 7.1.3.

To summarize, the Doob decomposition is a way of representing almost any stochastic process. We have made almost no probabilistic assumptions in the kind of modeling done thus far; parameterizing the predictable part of the process does not involve probabilistic assumptions. The innovation part of the decomposition allows us to use martingale theory, like the martingale central limit theorem and the law of large numbers for martingales [cf. Kurtz (1983)], to write out likelihoods, and to investigate issues of estimation. This facility is of particular value to those who subscribe to the frequentist point of view for inference and decision making.

The Doob decomposition of Equation (7.4) generalizes to continuous time stochastic processes as well. When such is the case, the decomposition is known as the *Doob–Meyer decomposition*, and is written as

$$X_t = U_t + M_t, \quad \text{for } t \geq 0. \tag{7.7}$$

The preceding process M_t is still a martingale, and by analogy with the U_t of Equation (7.4), the U_t here is an integral of "best" predictions; that is,

$$U_t = \int_0^t V_s \, ds.$$

The predictability of the V_t process is ensured by requiring that each V_t be known just before t.

A stochastic process X_t, be it in discrete or in continuous time, having the decomposition of Equation (7.4) or (7.7) is known as a *semimartingale*; the qualifier "semi" reflects the fact that one member of the decomposition, namely, U_t, is not a martingale. The material of this section is abstracted from Aalen's (1987) masterful exposition on dynamic modeling and causality.

7.1.2   The Doob–Meyer Decomposition of Counting Processes

We have seen that the Doob–Meyer decomposition, being based practically on no assumptions, is a very general construct. Thus the question arises as to whether we can meaningfully exploit this generality for addressing issues pertaining to the tracking of software performance. Such questions have been addressed, in one form or another, by investigators such as Koch and Spreij (1983), van Pul (1993), and Slud (1997). To appreciate how, we must first cast the problem of tracking software failures in a format that lends itself to a Doob–Meyer decomposition. This is done by looking at the software failure process as a continuous time counting process $N(t)$, $t \geq 0$. The precedent for doing so is in survival analysis, wherein $N(t)$ tracks the survival of a patient, with $N(t) = 0$ for $t < T$, and $N(t) = 1$, for $t \geq T$; T is, of course, the patient's lifelength. We return to this precedent later, but for now we note that in our context, $N(t)$ as a function of t is an integer-valued step function that is zero at time zero, with jumps of size +1. We suppose $N(t)$ to be right continuous (see Figure 2.5) so that $N(t)$ represents the number of times that the software experiences failure in the time interval $[0, t]$. In prescribing the foregoing, we are supposing that t is either the CPU time, or that the debugging and the re-initiation process are instantaneous.

Under some regularity conditions, which need not be of concern to us here, the process $N(t)$ has a random intensity process $\lambda^*(t)$, $t \geq 0$ (see Section 2.3.1), whose realization $\lambda(t)$ depends on $\mathcal{F}_{t^-}$, where $\mathcal{F}_{t^-}$ denotes *everything* that has happened until just before time t. That is, $\mathcal{F}_{t^-}$ encompasses a complete specification of the path of $N(t)$ on $[0, t]$, as well as other events and factors that have a bearing on the behavior of $N(t)$. Specifically,

$$\lambda(t)dt = \mathcal{P}[N(t) \text{ jumps in an interval } dt \mid \mathcal{F}_{t^-}]. \tag{7.8}$$

Observe that the preceding setup parallels that of the self-exciting point process of Section 2.3.3, with $\mathcal{F}_{t^-}$ being $(\mathcal{H} \cup \mathcal{H}_t)$.

The implication of Equation (7.8) is that in a small interval of time dt, $N(t)$ either jumps or does not, and so by analogy with the expected value of a Bernoulli random variable, the probability of a jump in dt is simply the expected number of jumps in dt. Thus

$$\lambda(t)dt = \mathrm{E}[dN(t) \mid \mathcal{F}_{t^-}],$$

and if we define

$$dM(t) = dN(t) - \lambda(t)dt, \tag{7.9}$$

then $\mathrm{E}(dM(t) \mid \mathcal{F}_{t^-}) = 0$, which is the continuous time analogue of Equation (7.3), a defining property of martingales. It now follows from Equation (7.9), that for $t \geq 0$,

$$M(t) = N(t) - \int_0^t \lambda(u)du$$

is a martingale with $\int_0^t \lambda(u)du$ as a compensator of $N(t)$, and $N(t)$, $t \geq 0$, is a semimartingale.

Thus to summarize, the counting process $N(t)$, $t \geq 0$, generated by software failures admits a Doob–Meyer decomposition of the type described. Since the compensator of $N(t)$ must be a predictable process, we need to explore parameterizations of $\lambda(t)$ that are meaningful and ensure predictivity. For this, it is instructive to look at a parameterization that is commonly used in survival analysis. To start with, suppose that $N(t)$, $t \geq 0$ tracks the survival of a single patient, so that if $Y(t)$ is defined as

$Y(t) = 1$, if the patient is under observation just before t, and

$= 0$, otherwise,

and if $\lambda(t)$ is parameterized as $\lambda(t) = \lambda_0(t) \exp(\beta_0 z(t))$, $t \geq 0$, then

$$\lambda(t)dt = Y(t) \lambda_0(t) \exp(\beta_0 z(t))\, dt. \tag{7.10}$$

The preceding reparameterization is the famous *Cox regression model* [cf. Gill (1984)]; $\lambda_0(t)$ is known as the *baseline failure rate*, and $z(t)$ is a known covariate; β_0 is a constant. Thus given the past, up to (but not including) time t, $Y(t)$ is predictable, and since $z(t)$ is known, $\lambda(t)$ is also predictable. A generalization of this setup is to consider the tracking of several, say n, patients so that $N(t)$ can take more than one jump, and to allow $Y(t)$ to take forms different from that given previously. Furthermore, the fixed covariate $z(t)$ can be replaced by a random covariate $Z(t)$, or by a collection of several fixed and/or random covariates. All that is required for the decomposition of $N(t)$, $t \geq 0$, is that $N(t)$, $Y(t)$, and $Z(t)$, $t \geq 0$ be observable, and that $Y(t)$ and $Z(t)$ be predictable. As an illustration of these generalizations, we may parameterize $\lambda(t)$ as

$$\lambda(t)dt = (n - N(t^-)) \lambda_0(t) \exp(\underset{\sim}{\beta}_0' \underset{\sim}{Z}(t))dt, \tag{7.11}$$

where now $Y(t)$ is replaced by $(n - N(t^-))$, the *risk set* at time t, and $\underset{\sim}{Z}(t)$ is a vector of random covariates; $\underset{\sim}{\beta}_0$ is a vector of parameters. To see why the risk set $(n - N(t^-))$ is a meaningful choice for $Y(t)$, we consider a special type of counting process, namely, the "order statistics process." Specifically, suppose that $T_{(1)} \leq T_{(2)} \leq \cdots \leq T_{(n)}$ are the order statistics (see Section 3.5.3) of a sample of size n from an absolutely continuous (predictive) distribution function $F(t \mid \mathcal{H})$ and a (predictive) failure rate $\lambda_0(t)$; the $\mathcal{H}$ has been suppressed. We could view the $T_{(i)}$s as the survival times of n items under observation starting from time 0. Let $I(A)$ be the indicator of a set A, and for $t \geq 0$, we define the counting process $N(t)$ as

$$N(t) = \sum_{i=1}^{n} I(T_{(i)} \leq t);$$

then $N(t)$, $t \geq 0$ is called the *order statistics process*. We next define a process $Y(t)$, $t \geq 0$ as

$$Y(t) = \sum_{i=1}^{n} I(T_{(i)} \geq t) = (n - N(t^-));$$

then $Y(t)$ represents the number of items that are at the risk of failure just before time t; thus the term "risk set." The intensity process $\lambda(t)$, $t \geq 0$ of the counting process $N(t)$ is the rate at which $N(t)$ jumps. Clearly, this will be of the form

$$\lambda(t)dt = (n - N(t^-))\, \lambda_0(t)\, dt, \tag{7.12}$$

and thus $Y(t) = (n - N(t^-))$, the risk set. Observe that $\lambda(t)$ is random, since it depends on $N(t)$; however, given $N(t^-)$, $\lambda(t)$ is known and thus $\lambda(t)$ is predictable, but $N(t)$ itself is not.

7.1.3 Incorporating the Operational Profile

The parameterization of Equation (7.12) also appears in the tracking of software failures by bug counting models like the model of Jelinski and Moranda; see Section 3.2.2. There, the Λ of Equation (3.4) is $\lambda_0(t)$ of Equation (7.12) and $(n-N(t^-))$ parallels $(N-i)$ of Equation (3.4). For the model by Goel and Okumoto (see Section 3.3.1) $\lambda_0(t)$ is to be identified with be^{-bt}, n identified with the constant a, and $\lambda(t)$ does not depend on $N(t^-)$.

Finally, the parameterizations (7.10) and (7.11) can be made pertinent to tracking software failures if the fixed covariate $z(t)$ can be identified with a nonrandom (or predetermined) operational profile, and the random covariate $Z(t)$ identified with a random operational profile. The random operational profile can be any stochastic process that is deemed meaningful.

Thus to conclude, when software failures are tracked by a counting process model, the Doob–Meyer decomposition results in the integral of its intensity process as a compensator, and if the counting process is modeled as an order statistics process, then the bug counting models of software reliability arise as special cases. Thus, in addition to some unification, the counting process set-up facilitates the incorporation of an operational profile, be it fixed or random. Finally, as mentioned before, the martingale theory facilitates asymptotic inference for those who wish to work in the frequentist paradigm.

7.2 Statistical Aspects of Software Testing: Experimental Designs

Software testing tends to consume a significant proportion of its development budget. It also tends to prolong the software development cycle raising the specter of its obsolescence. Thus the need to make the testing process more efficient and cost effective has been very germane. The literature in software engineering draws attention to two types of testing strategies, "random testing" and "partition testing." Both strategies raise issues of statistical inference. Also, the statistical technique of design of experiments (henceforth DOE) has been proposed as a way to implement partition testing. Thus the aim of this section is to highlight the statistical issues that the problem of software testing poses, and to place DOE in the broader context of software engineering. To do so, we start with the following preamble which introduces some terminology and which defines the terms mentioned previously.

The set of all possible inputs to a piece of software is known as (its) *input domain*. Typically, this set tends to be very large. Testing the software against its input domain serves two purposes: it weeds out the bugs in the software, and it enables us to ensure the software's overall quality. But a large input domain implies that exhaustive testing will be time consuming and expensive. Thus a compromise has been arrived at, wherein the software is tested against only a subset of the input domain; this subset is referred to as the set of test cases. It is hoped that the set of test cases is efficiently chosen, in the sense that it is representative of the inputs which the software is most likely to encounter. The selection of test cases can be done via random testing or via partition testing, strategies which have their underpinnings in the statistical theory of sample surveys; see, for example, Cochran (1977).

With random sampling, the test cases are selected from the input space using a random sampling scheme. This can be done in several ways, one of which is to assign a number to each member of the input space, and then to select those members whose assigned number appears in a table of random numbers. With partition testing, the input space is subdivided into "strata," and the test cases are selected at random from each stratum. According to Weyuker and Jeng (1991), the strata can be defined by considerations such as statement testing, data-flow testing, branch testing, path testing, mutation testing, and so on. The strata can also be defined using DOE techniques, as was done by Mandl (1985), Brownlie, Prowse, and Phadke (1992), and more recently by Cohen et al.(1994). Clearly, the efficacy of partition testing depends on the manner in which the strata are defined and their representativeness of the actual environment in which the software is to operate. Nair et al.(1998) provide a good discussion and a comprehensive treatment of the comparative advantages of partition testing over random testing.

7.2.1 Inferential Issues in Random and Partition Testing

In what follows, we have adopted the formulation of Nair et al. (1998) for describing the inferential issues that the problem of software testing poses. We start by supposing that the input domain consists of N members, where N is quite large. Our aim is to make some statements about the quality of a piece of software that will be subjected to inputs from this input domain. The metric that we use for expressing quality is "the expected failure rate," a notion that we make precise soon.

With random testing, we will want to select at random n inputs from the set of N possible inputs. This can be done on the basis of any probability distribution, say $p(j), j = 1, \ldots, N$, where $p(j)$ represents the probability that input j will be selected; clearly $\sum_j p(j) = 1$. By random selection, we mean a process of selection wherein the input to be selected is not based on what inputs have already been selected. The distribution $p(j), j = 1, \ldots, N$, can be specified in any manner, but for a realistic assessment of the software's quality, it is appropriate to choose $p(j)$ in a manner that reflects the software's operational profile. Recall that the operational profile describes the software's usage in the field environment. We now define a binary variable X_j, where $X_j = 1$ (0), if the jth input will result in a failure (success) of the software. Then, according to Nair et al. (1998), the software's *expected failure rate* under random testing is

$$\theta = \sum_j X_j p(j). \tag{7.13}$$

If the sampling is such that $p(j) = N^{-1}, j = 1, \ldots, N$ (i.e., if all the N inputs have an equal probability of being selected), then θ reduces to X/N, where $X = \sum_1^N X_j$, is the total number of inputs that would lead to software failure.

Clearly, θ will be known only if the disposition of each X_j is known. In actuality this, of course, is not the case; indeed an aim of software testing is to discover those X_js that take the value one. Thus, if θ is to be of any use, it is important to infer its value. In the absence of any prior assumptions about the X_js, inference about θ can only be made on the basis of the results of the n test cases. Let x_i denote the revealed value of X_i, $i = 1, \ldots, n$, and suppose that bugs that are the cause of an X_i taking the value one are not eliminated. We are mimicking here the scenario of sampling with replacement. Then, if N is large, an estimator of θ could be $\widehat{\theta} = \sum_1^n p(i)x_i$; if $p(j) = N^{-1}$, for all j, then $\widehat{\theta}$ is simply the sample average $\sum_1^n (x_i/n)$.

There are certain aspects of Equation (7.13) that warrant some discussion. The first is that θ is a proxy for the software's quality prior to any testing, and is meant to be the probability of the software's experiencing a failure when it is subject to the next randomly selected input. Once the software experiences a failure, the cause of failure will be eliminated so that the corresponding X_j will make the transition from a one to a zero. Consequently, the $X = \sum_1^N X_j$ will decrease by one after the software experiences a failure. The second aspect of Equation (7.13) is that θ, as defined, is really a propensity, rather than a probability. A measure of software quality that is more in keeping with our subjective viewpoint requires that we specify, for each input j, our personal probability that the input will lead to a failure. That is, we are required to assess $P(X_j = 1) \stackrel{\text{def}}{=} \pi(j)$, for $j = 1, \ldots, N$. Then, if we assume that the two events $\{X_j = 1\}$ and $\{j$ is the next input to the software$\}$ are independent, an analogue for θ based on the $\pi(j)$s would be

$$\theta^* = \sum_1^N p(j)\, \pi(j)\,. \tag{7.14}$$

Observe that θ^* will reduce to θ if every $\pi(j)$ is either a one or a zero, and that $\theta^* = \pi$, if $\pi(j) = \pi$ and $p(j) = 1/N$, for all values of j. The requirement of event independence mentioned previously is implicit in the definition of θ. In defining θ^* we have not been explicit about our assumptions about the dependence or the independence of the sequence $\{X_j;\ j = 1, \ldots, N\}$. Such assumptions are reflected in our choice of $\pi(j)$, the marginal distribution of X_j. Since θ^* involves a use of personal probabilities, it would be natural to conduct inference about θ^* using a normative approach wherein the $\pi(j)$s will be updated on the basis of $x_1, \ldots, x_n$. This is a topic that needs to be investigated and which depends on the nature of assumptions about the sequence $\{X_j; j = 1, \ldots, N\}$.

We now discuss the scenario of partition testing. Here the input domain is decomposed into, say K, strata (or cells), with cell i consisting of N_i inputs; $i = 1, \ldots, K$, and $\sum_1^K N_i = N$. Suppose that n is the number of test cases that are allocated among the K strata in such a way that stratum i receives an allocation n_i, where $\sum_1^K n_i = n$. Within each stratum, the n_i test cases are chosen according to some distribution, say $p^{(i)}$, $i = 1, \ldots, K$, where $p^{(i)}(j)$ is the probability that the jth input of stratum i is selected; $j = 1, \ldots, N_i$. Thus within each stratum the testing protocol parallels that of random testing. Analogous to the θ of Equation (7.13), we define θ_i as the *expected partition failure rate* of stratum i as

$$\theta_i = \sum_{j=1}^{N_i} p^{(i)}(j)\, X_{ij}, \quad i = 1, \ldots, K, \tag{7.15}$$

where $X_{ij} = 1(0)$ if the jth input of the ith partition will result in the software's failure (success). Verify that when $p^{(i)}(j) = (N_i)^{-1}$, for $j = 1, \ldots, N_i$, and for each i, $i = 1, \ldots, K$, then $\theta = \sum_1^K ((\theta_i N_i)/N)$. Thus the expected partition failure rate bears a relationship to the expected failure rate when simple random sampling is used for each stratum. Inference for θ_i follows along lines parallel to that for θ, with $\widehat{\theta}_i = \sum_{j=1}^{n_i} p^{(i)}(j) \, x_{ij}$, where x_{ij} is the revealed value of X_{ij}. Finally, if $\pi^{(i)}(j)$ is our personal probability that input j in the ith stratum will lead to failure, then analogous to θ^* of Equation (7.14) we have, for the ith stratum,

$$\theta_i^* = \sum_{j=1}^{N_i} p^{(i)}(j) \, \pi^{(i)}(j), \quad i = 1, \ldots, K. \tag{7.16}$$

Here again, normative inference about θ_i^* is a topic that remains to be explored.

7.2.2 Comparison of Random and Partition Testing

There has been some debate in the software testing literature on the merits of partition testing over random testing. The paper by Nair et al.(1998) is signal, because it settles this debate in a formal manner using the quality metrics θ and θ_i, $i = 1, \ldots, K$, defined before. Their conclusion is that partition testing can produce gains in efficiency over random testing if certain guidelines about defining the strata, and about allocating the test cases to each stratum, are followed. To appreciate this we consider the *failure detection probability* (i.e., the probability of observing at least one failure) as a criterion for comparing the two strategies. Then it is easy to see that the probability of detecting at least one failure in a sample of n test cases which are selected using the random testing strategy, and conditional on knowing θ, is

$$\beta_{R,n}(\theta) = 1 - (1-\theta)^n. \tag{7.17}$$

The corresponding detection probability under the partition testing scheme, and conditional on $\theta_1, \ldots, \theta_K$, is

$$\beta_{P,n}(\theta_1, \ldots, \theta_K) = 1 - \prod_{i=1}^{K}(1-\theta_i)^{n_i}. \tag{7.18}$$

If we let $\eta_p = 1 - \prod_{i=1}^{K}(1-\theta_i)^{\alpha_i}$, where $\alpha_i = n_i/n$, $i = 1, \ldots, K$, then Equation (7.18) becomes

$$\begin{aligned}\beta_{P,n}(\theta_1, \ldots, \theta_K) &= 1 - (1-\eta_P)^n \\ &= \beta_{R,n}(\eta_P),\end{aligned}$$

because of Equation (7.17). Thus partition testing will be more effective than random testing if $\eta_P > \theta$, and vice versa otherwise. This conclusion is independent of the sample size.

Since η_P depends on the θ_is and the α_is, the effectiveness of partition testing over random testing depends on the partitioning and the sample allocation strategy that is used. For example, it is easy to see that since η_P is maximized when all the test runs are allocated to the stratum for which θ_i is the largest, a partition testing strategy with sampling concentrated in a cell can be more efficient than a random testing strategy. However, there is a caveat to such a proposal. This is because the Equations (7.17) and (7.18) upon which our claims are based require a sure knowledge of θ and $\theta_1, \ldots, \theta_K$. Thus in order to select a testing strategy (random or partition), and to implement a sample allocation scheme, the θ and the θ_is should be replaced by the θ^* and θ_i^*s of Equations (7.15) and (7.16), respectively. If the testing is to be done in several phases (i.e., the software is first tested against a run of, say $n^{(1)}$ inputs, and then based on the results of this run, a second run involving, say $n^{(2)}$ inputs, is done, etc.), then a preposterior analysis of the decision problem involving the selection of a testing strategy needs to be conducted; see Section 6.4. This too is a topic that remains to be addressed.

Finally, criteria other than the failure detection probability can also be used to compare testing strategies. Examples of these are: the expected number of detected failures, the precision in estimating θ, the upper confidence bound for θ, the cost of testing, and so on. However, in the final analysis, what seems to matter most appears to be the manner in which the partitions are defined. In many cases, a knowledge about the software development process, or the logic of the software code, will suggest partitions that are natural. Such partitionings will also suggest those partitions that are likely to experience high failure rates so that sample size allocations can be judiciously made. In other cases a knowledge about the software's requirements and other features such as the nature of the fields in its input screen can be used to define the partitions. When such is the case, DOE techniques can be used to construct partitions that have a good coverage of the input domain, and to ensure sample size allocations to each partition are in some sense balanced. In what follows, we give a brief overview of the DOE techniques and motivate their use in software testing.

7.2.3   Design of Experiments in Software Testing

Design of experiment techniques are statistical procedures that are used for planning experiments. Such techniques endeavor to ensure that the maximum possible information can be gleaned from as few experiments as possible. DOE techniques have been widely used in agriculture, industry, and medicine for quite some time. However, with the growing emphasis on quality engineering and robust design, the DOE approach has of late received enhanced visibility [cf. Phadke (1989)]. In the arena of software engineering, Mandl (1985) has used DOE for compiler testing, Brownlie, Prowse, and Phadke (1992) for software testing, and Cohen et al. (1994) for the "screen testing" of input data to an inventory control system.

7.2 Statistical Aspects of Software Testing: Experimental Designs   233

For purposes of motivation and discussion, we focus on the screen testing scenario of Cohen et al. (1994). By screen testing, it is meant the checking of user inputs [to the (several) data fields of a computer screen], prior to performing any operations on the data. This checking is often done by a piece of software; its purpose is to ensure consistency and admissibility of the input data. Of interest here is an assessment of the quality of this screen testing software. Screen testing can be very time consuming since it is not uncommon for a large system to have hundreds of screens, with each screen having as many as 100 data fields. Even if each field can take only two possible input values, say (+) or (−), then a typical screen could have 2^{100} possible combinations of inputs. Thus exhaustive testing of all these inputs in the input domain is expensive and time consuming, and this is so for just a single screen. An alternative to exhaustive testing is therefore clearly in order, and one such alternative is random testing. Another alternative is what is known as "default testing" wherein all the fields, save one, are set at their default values and the software is tested against all the values that the excluded field can take. The third alternative is to use DOE techniques, and as mentioned before, this is also a strategy for implementing partition testing.

To appreciate the value of DOE, consider a simple situation involving three fields, with each field having two inputs, say 1 and 2. An exhaustive test set for this scenario would have the eight possible combinations {(1, 1, 1), (1, 1, 2), (1, 2, 1), (2, 1, 1), (1, 2, 2), (2, 1, 2), (2, 2, 1), (2, 2, 2)}, where, for example, (2, 1, 2) denotes the settings of field 1 at level 2, field 2 at level 1, and field 3 at level 2. A 50% reduction of these eight test cases is provided by an *orthogonal array*, in which every pair of inputs occurs exactly once. This turns out to be {(1, 1, 1), (1, 2, 2), (2, 1, 2), (2, 2, 1)}. Orthogonal array designs are test sets such that, for any pair of fields, all combinations of input values occur, and every pair occurs the same number of times. Thus, for example, if an input screen consists of seven fields, with each field having two inputs, an exhaustive test would entail $2^7 = 128$ test cases, whereas an orthogonal array would entail the eight test cases {(1, 1, 1, 1, 1, 1, 1), (1, 1, 1, 2, 2, 2, 2), (1, 2, 2, 1, 1, 2, 2), (1, 2, 2, 2, 2, 1, 1), (2, 1, 2, 1, 2, 1, 2), (2, 1, 2, 2, 1, 2, 1), (2, 2, 1, 1, 2, 2, 1), and (2, 2, 1, 2, 1, 1, 2)}; see Phadke (1989), p. 286.

Orthogonal arrays have been used for software testing by Brownlie, Prowse, and Phadke (1992). Whereas such arrays give test sets that cover every pair of inputs with fewer test cases than the exhaustive test set, they do have their limitations. As pointed out by Cohen et al. (1994), such arrays are difficult to construct (there does not exist a unified approach for doing so), and do not always exist. For example, there does not exist an orthogonal array when the number of fields is six, and each field has seven possible inputs. More important, in the context of software testing, orthogonal arrays can be wasteful. This is because orthogonal arrays are required to be "balanced." That is, each pair of inputs must occur exactly the same number of times. Observe, that in the seven-field example given before, for fields one and two, the pair (1, 1) occurs two times; similarly, for fields six and seven, the pair (2, 2) also occurs two times. In industrial experimentation the replication of test cases is desirable, since replication is the basis for learning about precision (variance), and the precisions should be based on an equal number of replicates. But with software testing replication is wasteful because replicated tests give identical results. In response to such concerns about orthogonal arrays, Cohen et al. (1994) have proposed an alternate design which they label the "AETG Design," Due to the proprietary nature of this design, details are unavailable. However, the design does not suffer from the limitations of orthogonal arrays, and in the seven-field example, the number of test cases drops down to six. More details about the use of the AETG design can be found in the aforementioned reference.

The screen testing scenario has also been considered by Nair et al. (1998) in their comparison of random and partition testing. Here, based on knowledge about the requirements of the screen field and the software development process, four factors were identified as being relevant. These are:

> A — the number of unique tasks,
> B — the replicates per task,
> C — the replicate type, and
> D — the mode of user-input.

Each of these factors was further broken down into categories, the categories being based on subject matter knowledge. Specifically, each of the factors A and D had four categories, and each of the factors B and C had two. With such a decomposition, the total number of combinations (i.e., partitions) was $4 \times 2 \times 2 \times 4 = 64$. For conducting the software test, one test case was selected at random from each partition. Thus the total number of test cases with this type of a partition testing strategy was 64, a significant saving as compared to the 64,746 test cases that would have resulted an exhaustive testing scheme. The possible values that factor A alone can take is 162. A design such as the one described, namely, partitioning the input domain into factors, and then creating categories (or levels) within each factor is known as a *factorial design*. A further reduction in the number of test cases is possible if, instead of testing at all the combinations of a factorial design (64 in our example), we test at only a sample from this set of combinations. Such a design is aptly termed a *fractional factorial design*. That is, a fractional factorial design is a sampling scheme on the set of all possible factor combinations. The orthogonal arrays described before are examples of fractional factorial designs. It is important to note that sampling on the set of all possible factor combinations is not a random sample; for example, the orthogonal arrays are constrained in the sense that all the pairwise combinations appear at least once.

7.2 Statistical Aspects of Software Testing: Experimental Designs

		Levels of Factor 1			
		ℓ_{11}	ℓ_{12}	ℓ_{13}	ℓ_{14}
Levels of Factor 2	ℓ_{21}	A	B	C	D
	ℓ_{22}	C	D	A	B
	ℓ_{23}	D	C	B	A
	ℓ_{24}	B	A	D	C

FIGURE 7.1. An Orthogonal Latin Square Design.

Another example of a fractional factorial design is what is known as an *orthogonal Latin square*. This design was used by Mandl (1985) for compiler testing and validation of a piece of software written in Ada. To appreciate the nature of a Latin square design, suppose that our input domain can be partitioned into three factors, with each factor having four levels. The total number of possible combinations is therefore $4 \times 4 \times 4 = 64$. A Latin square design will reduce the number of test cases to 16, and yield much of the same information as the full set of 64 tests. To see the structure of the Latin square design, suppose that $\mathcal{L}_{ij}$ denotes the *j*th level of factor *i*, $i = 1, 2$, and $j = 1, 2, 3$, and 4. The four levels of factor three are denoted as A, B, C, and D. This notation may seem idiosyncratic, but it is in keeping with the DOE convention. The name Latin square derives from the fact that the four levels of factor three are denoted by the Latin alphabet. In Figure 7.1 we show an orthogonal Latin square design as a balanced two-way classification scheme in which every level of every factor appears at least once. The design is depicted by a square matrix in which each of the four levels A, B, C, and D appears precisely once in each row and once in each column of the matrix.

The generalization from four levels to *n* levels for each of the three factors is immediate. The total number of possible combinations is now n^3, whereas an orthogonal Latin square based design would entail n^2 tests.

Suppose now that it is desirable to partition the input domain into four factors, with each factor having four levels. The total number of possible combinations is $4^4 = 256$. Suppose that the four levels of the (new) fourth factor are denoted by the Greek letters, α, β, γ, and δ. The levels of the other three factors are denoted, as before, by the $\mathcal{L}_{ij}$s, and by Latin letters. Here, an analogue of the Latin square design is a Latin hypercube design, known as a *Greco-Latin square*. This name derives from the fact that the levels of the third and the fourth

		Levels of Factor 1			
		ℓ_{11}	ℓ_{12}	ℓ_{13}	ℓ_{14}
Levels of Factor 2	ℓ_{21}	A-α	B-β	C-γ	D-δ
	ℓ_{22}	C-δ	D-γ	A-β	B-α
	ℓ_{23}	D-β	C-α	B-δ	A-γ
	ℓ_{24}	B-γ	A-δ	D-α	C-β

FIGURE 7.2. An Orthogonal Greco-Latin Square Design.

factors are denoted by the Latin and Greek alphabets, respectively. In Figure 7.2 we show an orthogonal Greco-Latin square design for the case of four factors, each at four levels. Interestingly, despite the addition of a new factor, the total number of test cases remains at 16. Observe that each of the 16 Greek–Latin alphabet combinations occurs exactly once, and that each level of every factor appears exactly once in each row and in each column.

Tables of Latin square designs, such as those of Figures 7.1 and 7.2 are given in Fisher and Yates (1953). It is important to note that Latin square designs are only possible when the number of levels of all the factors is the same. That is, the design in a two-way classification results in a square. However, it can sometimes happen that an entire row, or an entire column, of a Latin square can be missing. When such is the case, the resulting incomplete Latin square is called a *Youden square*; see, for example, Hicks (1982), p. 80.

7.2.4 Design of Experiments in Multiversion Programming

The literature on DOE describes another commonly used design, namely, the "randomized complete block design," that can be seen as a precursor to the Latin square design. The role of this design can be appreciated via the scenario of evaluating n-version programming by several evaluation teams, say also n. By n-version programming, we mean n typically nonidentical copies of a program that are developed by n separate teams using a common set of requirements and specifications. Conceivably, such programs can be used for ensuring high reliability through fault tolerance; see, for example, Knight and Levenson (1986). Suppose that n is four, and let the four versions of the program be denoted by the Latin letters, A, B, C, and D. Suppose also, that there are four testing teams whose role is to test and to evaluate the four versions of the same functional program. Let the testing teams be denoted by the Roman numerals, I,

	Testing Team		
I	II	III	IV
B	D	A	C
C	C	B	D
A	B	D	B
D	A	C	A

FIGURE 7.3. A Randomized Complete Block Design.

II, III, and IV, and suppose that each team is required to conduct four tests. This latter requirement may make sense if the input domain is partitioned into four strata, and the test team is required to choose, at random, any one of the four strata for its test. The scenario described previously is for illustrative purposes only; it is not, in any way, intended to be realistic. How should we allocate the four versions of the program to the four testing teams, so that each team conducts four tests?

A naive solution is to allocate a version to a team, and require that it do so four times. For example, we may require that team I test version A four times, team II test version B four times, team III test version C, and team IV test version D, four times each. Such an approach is fallible since we are unable to distinguish, in our analysis, between teams and versions. Such designs are called *completely confounded* because averages for teams are also averages for the versions.

An improvement over the completely confounded design is the *completely randomized design* wherein the assignment of a version to a team is random. However, such an assignment is also fallible, because it could result in the situation wherein a version, say A, is never tested by a team, say III. By contrast, in a *randomized complete block design*, every version is tested exactly once by every team. Figure 7.3 shows such an assignment.

Finally, if it so happens that a particular version of the program, say B, cannot be assigned to a particular test team, say I, then the resulting design is known as an *incomplete block design*; see, for example, Hicks (1982), p. 80.

7.2.5 Concluding Remarks

The subject of experimental design is vast and specialized. We have attempted to give merely an overview of this topic, keeping in mind the intended applications. There are several excellent books on this subject, the one by Hicks (1982) offering a relaxed introduction.

The topic of software testing offers much opportunity for using some well-known techniques of applied statistics. We have highlighted two of these, namely, sampling and the design of experiments. Special features of the software testing problem call for modifications of the available methodologies. In the context of random and partition testing, we have alluded to a few open problems. In the context of DOE, the need for new designs has been recognized, and some work involving constrained arrays, vis-à-vis the AETG design has been reported. However work in this arena seems to be continuing, a recent contribution being the paper of Dalal and Mallows (1998).

7.3   The Integration of Module and System Performance

It is often the case that a large software system can be decomposed into *modules*, where each module is a subset of the entire code. The modules are designed to perform a well-defined subtask, and a part of the code may be common to several modules. The output of a module could be an input to another module, or the output of the entire system itself. In the interest of clarity, we find it useful to define the *input specific reliability* of a module as the probability that the module produces a correct output against a given input. Since the number of distinct inputs to a module can be very large — — conceptually infinite — — it is useful to think in terms of the overall or *composite reliability* of a module as the probability that the module produces a correct output for any randomly chosen input from the input domain.

It is often the case that each module is tested individually to assess its input-specific reliability with respect to the subtask that it is required to perform. When such is the case, the input-specific reliability will be one or zero, depending on whether the module's observed output is correct. Typically, the causes of an incorrect output are identified and eliminated through debugging. However, it is not feasible to test a module against all its possible inputs. Thus the best that one can hope for is to estimate the module's composite reliability. This estimate will depend on the size of the sample of inputs against which it is tested. Clearly, because of debugging (which is often assumed to be perfect), the estimated composite reliability will increase with the number of inputs against which the module is tested. The purpose of this section is to propose a framework by which a software system's composite reliability can be assessed via the (estimated) composite reliability of each module. For this, we need to know the relationships among the various modules of the software system, that is, the manner in which the modules are linked with each other. In what follows, by the term reliability we mean composite reliability.

7.3 The Integration of Module and System Performance 239

FIGURE 7.4a. The "Sequence" Control Flow.

7.3.1 The Protocols of Control Flow and Data Flow

The modules of a software system may or may not be linked with each other. When linked, there are two protocols that describe the nature of linkage. The first is *control flow* which specifies the *time sequence* of events and operations. The second is *data flow* which describes how data (or information) is transmitted to, from, and between modules. By convention, we use solid lines to indicate control flow, whereas dotted lines indicate data flow; see Figures 7.4 and 7.5. There are three types of control flow that need to be considered: "sequence," "selective," and "iteration;" we describe these later. In what follows, we assume that these three control flows can be used to represent any software system with any number of modules.

For purposes of exposition, consider a software system with only two modules M_1 and M_2 and a "condition gate," which is denoted by the letter C inscribed within a diamond. The condition gate is a binary logic gate with two outputs, "*t*" for truth, and "*f*" for false. Figures 7.4a), b), and c) show how M_1 and M_2 are linked via the sequence, the selective, and the iteration flows, respectively.

When the control flow is a sequence [Figure 7.4a)] the data flow can take various possibilities. Five of these are shown in Figure 7.5. With the possibility labeled $\mathcal{P}_1$, a user inputs data to M_1 which then processes it and delivers its output to M_2 which in turn processes this input and delivers to the user its output. This flow of data (or information) is indicated by the dotted lines of Figure 7.5. With the possibility labeled $\mathcal{P}_2$ a user may input data either to M_1 or to M_2 directly; in the latter case M_1 is completely bypassed. Possibility $\mathcal{P}_3$ is a combination of $\mathcal{P}_1$ and $\mathcal{P}_2$. With possibility $\mathcal{P}_4$ a user inputs data to M_1, but the output of M_1 consists of two parts, one part going directly to the user and the other going to M_2 as an input; M_2 processes its input and delivers its output to the user. Possibility $\mathcal{P}_5$ is $\mathcal{P}_4$ with the added feature that a user may bypass M_1 and feed the data directly to M_2. The key aspect of Figure 7.5 is that whenever data need to go from one module to the other, it is always from M_1 to M_2 and not vice versa. Clearly, an erroneous output of either M_1 or of M_2 would result in a failure of the software system.

When the control flow is selective [Figure 7.4b)] the data flow has only one possibility. A user first inputs data to the condition gate C which then classifies

240   7. Other Developments: Open Problems

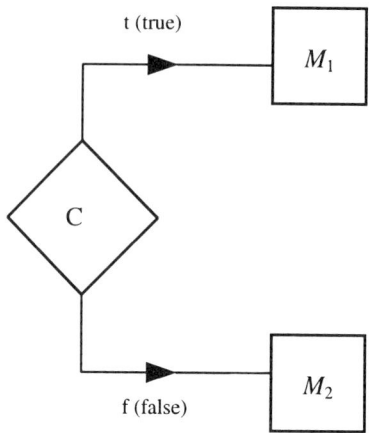

FIGURE 7.4b. The "Selective" Control Flow.

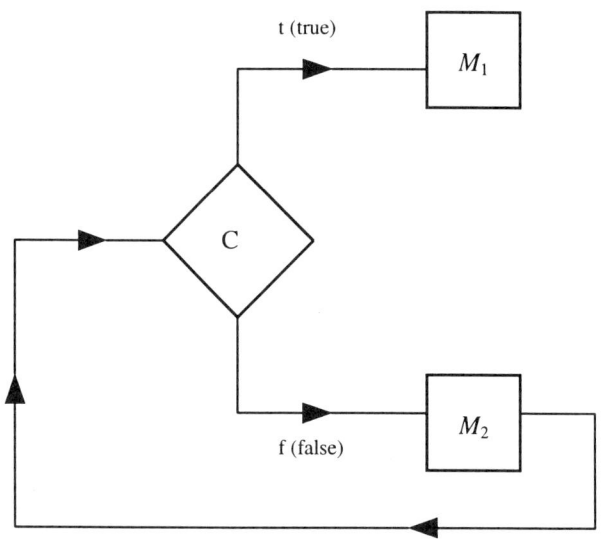

FIGURE 7.4c. The "Iteration" Control Flow.

7.3 The Integration of Module and System Performance 241

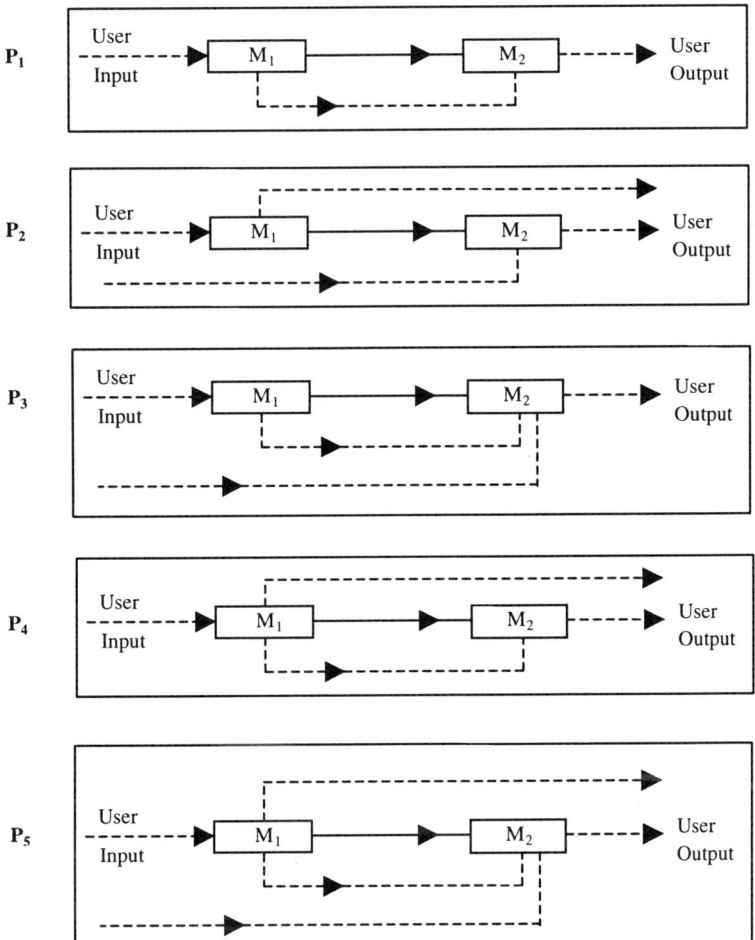

FIGURE 7.5. The Sequence Control Flow and Five Possibilities for Its Data Flow.

them as either t (for true) or f (for false). All inputs that are classified as t become inputs to M_1 which then processes them and delivers to the user its outputs. Similarly, all inputs classified as f become inputs to M_2. Here, the data flow diagram mimics the control flow diagram; see Figure 7.6.

An erroneous output of either M_1 or of M_2 or a misclassification by the condition gate C will result in a failure of the software system. It is easy to see that the two-module software system can be expanded to incorporate additional modules, either in a sequence or a selective flow, by the introduction of additional condition gates.

The third control flow protocol, namely "iteration" [see Figure 7.4c)] is de facto a sequence flow with an intervening condition gate. Here the user inputs a datum to the condition gate C which classifies it as either t or f. If t, then the datum becomes an input to M_1 which processes it and delivers the output to the user. If a datum is classified f, then it becomes an input to M_2 which processes it and provides as output an input to the condition gate for reclassification as t or f. This process (referred to by programmers as a "loop") repeats itself zero or more times, and thus the term "iteration;" see Figure 7.7. In Figure 7.7, the module M_2 has an additional index j, $j = 1, 2, \ldots$, to indicate the jth iteration of M_2 for a particular input datum.

It is sometimes true that input data of a certain kind can affect M_2 in such a way that it provides correct outputs for the first k iterations, and an incorrect output at the $(k + 1)$th iteration. Thus M_2 could, de facto, be viewed as a collection of submodules $M_2(1), M_2(2), \ldots, M_2(j), \ldots$, that are linked in a sequence flow. Often, there may be an upper limit, say J, to the number of iterations per input that M_2 is allowed to perform; in such cases, $j = 1, \ldots, J$, so that M_2 is essentially a maximum of J submodules linked in a sequence flow with an intervening condition gate between each iteration. Clearly, an erroneous output, be it M_1 or any one of the $M_2(j)$s, or a misclassification by the condition gate, might result in a failure of the software system.

The reliability of a modularized software system is the probability that the system provides a correct user output given that the user input data conform to specifications. The following structure function calculus enables us to obtain the reliability of the software system given the reliabilities of each of its modules and its condition gates. That is, it facilitates an integration of module and system performance. The reliability of a condition gate is the probability of its correct classification.

7.3.2   The Structure Function of Modularized Software

Let $I_1, I_2, \ldots, I_k, \ldots$, denote the possible distinct user inputs to the software system. For purposes of discussion, we focus attention on a single input, say input I_k. With respect to I_k, each module of the software, its condition gates, and the software system itself, will be in one of two states, functioning correctly or not. Also, each condition will make a binary classification, true or

7.3 The Integration of Module and System Performance   243

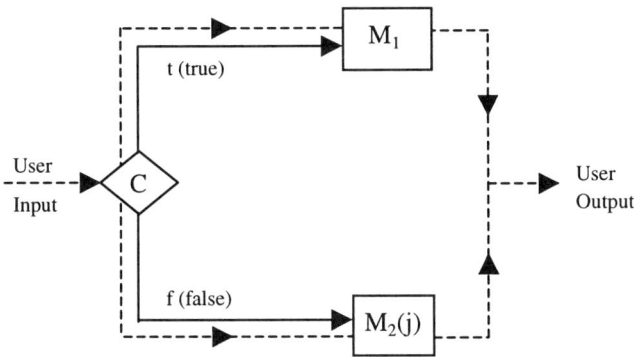

FIGURE 7.6. The Selective Control and Its Associated Data Flow.

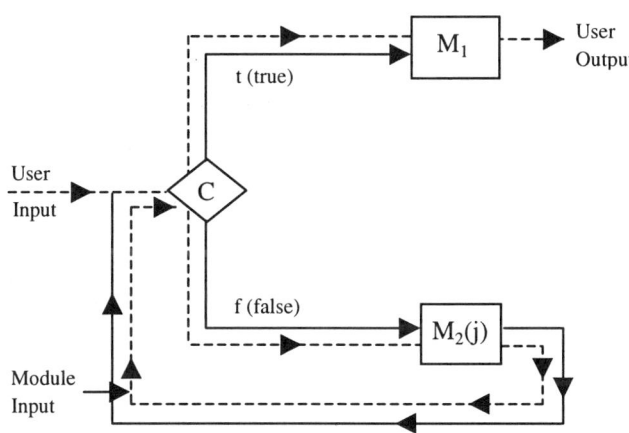

FIGURE 7.7. The Iteration Control and Its Data Flow.

false. These binary states will be represented by indicator variables, and the *structure function* is a binary function that describes the relationship between the state of a software system and the states of its modules and its condition gates. We next introduce some notation that helps us to describe the preceding relationships.

For a two-module system having at most one condition gate C, let

$X_i = 1(0)$ if module M_i, $i = 1, 2$, functions correctly (incorrectly) for input I_k;

$Y = 1(0)$ if the condition gate C makes a correct (incorrect) classification when the user input is I_k;

$C = 1(0)$ if the condition gate C makes a correct classification and declares $t(f)$.

When there is an iteration flow associated with module M_i, $i = 1, 2$, let the indicator variable $X_i(j)$, $j = 1, 2, \ldots$, be such that

$X_i(j) = 1(0)$ if the jth iteration of M_i produces a correct (incorrect) output given that its input generated via I_k is correct.

The possibilities labeled $\mathcal{P}_4$ and $\mathcal{P}_5$ of Figure 7.5 show that module M_1 can generate two types of output: one that goes directly to the user, and another that becomes an input to M_2. To account for these types of outputs, one needs to introduce an additional index to X, so that

$X_{11} = 1(0)$ if the output of M_1 which goes to the user is correct (incorrect) for input data I_k; similarly

$X_{12} = 1(0)$ if the output of M_1 which goes to M_2 is correct (incorrect).

Finally, the binary state of the entire software system is described by the indicator variable X, where $X = 1(0)$ if the entire software system performs correctly (incorrectly), under user input I_k.

Clearly, X is a function of some or all of the indicator variables previously defined. Let $\underline{X} = (X_1, X_2, Y, C, X_1(j), X_2(j), X_{11}, X_{12})$; $\underline{X}$ denotes the states of the modules and the condition gate(s) of the software system for a user input I_k. Then

$$X = \phi(\underline{X}),$$

where the binary function ϕ is known as the *structure function* of the software system. The form of ϕ is dictated by the control and the data flow protocols of the software system. Since both $\underline{X}$ and X depend on the input data I_k, one may replace $\underline{X}$ with $\underline{X}(I_k)$ and X with $X(I_k)$, so that the preceding relationship takes the general form

$$X(I_k) = \phi(\underline{X}(I_k)).$$

The easiest cases to consider are those involving a sequence flow. Specifically, under the possibilities $\mathcal{P}_1$, $\mathcal{P}_2$, and $\mathcal{P}_3$ of Figure 7.5, $X = \phi(X_1, X_2) = X_1 \bullet X_2$, and under the possibilities $\mathcal{P}_4$ and $\mathcal{P}_5$ of Figure 7.5, $X = \phi(X_{11}, X_{12}, X_2) = X_{11} \bullet X_{12} \bullet X_2$. The next case to consider is the selective flow of Figure 7.6, for which $X = \phi(X_1, X_2, Y, \mathcal{C}) = Y \bullet [\mathcal{C} \bullet X_1 + (1 - \mathcal{C}) \bullet X_2]$. When the control flow is an iteration, Figure 7.7, then it is easy to verify that, for $j = 1, 2, \ldots, X = \phi(X_1, X_2(j), Y, \mathcal{C}) = Y \bullet [\mathcal{C} \bullet X_1 + (1 - \mathcal{C}) \prod_1^\infty X_2(j)]$, or if there is an upper limit of J on the number of iterations that M_2 is allowed to perform, $X = Y \bullet [\mathcal{C} \bullet X_1 + (1 - \mathcal{C}) \prod_1^J X_2(j)]$. A special case of the preceding arises if the internal code of M_2 is not affected by the input data of each iteration; that is, module M_2 remains unchanged from iteration to iteration. In such cases $X_2(j) = X_2$, for $j = 1, 2, \ldots$, so that

$$X = Y \bullet [\mathcal{C} \bullet X_1 + (1 - \mathcal{C})(X_2)^j] = Y \bullet [\mathcal{C} \bullet X_1 + (1 - \mathcal{C}) \bullet (X_2)],$$

since X_2 is either 0 or 1.

For software systems with more than two modules, we can decompose the system into pairs of modules (this is called *modular decomposition*), and view each pair as a module. Thus, in principle, the structure function of any software system will take a form that is a composition of one or more of the preceding forms. Once the foregoing is done, the task of integrating module and system performance is complete. However, there are many other issues that still remain to be addressed. For one, how do we incorporate the effect of fault tolerance into the structure function? For another, how do estimates of module reliabilities propagate to estimates of system reliability, given the nature of our structure functions? How must we incorporate dependencies between the indicator variables that describe the performance of each module? These and other issues are potential candidates for further research.

Appendices

A	Statistical Computations using the Gibbs Sampler		249
	A.1 An Overview of the Gibbs Sampler		250
	A.2 Generating Random Variates— The Rejection Method		253
	A.3 Examples : Using the Gibbs Sampler		254
		A.3.1 Gibbs Sampling the Jelinski–Moranda Model	254
		A.3.2 Gibbs Sampling the Hierarchical Model	255
		A.3.3 Gibbs Sampling the Adaptive Kalman Filter Model	256
		A.3.4 Gibbs Sampling the Non-Gaussian Kalman Filter Model	258
B	The Maturity Questionnaire and Responses		261
	B.1 The Maturity Questionnaire		261
	B.2 Binary (Yes, No) Responses to the Maturity Questionnaire		265
	B.3 Prior Probabilities and Likelihood		266
		B.3.1 The Maturity Levels $\mathcal{P}(M_i\|M_{i-1})$	266
		B.3.2 The Key Process Areas $\mathcal{P}(K_{ij})$ and $\mathcal{P}(K_{ij} \| M_i)$	266
		B.3.3 The Likelihoods $\mathcal{L}(K_{ij}; \underline{R}_{ij})$	268

APPENDIX A
STATISTICAL COMPUTATIONS USING THE GIBBS SAMPLER

Markov Chain Monte Carlo (MCMC) methods are computer-intensive techniques that have greatly facilitated statistical computations, both Bayesian and frequentist. As is evident from the material of Chapter 4, the role of integration is central to Bayesian inference. However, integration is often a difficult task, especially when one has to deal with complicated kernels involving multiple variables. Bayesian inference for the concatenated failure rate model is a case in point; see Section 4.7. Sometimes numerical integration or analytical approximations can be used, but these too may pose formidable difficulties. The MCMC method is an alternative to these. Here we are able to indirectly generate random samples from the distributions of interest (univariate or multivariate), and obtain sample estimates of the desired quantities. In so doing, we have implicitly performed the required integration.

There are several MCMC methods that have been proposed in the literature, one of which is the "Metropolis–Hastings Algorithm" [cf. Chib and Greenberg (1995)]. A special case of this algorithm is the "Gibbs sampler," which has now become a popular statistical tool [cf. Brooks (1998)]. The purpose of this Appendix is to overview the Gibbs sampler, and to describe how it may be used to address the computational issues that arise in the context of the models of Chapter 3. Since the technique is quite general, its scope of application is wide, and thus it behooves us to devote some effort to understand its workings. One of the best descriptions of the Gibbs sampler (that we have encountered) is the paper by Casella and George (1992); the material that follows is largely based on their exposition.

A.1 An Overview of the Gibbs Sampler

An archetypical problem in Bayesian inference involves a *known* joint density function, say $f(t, \theta_1, \ldots, \theta_p \mid \underline{t}, \bullet)$, where $\underline{t}$ denotes the observed data, $\theta_1, \ldots, \theta_p$ the unknown parameters, and t an unknown observable of interest, say the time to next failure; the $\bullet$ represents the specified hyperparameters of the underlying prior distributions. Our interest is in obtaining the marginal (predictive) density

$$f(t \mid \underline{t}, \bullet) = \int_{\theta_1} \ldots \int_{\theta_p} f(t, \theta_1, \ldots, \theta_p \mid \underline{t}, \bullet) d\theta_1 \ldots d\theta_p,$$

or its characteristics such as its mean, its variance, and so on. An example is Equation (4.56) of Section 4.7.2. The straightforward approach would be to perform the preceding multiple integration, and then obtain the desired characteristics. The Gibbs sampler provides an easy alternative for obtaining $f(t \mid \underline{t})$. It does this by generating a random sample from $f(t \mid \underline{t}, \bullet)$. The novelty of the approach is that the random sample can be generated *without* an explicit knowledge of $f(t \mid \underline{t}, \bullet)$. The random sample can be used to obtain an estimate of $f(t \mid \underline{t}, \bullet)$ itself, or to obtain estimates of its characteristics of interest. The accuracy of our estimates would depend on m; in general, the larger the m, the better the estimate.

To describe the workings of the Gibbs sampler, we let $p = 1$, and for convenience, suppress the conditioning arguments $\bullet$ and $\underline{t}$, and also the index 1 of θ_1. Thus, to obtain

$$f(t) = \int_\theta f(t, \theta) d\theta,$$

using the Gibbs sampling algorithm we proceed as follows.

First, we select a starting value of θ, say $\theta_0^{(1)}$, and then generate (i.e., simulate) a value from $f(t \mid \theta_0^{(1)})$; we denote this value as $t_0^{(1)}$. Next, we use $t_0^{(1)}$ to generate a value $\theta_1^{(1)}$ from $f(\theta \mid t_0^{(1)})$. We then use $\theta_1^{(1)}$ to generate $t_1^{(1)}$ from $f(t \mid \theta_1^{(1)})$, and so on, so that in general, for $j = 0, 1, 2, \ldots$,

$$t_j^{(1)} \sim f(t \mid \theta_j^{(1)}), \quad \text{and}$$

$$\theta_{j+1}^{(1)} \sim f(\theta \mid t_j^{(1)}).$$

This procedure of iteratively generating values of θ and t by alternating between the conditional densities $f(\theta \mid t)$ and $f(t \mid \theta)$ is called *Gibbs sampling*,

Appendix A.1 An Overview of the Gibbs Sampler 251

and the generated pairs $(\theta_0^{(1)}, t_0^{(1)})$, $(\theta_1^{(1)}, t_1^{(1)})$, ..., $(\theta_k^{(1)}, t_k^{(1)})$, $k = 0, 1, 2, \ldots$, is called the *Gibbs sequence*. It can be shown that when k is large, the value $t_k^{(1)}$ can be regarded as a realization from the density function $f(t)$; similarly $\theta_k^{(1)}$ a realization from $f(\theta)$, the marginal density function of θ. Thus to generate a sample of size m from $f(t)$ one repeats the foregoing iterative procedure m times, starting each of the m cycles with suitable choices of initial values $\theta_0^{(1)}, \theta_0^{(2)}, \ldots, \theta_0^{(m)}$. Clearly, to generate the Gibbs sequence a knowledge of the conditionals $f(t \mid \theta)$ and $f(\theta \mid t)$ is necessary. Furthermore, we should be able to generate values from $f(t \mid \theta)$ and $f(\theta \mid t)$. Since $t_k^{(i)}$ and $\theta_k^{(i)}$ depend on the starting value $\theta_0^{(i)}$, $i = 1, 2, \ldots, m$, it is important to ensure that the starting values constitute an independent sequence.

Once the Gibbs sequence is generated, the densities $f(t)$ and $f(\theta)$ can be estimated by constructing a histogram of $t_k^{(1)}, t_k^{(2)}, \ldots, t_k^{(m)}$, and $\theta_k^{(1)}, \theta_k^{(2)}, \ldots, \theta_k^{(m)}$, respectively, for large values of k and m. However, as was pointed out by Gelfand and Smith (1990), better estimates of the densities can be obtained by averaging the conditional densities. Specifically, $\hat{f}(t)$, a Gibbs sequence estimate of $f(t)$, is given by the average

$$\hat{f}(t) = \frac{1}{m} \sum_{i=1}^{m} f(t \mid \theta_k^{(i)}),$$

and a Gibbs sequence estimate of $f(\theta)$ by the average

$$\hat{f}(\theta) = \frac{1}{m} \sum_{i=1}^{m} f(\theta \mid t_k^{(i)}).$$

Similarly, estimates of the means of $f(t)$ and $f(\theta)$ can be obtained by the sample averages $1/m \sum_{i=1}^{m} t_k^{(i)}$ and $1/m \sum_{i=1}^{m} \theta_k^{(i)}$, respectively.

In the case of three variables, say t, θ_1, and θ_2, we choose the starting values $\theta_{10}^{(1)}$ and $\theta_{20}^{(1)}$, and then sample iteratively from the three full conditional densities $f(t \mid \theta_1, \theta_2)$, $f(\theta_1 \mid t, \theta_2)$, and $f(\theta_2 \mid t, \theta_1)$. After k iterations we produce the Gibbs sequence $(\theta_{1k}^{(1)}, \theta_{2k}^{(1)}, t_k^{(1)})$; $t_k^{(1)}$ is then a realization from the marginal density $f(t)$, similarly, $\theta_{1k}^{(1)}$ and $\theta_{2k}^{(1)}$. As before, we repeat this procedure m times to generate samples of size m from the required densities. The procedure generalizes to several variables.

To conclude, the Gibbs sampling algorithm can be thought of as a practical implementation of the fact that a knowledge of the conditional distributions is sufficient to determine a joint distribution, *should* the joint distribution exist. Note that it is not always true that the existence of proper conditional distributions ensures the existence of a proper joint distribution. If a proper joint distribution does not exist, then a marginal distribution will not exist. When such is the case the outputs from a Gibbs sampler will be misleading. Thus before invoking the Gibbs sampler one should ensure the existence of a proper joint distribution. One way to do this is to solve a certain fixed point integral equation and see if the solution is the required marginal density. Another way is to restrict all conditional densities to lie on compact intervals. Since the situation described is rare, we may for all practical purposes ignore it and proceed with the iterative scheme.

A.2 Generating Random Variates—The Rejection Method

For a successful implementation of the Gibbs sampling algorithm, it is important that we are able to efficiently generate realizations from the full conditional distributions. Often, these conditional distributions are not of a standard form, being compositions of priors and likelihoods. Thus, for example, to generate a realization from, say $f(\theta_2 \mid \theta_1; t)$, where θ_1 and θ_2 are unknown quantities and t are the observed data, we may find it convenient to express $f(\theta_2 \mid \theta_1; t)$ as

$$f(\theta_2 \mid \theta_1; t) \propto \mathcal{L}(\theta_2 \mid \theta_1; t)\, \mathcal{P}(\theta_2 \mid \theta_1),$$

by Bayes' Law where the first term on the right-hand side of the preceding is the likelihood and the second term is the prior of θ_2 conditional on θ_1. If the prior happens to be a standard (well-known) form, then generating samples from $\mathcal{P}(\theta_2 \mid \theta_1)$ may be relatively straightforward. The method of *rejection sampling* enables us to generate samples from $f(\theta_2 \mid \theta_1; t)$ by modifying the samples generated from $\mathcal{P}(\theta_2 \mid \theta_1)$ via the likelihood $\mathcal{L}(\theta_2 \mid \theta_1; t)$. The method of rejection sampling proceeds as follows.

(a) First we generate a realization, say $\theta_2^{(a)}$, from the prior distribution $\mathcal{P}(\theta_2 \mid \theta_1)$, with θ_1 specified.

(b) We then generate a realization, say $u^{(a)}$, from a uniform distribution on $(0,1)$.

(c) We then compute the *rejection kernel* (also known as a *blanketing function*) $(\mathcal{L}(\theta_2^{(a)} \mid \theta_1; t))/(\mathcal{L}(\widehat{\theta}_2 \mid \theta_1; t))$, where $\widehat{\theta}_2$ is that value of θ_2 which maximizes the likelihood function $\mathcal{L}(\theta_2 \mid \theta_1; t)$.

(d) If $u^{(a)} \leq (\mathcal{L}(\theta_2^{(a)} \mid \theta_1; t))/(\mathcal{L}(\widehat{\theta}_2 \mid \theta_1; t))$, then $\theta_2^{(a)}$ is a realization from $\mathcal{P}(\theta_2 \mid \theta_1; t)$, otherwise $\theta^{(a)}$ is discarded (rejected).

(e) We repeat the steps (a) to (d) until the desired number of $\theta_2^{(a)}$'s have been obtained.

Rejection sampling is one of several approaches for generating samples from one distribution by modifying the samples generated by another. An overview of some of the alternatives may be found in Smith and Gelfand (1992).

254   Appendix A

A.3   Examples: Using the Gibbs Sampler

Whereas the general methodology for implementing the Gibbs sampling algorithm is relatively straightforward, some initial preparation to get all the full conditional densities in a workable form is necessary. This is illustrated by Equation (4.54) of Section 4.7.2 where Bayesian inference for the concatenated failure rate model was discussed. The purpose of this section is to show how the Gibbs sampling algorithm can be used in place of numerical integration and approximations that were the mainstay of many of the examples of Chapter 4. We start with the simplest.

A.3.1   Gibbs Sampling the Jelinski–Moranda Model

Recall, that for this model, the predictive density of T_{n+1} at t, given the data $\underline{t} = (t_1, \ldots, t_n)$ and the prior parameters θ, μ, and α (henceforth $\bullet$), had to be numerically obtained; see Section 4.2.3. Thus to obtain $f(t \mid \underline{t}, \bullet)$ we need to generate the Gibbs sequence $(t_k^{(i)}, N_k^{(i)}, \Lambda_k^{(i)} \mid \underline{t}, \bullet)$, for $i = 1, 2, \ldots, m$. For this we need to know the full conditionals $f(t \mid N, \Lambda, \underline{t})$, $f(N \mid t, \Lambda, \underline{t})$, and $f(\Lambda \mid t, N, \underline{t})$; for convenience, the $\bullet$ has been suppressed in this and subsequent sections.

But from Equation (3.5), we know that $f(t \mid N, \Lambda, \underline{t}) = f(t \mid N, \Lambda) = \Lambda \exp(-\Lambda t(N - n))$, an exponential distribution, so that given the starting values N_0 and Λ_0 we can easily generate a t_0. For the full conditional $f(N \mid t, \Lambda_0, \underline{t})$, we use Bayes' Law whereby

$$f(N \mid t, \Lambda_0, \underline{t}) = f(N \mid \Lambda_0, \underline{t})$$

$$\propto \mathcal{L}_n(N \mid \Lambda_0; \underline{t}) \, \mathcal{P}(N \mid \Lambda_0)$$

$$= \mathcal{L}_n(N \mid \Lambda_0; \underline{t}) \, \mathcal{P}(N), \quad \text{for } N \geq n,$$

where $\mathcal{P}(N)$, our prior for N, is assumed to be independent of Λ_0. The likelihood of N, for fixed values of Λ_0 and $\underline{t}$, is of the form

$$\mathcal{L}_n(N \mid \Lambda_0; \underline{t}) = \prod_{i=1}^{n} \Lambda_0 \exp(-\Lambda_0 t_i (N - i + 1)).$$

In writing the preceding, we have assumed that given Λ_0 and $\underline{t}$, the distribution of N does not depend on T_{n+1}. Thus given Λ_0 and $\underline{t}$, we may generate a realization N_1 from $f(N \mid \Lambda_0, \underline{t})$ by using rejection sampling on samples generated from $\mathcal{P}(N)$, which we recall was assumed to be a Poisson distribution. Similarly, for the full conditional $f(\Lambda \mid t, N_1, \underline{t})$, we have

$$f(\Lambda \mid t, N_1, \underline{t}) = f(\Lambda \mid N_1, \underline{t}) \propto \mathcal{L}_n(\Lambda \mid N_1; \underline{t}) \mathcal{P}(\Lambda),$$

with $\mathcal{L}_n(\Lambda \mid N_1; \underline{t})$ taking the same form as $\mathcal{L}_n(N \mid \Lambda_0; \underline{t})$. Thus given N_1 and $\underline{t}$ we may use rejection sampling on samples generated from $\mathcal{P}(\Lambda)$—a gamma distribution—to generate Λ_1. The iterative scheme is now in place.

A.3.2 Gibbs Sampling the Hierarchical Model

The hierarchical model of Section 4.4 involved the parameters α, β_0, β_1, and Λ_i, $i = 1, 2, \ldots$, and the predictive distribution of T_{n+1} given $\underline{t} = (t_1, \ldots, t_n)$; see Equations (4.21) and (4.22). Thus to Gibbs sample this model, the number of full conditionals that we need to consider is five. As a start, consider the full conditional $f(t \mid \Lambda_i, \alpha, \beta_0, \beta_1, \underline{t})$, which, because of the obvious independence considerations, is in fact $f(t \mid \alpha, \beta_0, \beta_1)$. But from Equation (4.17), T_{n+1} has a Pareto density at t of the form $(\alpha(\beta_0 + \beta_1(N+1))^\alpha)/((t + \beta_0 + \beta_1(N+1))^{\alpha+1})$. Thus given the starting values for α, β_0, and β_1, a realization t from $f(t \mid \alpha, \beta_0, \beta_1)$ can be generated using this Pareto density. Similarly, from the same starting values, we can also generate a realization from $f(\Lambda_i \mid \alpha, \beta_0, \beta_1, \underline{t})$, using the density of Equation (4.20); note that to generate a realization from the preceding density of Λ_i we only need to use the t_i from the collection $(t_1, \ldots, t_i, \ldots, t_n)$. To generate realizations from the remaining three densities, we must decompose them. Specifically, $f(\alpha \mid t, \Lambda_i, \beta_0, \beta_1, \underline{t}) = f(\alpha \mid \Lambda_i, \beta_0, \beta_1, \underline{t})$, and by Bayes' Law

$$f(\alpha \mid \Lambda_i, \beta_0, \beta_1, \underline{t}) \propto \mathcal{L}_n(\alpha \mid \Lambda_i, \beta_0, \beta_1; \underline{t}) \, \mathcal{P}(\alpha \mid \Lambda_i, \beta_0, \beta_1),$$

where $\mathcal{L}_n(\alpha \mid \Lambda_i, \beta_0, \beta_1, \underline{t})$ is the likelihood of α for fixed values of the other arguments. To specify this likelihood we use Equation (4.20); accordingly,

$$\mathcal{L}_n(\alpha \mid \Lambda_i, \beta_0, \beta_1; \underline{t}) = \prod_{i=1}^{n} \frac{\Lambda_i^\alpha (t_i + \beta_0 + \beta_1 i)^{\alpha+1}}{\Gamma(\alpha+1)} \, e^{-\Lambda_i(t_i + \beta_0 + \beta_1 i)}.$$

For the prior $\mathcal{P}(\alpha \mid \Lambda_i, \beta_0, \beta_1)$ we observe that, by the multiplication rule,

$$\mathcal{P}(\alpha \mid \Lambda_i, \beta_0, \beta_1) \propto \mathcal{P}(\Lambda_i \mid \alpha, \beta_0, \beta_1) \, \pi(\alpha \mid \beta_0, \beta_1),$$

which because of Equation (4.18) is of the form

$$\mathcal{P}(\alpha \mid \Lambda_i, \beta_0, \beta_1) \propto \mathcal{P}(\Lambda_i \mid \alpha, \beta_0, \beta_1) \, \pi(\alpha \mid \omega).$$

But by our model construction, Λ_i has a gamma distribution with a shape parameter α and a scale parameter $\beta_0 + \beta_1 i$, and $\pi(\alpha \mid \omega)$ is a uniform distribution on $(0, \omega)$. Generating random variates from $\pi(\alpha \mid \omega)$ is therefore very straightforward. In order to generate realizations from $f(\alpha \mid \Lambda_i, \beta_0, \beta_1, \underline{t})$ we may use rejection sampling on samples generated from $\pi(\alpha \mid \omega)$ using

$$\frac{\mathcal{L}_n(\alpha|\Lambda_i, \beta_0, \beta_1, \underline{t}) \times \mathcal{P}(\Lambda_i|\alpha, \beta_0, \beta_1)}{\mathcal{L}_n(\widehat{\alpha}|\Lambda_i, \beta_0, \beta_1, \underline{t}) \times \mathcal{P}(\Lambda_i|\widehat{\alpha}, \beta_0, \beta_1)}$$

as the rejection kernel, or do rejection sampling in two stages, first using $\mathcal{P}(\Lambda_i \mid \alpha, \beta_0, \beta_1)/\mathcal{P}(\Lambda_i|\widetilde{\alpha}, \beta_0, \beta_1)$ as the rejection kernel, and then using $\mathcal{L}_n(\alpha \mid \Lambda_i, \beta_0, \beta_1, \underline{t})/\mathcal{L}_n(\widehat{\widetilde{\alpha}} \mid \Lambda_i, \beta_0, \beta_1, \underline{t})$ as a rejection kernel on those samples that have been accepted by the first rejection kernel. Note that $\widehat{\alpha}$ ($\widetilde{\alpha}$) [$\widehat{\widetilde{\alpha}}$] is that value of α that maximizes the numerator terms of their respective rejection kernels. Thus given the starting values of β_0 and β_1, and the previously generated value of Λ_i, we can generate realizations from $f(\alpha \mid \Lambda_i, \beta_0, \beta_1, \underline{t})$.

Generating samples from the full conditionals $f(\beta_0 \mid t, \alpha, \beta_0, \beta_1, \underline{t})$ and $f(\beta_1 \mid t, \alpha, \beta_0, \beta_1, \underline{t})$ proceeds along similar lines, except that now $\pi(\alpha \mid w)$ gets replaced by $\pi(\beta_0 \mid \beta_1, a, b)$ and $\pi(\beta_1 \mid c, d)$, respectively; see Equation (4.18).

A.3.3 Gibbs Sampling the Adaptive Kalman Filter Model

In what follows we use the notation of Section 4.5. With dynamic models, interest centers around the state of nature θ_i, and predictions about future observables $Y_{i+1}, Y_{i+2}, \ldots$, given the observed data $\underline{y}^{(i)} = (y_1, \ldots, y_i)$, for $i = 1, 2, \ldots$. Interest may also center around other parameters such as the α of Equation (4.26), or the C of Equation (4.28).

Recall that for the adaptive Kalman filter model,

$$(y_i \mid \theta_i) \sim \mathcal{N}(\theta_i, \sigma_1^2),$$

$$(\theta_i \mid \alpha_1, \theta_{i-1}) \sim \mathcal{N}(\alpha\theta_{i-1}, W_i^2), \quad \text{and}$$

$$\alpha \sim \mathcal{U}(-2, +2), \quad \text{for } i = 1, 2, \ldots;$$

$\mathcal{U}(-2, +2)$ denotes a uniform distribution over the interval $(-2, +2)$.

For inference about $(\theta_i; \underline{y}^{(i)})$, $(\alpha; \underline{y}^{(i)})$, and the predictive density at y of $(Y_{i+1}; \underline{y}^{(i)})$, we consider the full conditionals generated by the 5-tuple $(\theta_i, \theta_{i+1}, Y_{i+1}, \alpha, \underline{y}^{(i)})$. For purposes of discussion we focus attention on the case $i = 1$, so that our 5-tuple is $(\theta_1, \theta_2, Y_2, \alpha, y_1)$; also note that the starting value θ_0 has been specified in advance. Because of the underlying distributional assumptions given previously, we see that the full conditionals of the 5-tuple are distributed as

$$(\theta_1 \mid \theta_2, \alpha, y_1, Y_2) \sim (\theta_1 \mid \theta_2, \alpha, y_1),$$

$$(\theta_2 \mid \theta_1, \alpha, y_1, Y_2) \sim (\theta_2 \mid \theta_1, \alpha, Y_2),$$

$$(Y_2 \mid \theta_1, \theta_2, \alpha, y_1) \sim (Y_2 \mid \theta_2), \quad \text{and}$$

Appendix A.3 Examples: Using the Gibbs Sampler

$$(\alpha \mid \theta_1, \theta_2, y_1, Y_2) \sim (\alpha \mid \theta_1, \theta_2).$$

To generate realizations from the preceding set of conditionals we choose θ_{10} and θ_{20} as starting values of θ_1 and θ_2, respectively. We can then generate a Y_{20} from the distribution of $(Y_2 \mid \theta_{20})$, which is a Gaussian with mean θ_{20} and variance σ_1^2. To generate a realization from $(\alpha \mid \theta_1, \theta_2)$, we use the multiplication rule whereby

$$\mathcal{P}(\alpha \mid \theta_1, \theta_2) \propto \mathcal{P}(\theta_2 \mid \theta_1, \alpha) \, \mathcal{P}(\theta_1 \mid \alpha) \, \mathcal{P}(\alpha),$$

and observe that $(\theta_2 \mid \theta_1, \alpha)$ has a Gaussian distribution with mean $\alpha\theta_1$ and variance W_2^2, and that $(\theta_1 \mid \alpha)$ has a Gaussian distribution with mean $\alpha\theta_0$ and variance W_1^2. These relationships with θ_{20} replacing θ_2 and θ_{10} replacing θ_1 define the rejection kernel for samples generated from the uniform $(-2, +2)$ distribution of α. Consequently, we are able to generate α_0, a realization from the distribution of $(\alpha \mid \theta_{10}, \theta_{20})$.

Having chosen the starting values θ_{10} and θ_{20} (note that the starting value θ_0 is external to the Gibbs sampling algorithm), and having generated the values Y_{20} and α_0, we now proceed to generate a realization from the distribution of $(\theta_2 \mid \theta_{10}, \alpha_0, Y_{20})$. Invoking the multiplication rule, we observe that

$$\mathcal{P}(\theta_2 \mid \theta_{10}, \alpha_0, Y_{20}) \propto \mathcal{P}(Y_{20} \mid \theta_2, \theta_{10}, \alpha_0) \, \mathcal{P}(\theta_2 \mid \theta_{10}, \alpha_0)$$

$$= \mathcal{P}(Y_{20} \mid \theta_2) \, \mathcal{P}(\theta_2 \mid \theta_{10}, \alpha_0),$$

since Y_{20} is independent of θ_{10} and α_0, given θ_2.

But $(\theta_2 \mid \theta_{10}, \alpha_0)$ has a Gaussian distribution with mean $\alpha_0\theta_{10}$ and variance W_2^2, and so samples from this distribution can be easily generated. Also, $(Y_{20} \mid \theta_2)$ has a Gaussian distribution with mean θ_2 and variance σ_1^2, and this forms the basis of a rejection kernel for samples generated from $(\theta_2 \mid \theta_{10}, \alpha_0)$. Consequently, we are able to update θ_{20} to θ_{21}.

Our final task is to generate samples from the distribution of $(\theta_1 \mid \theta_{21}, \alpha_0, y_1)$, which because of the multiplication law is

$$\mathcal{P}(\theta_1 \mid \theta_{21}, \alpha_0, y_1) \propto \mathcal{P}(\theta_{21} \mid \alpha_0, \theta_1, y_1) \, \mathcal{P}(y_1 \mid \theta_1, \alpha_0) \, \mathcal{P}(\theta_1 \mid \alpha_0)$$

$$= \mathcal{P}(\theta_{21} \mid \alpha_0, \theta_1) \, \mathcal{P}(y_1 \mid \theta_1) \, \mathcal{P}(\theta_1 \mid \alpha_0)$$

since θ_{21} is independent of y_1 given α_0 and θ_1.

But with θ_0 specified (it is the starting value for filtering), $\mathcal{P}(\theta_1 \mid \alpha_0)$ is Gaussian with mean $\alpha_0\theta_0$ and variance W_1^2; thus samples from this distribution can be easily generated. To construct the rejection kernel, we first observe that $(\theta_{20} \mid \alpha_0, \theta_1)$ has a Gaussian distribution with mean $\alpha_0\theta_1$ and variance W_2^2, and that since y_1 has been observed as y_1, $\mathcal{P}(y_1 \mid \theta_1)$ is the likelihood $\mathcal{L}(\theta_1; y_1)$. This likelihood can be assessed from the assumption that $(y_1 \mid \theta_1) \sim \mathcal{N}(\theta_1; \sigma_1^2)$. Thus the rejection kernel for samples generated from $\mathcal{P}(\theta_1 \mid \alpha_0)$ is provided by the function $\mathcal{L}(\theta_1; y_1) \times \mathcal{P}(\theta_{21} \mid \alpha_0, \theta_1)$. We can now generate a realization θ_{11} from $\mathcal{P}(\theta_1 \mid \theta_{21}, \alpha_0; y_1)$; θ_{10} is thus updated to θ_{11}.

The process repeats itself, so that after k iterations we are able to produce (given θ_0 and y_1) the realizations $\theta_{1k}^{(1)}$, $\theta_{2k}^{(1)}$, $Y_{2k}^{(1)}$, and $\alpha_k^{(1)}$, based on the starting values θ_{10} and θ_{20}. Repeating this m times, each time using a new pair of starting values, we can produce the realizations $\theta_{1k}^{(m)}$, $\theta_{2k}^{(m)}$, $Y_{2k}^{(m)}$, and $\alpha_k^{(m)}$. The Gibbs sample based estimate of $\mathcal{P}(\theta_j; y_1, \theta_0)$, $j = 1, 2$, is the histogram of $\theta_{jk}^{(\ell)}$, $\ell = 1, \ldots, m$; similarly, an estimate of $\mathcal{P}(\alpha; y_1, \theta_0)$, based on its uniform (-2, +2) prior, is the histogram of $\alpha_k^{(\ell)}$, and an estimate of $\mathcal{P}(Y_2; y_1, \theta_0)$ is the histogram of $Y_{2k}^{(\ell)}$, $\ell = 1, \ldots, m$.

For the case $i = 2$, we need to consider the 6-tuple $(\theta_2, \theta_3, Y_3, \alpha, y_1, Y_2)$, and the histograms mentioned above provide the starting values, and also the sampling distributions for θ_2 and α.

A.3.4 Gibbs Sampling the Non-Gaussian Kalman Filter Model

The non-Gaussian Kalman filter model, defined by Equations (4.27) and (4.28), leads to the relationships:

$$(T_1 \mid \theta_1, \omega_1) \sim \mathcal{G}(\theta_1, \omega_1),$$

$$(\theta_1 \mid \theta_0, C) \sim \frac{\theta_0}{C} \epsilon_1, \quad \text{and}$$

$$(\theta_0 \mid \sigma_0, \nu_0) \sim \mathcal{G}(\sigma_0 + \nu_0, u_0),$$

for ϵ_1 having a beta distribution on (0, 1) with parameters σ_0 and ν_0; θ_1 is the scale parameter of the gamma distribution of T_1. The prior distribution of C was assumed uniform on the interval (0, 1).

Suppose that T_1 has been observed as t_1, and that inference about $(\theta_1; t_1)$, $(\theta_2; t_1)$, $(C; t_1)$, and $(T_2; t_1)$ is desired. For Gibbs sampling under the preceding setup, we need to consider the full conditionals generated by the 6-tuple $(\theta_0, \theta_1, \theta_2, T_2, C, T_1)$. The incorporation of θ_0 in the tuple is necessary because of the fact that unlike the fixed θ_0 of the adaptive Kalman filter model, the θ_0 here has a gamma distribution. The full conditionals of the 6-tuple have distributions determined by quantities such as $(C \mid \theta_2, \theta_1, \theta_0)$. Generating realizations from such conditionals poses a difficulty. This is because the multiplicative term $\theta_0 \epsilon_1 / C$, of the second relationship given previously, makes it difficult to obtain a rejection kernel. Whereas such difficulties can be overcome using the Metropolis–Hastings algorithm [cf. Chib and Greenberg (1995)], the fact that closed form inference when we condition on C is available makes a more direct approach feasible. To see how, consider all the full conditionals of the 5-tuple $(\theta_1, \theta_2, T_2, C, T_1)$, and observe, in the light of Equations (4.37) through (4.39), that generating realizations from these conditionals involves generating realizations from the conditionals $(\theta_1 \mid C, T_1)$, $(\theta_2 \mid C, T_2)$, $(T_2 \mid C, T_1)$, and $(C \mid t_1, T_2)$. For a starting value C_0 of C, generating the values θ_{10}, θ_{20}, and T_{20}, given an observed value t_1 of T_1, follows from Equations (4.37) through (4.39). To update C_0 to C_1, via the generation of a realization from the distribution of $(C \mid t_1, T_2)$, it is necessary that T_2 be observed. This is because T_1 alone does not provide information about C. Suppose then, that T_2 has been observed as t_2. Then, Equation (4.39) can be used to construct a likelihood for C, and this likelihood facilitates the formation of a rejection kernel. Specifically,

$$\mathcal{P}(C; t_1, t_2) \propto \mathcal{L}(C; t_2, t_1)\, \mathcal{P}(C; t_1)$$

$$= \mathcal{L}(C; t_2, t_1)\, \mathcal{P}(C)\,,$$

where $\mathcal{P}(C)$ is our prior for C. The rest proceeds in the usual manner.

APPENDIX B
THE MATURITY QUESTIONNAIRE AND RESPONSES

B.1 The Maturity Questionnaire

Maturity Level 2

Key Process Area 1 (K_{21})—Requirements Management

1. For each project involving software development, is there a designated software manager?

2. Does the project software manager report directly to the project (or project development) manager?

3. Does the Software Quality Assurance (SQA) function have a management reporting channel separate from the software development project management?

4. Is there a designated individual or team responsible for the control of software interfaces?

5. Is there a software configuration control function for each project that involves software development?

Key Process Area 2 (K_{22})—Software Quality Assurance

6. Does senior management have a mechanism for the regular review of the status of software development projects?

7. Is a mechanism used for regular technical interchanges with the customer?

8. Do software development first-line managers sign off on their schedules and cost estimates?

9. Is a mechanism used for controlling changes to the software requirements?

10. Is a mechanism used for controlling changes to the code? (Who can make changes and under what circumstances?)

Key Process Area 3 (K_{23})—Software Project Planning

11. Is there a required training program for all newly appointed development managers designed to familiarize them with software project management?

12. Is a formal procedure used to make estimates of software size?

13. Is a formal procedure used to produce software development schedules?

14. Are formal procedures applied to estimating software development cost?

15. Is a formal procedure used in the management review of each software development prior to making contractual commitments?

Maturity Level 3

Key Process Area 1 (K_{31})—Integrated Software Management

16. Is a mechanism used for identifying and resolving system engineering issues that affect software?

17. Is a mechanism used for independently calling integration and test issues to the attention of the project manager?

18. Are the action items resulting from testing tracked to closure?

19. Is a mechanism used for ensuring compliance with the software engineering standards?

20. Is a mechanism used for ensuring traceability between the software requirements and top-level design?

Appendix B.1 The Maturity Questionnaire 263

Key Process Area 2 (K_{32})—Organization Process Definition

21. Are statistics on software design errors gathered?

22. Are the action items resulting from design reviews tracked to closure?

23. Is a mechanism used for ensuring traceability between the software top-level and detailed designs?

24. Is a mechanism used for verifying that the samples examined by Software Quality Assurance are representative of the work performed?

25. Is there a mechanism for ensuring the adequacy of regression testing?

Key Process Area 3 (K_{33})—Peer Review

26. Are internal software design reviews conducted?

27. Is a mechanism used for controlling changes to the software design?

28. Is a mechanism used for ensuring traceability between software detailed design and the code?

29. Are software code reviews conducted?

30. Is a mechanism used for configuration management of the software tools used in the development process?

Maturity Level 4

Key Process Area 1 (K_{41})—Quantitative Process Management

31. Is a mechanism used for periodically assessing the software engineering process and implementing indicated improvements?

32. Is there a formal management process for determining if the prototyping of software functions is an appropriate part of the design process?

33. Are design and code review coverage measured and recorded?

34. Is test coverage measured and recorded for each phase of functional testing?

35. Are internal design review standards applied?

Key Process Area 2 (K_{42})—Software Quality Management

36. Has a managed and controlled process database been established for process metrics data across all projects?

37. Are the review data gathered during design reviews analyzed?

38. Are the error data from code reviews and tests analyzed to determine the likely distribution and characteristics of the errors remaining in the product?

39. Are analyses of errors conducted to determine their process-related causes?

40. Is review efficiency analyzed for each project?

Maturity Level 5

Key Process Area 1 (K_{51})—Defect Prevention

41. Is software system engineering represented on the system design team?

42. Is a formal procedure used to ensure periodic management review of the status of each software development project?

43. Is a mechanism used for initiating error prevention actions?

44. Is a mechanism used for identifying and replacing obsolete technologies?

45. Is software productivity analyzed for major process steps?

B.2 Binary (Yes, No) Responses to the Maturity Questionnaire

K_{21}	K_{31}	K_{41}
1. N	16. N	31. Y
2. N	17. N	32. N
3. Y	18. Y	33. Y
4. Y	19. Y	34. Y
5. N	20. N	35. N

K_{22}	K_{32}	K_{42}
6. Y	21. Y	36. Y
7. N	22. Y	37. Y
8. Y	23. N	38. Y
9. Y	24. Y	39. Y
10. Y	25. Y	40. Y

K_{23}	K_{33}	K_{51}
11. N	26. Y	41. Y
12. Y	27. Y	42. Y
13. Y	28. N	43. Y
14. Y	29. Y	44. Y
15. Y	30. N	45. N

B.3 Prior Probabilities and Likelihoods

B.3.1 *The Maturity Levels* $\mathcal{P}(M_i \mid M_{i-1})$

The first column is based on common knowledge of maturity levels of U.S. companies. The second column is true by requirements of the hierarchical model.

$\mathcal{P}(M_2 = 1 \mid M_1 = 1) = 0.50$ $\mathcal{P}(M_2 = 1 \mid M_1 = 0) = 0$
$\mathcal{P}(M_2 = 0 \mid M_1 = 1) = 0.50$ $\mathcal{P}(M_2 = 0 \mid M_1 = 0) = 1$

$\mathcal{P}(M_3 = 1 \mid M_2 = 1) = 0.15$ $\mathcal{P}(M_3 = 1 \mid M_2 = 0) = 0$
$\mathcal{P}(M_3 = 0 \mid M_2 = 1) = 0.85$ $\mathcal{P}(M_3 = 0 \mid M_2 = 0) = 1$

$\mathcal{P}(M_4 = 1 \mid M_3 = 1) = 0.05$ $\mathcal{P}(M_4 = 1 \mid M_3 = 0) = 0$
$\mathcal{P}(M_4 = 0 \mid M_3 = 1) = 0.95$ $\mathcal{P}(M_4 = 0 \mid M_3 = 0) = 1$

$\mathcal{P}(M_5 = 1 \mid M_4 = 1) = 0.01$ $\mathcal{P}(M_5 = 1 \mid M_4 = 0) = 0$
$\mathcal{P}(M_5 = 0 \mid M_4 = 1) = 0.99$ $\mathcal{P}(M_5 = 0 \mid M_4 = 0) = 1$

B.3.2 *The Key Process Areas* $\mathcal{P}(K_{ij})$ *and* $\mathcal{P}(K_{ij} \mid M_i)$

These priors were specified according to expert opinion.

Maturity Level 2	Maturity Level 3	Maturity Level 4	Maturity Level 5
$\mathcal{P}(K_{21} = 1) = 0.8$	$\mathcal{P}(K_{31} = 1) = 0.5$	$\mathcal{P}(K_{41} = 1) = 0.2$	$\mathcal{P}(K_{51} = 1) = 0.02$
$\mathcal{P}(K_{21} = 0) = 0.2$	$\mathcal{P}(K_{31} = 0) = 0.5$	$\mathcal{P}(K_{41} = 0) = 0.8$	$\mathcal{P}(K_{51} = 0) = 0.98$
$\mathcal{P}(K_{22} = 1) = 0.9$	$\mathcal{P}(K_{32} = 1) = 0.6$	$\mathcal{P}(K_{42} = 1) = 0.3$	
$\mathcal{P}(K_{22} = 0) = 0.1$	$\mathcal{P}(K_{32} = 0) = 0.4$	$\mathcal{P}(K_{42} = 0) = 0.7$	
$\mathcal{P}(K_{23} = 1) = 0.9$	$\mathcal{P}(K_{33} = 1) = 0.6$		
$\mathcal{P}(K_{23} = 0) = 0.1$	$\mathcal{P}(K_{33} = 0) = 0.4$		

Appendix B.3 Prior Probabilities and Likelihood

The entries in the following table give $\mathcal{P}(K_{ij} \mid M_i)$.

Key Process Areas		M_2 1	M_2 0	M_3 1	M_3 0	M_4 1	M_4 0	M_5 1	M_5 0
K_{21}	1	0.95	0.30	-	-	-	-	-	-
	0	0.05	0.70	-	-	-	-	-	-
K_{22}	1	0.90	0.40	-	-	-	-	-	-
	0	0.10	0.60	-	-	-	-	-	-
K_{23}	1	0.90	0.40	-	-	-	-	-	-
	0	0.10	0.60	-	-	-	-	-	-
K_{31}	1	-	-	0.95	0.25	-	-	-	-
	0	-	-	0.05	0.75	-	-	-	-
K_{32}	1	-	-	0.90	0.35	-	-	-	-
	0	-	-	0.10	0.65	-	-	-	-
K_{33}	1	-	-	0.90	0.35	-	-	-	-
	0	-	-	0.10	0.65	-	-	-	-
K_{41}	1	-	-	-	-	0.98	0.20	-	-
	0	-	-	-	-	0.02	0.80	-	-
K_{42}	1	-	-	-	-	0.95	0.25	-	-
	0	-	-	-	-	0.05	0.75	-	-
K_{51}	1	-	-	-	-	-	-	1.0	0.0
	0	-	-	-	-	-	-	0.0	1.0

B.3.3 The Likelihoods $\mathcal{L}(K_{ij}; \underline{R}_{ij})$

These likelihoods are based on independence of the responses, and for simplicity are assumed to be the same for all the key process areas.

Responses $\underline{R}_{ij}$	Likelihood $L(K_{ij}=1; \underline{R}_{ij})$	Likelihood $L(K_{ij}=0; \underline{R}_{ij})$
0	0.0025	0.1200
00001	0.0035	0.0820
00010	0.0035	0.0820
00100	0.0035	0.0820
01000	0.0035	0.0820
10000	0.0035	0.0820
00011	0.0100	0.0350
00101	0.0100	0.0350
01001	0.0100	0.0350
10001	0.0100	0.0350
00110	0.0100	0.0350
01010	0.0100	0.0350
10010	0.0100	0.0350
01100	0.0100	0.0350
10100	0.0100	0.0350
11000	0.0100	0.0350
00111	0.0350	0.0100
01011	0.0350	0.0100
10011	0.0350	0.0100
01101	0.0350	0.0100
10101	0.0350	0.0100
11001	0.0350	0.0100
01110	0.0350	0.0100
10110	0.0350	0.0100
11010	0.0350	0.0100
11100	0.0350	0.0100
01111	0.0820	0.0035
10111	0.0820	0.0035
11011	0.0820	0.0035
11101	0.0820	0.0035
01110	0.0820	0.0035
11111	0.1200	0.0025

References

Aalen, O. O. (1987) Dynamic Modeling and Causality. *Scand. Actuarial J.*, 177–190.

Achcar, J. A., D. Dey, and M. Niverthy (1998) A Bayesian Approach Using Nonhomogeneous Poisson Process for Software Reliability Models in Frontiers in Reliability. *Series on Quality, Reliability and Engineering Statistics* (S. K. Basu and S. Mukhopadhyay, Eds.), 4: Calcutta University, India.

Al-Mutairi, D., Y. Chen, and N. D. Singpurwalla (1998) An Adaptive Concatenated Failure Rate Model for Software Reliability. *J. Amer. Statist. Assoc.*, **93** 443: 1150–1163.

Andersen, P. K. and Ø. Borgan (1985) Counting Process Models for Life History Data: A Review (with Discussion). *Scand. J. Statist.*, **12**: 97–158.

Andreatta, G. and G. M. Kaufman (1986) Estimation of Finite Population Properties When Sampling is Without Replacement and Proportional to Magnitude. *J. Amer. Statist. Assoc.*, **81** 395: 657–666.

Arjas, E. and P. Haara (1984) A Marked Point Process Approach to Censored Failure Data with Complicated Covariates. *Scand. J. Statist.*, **11**: 193–209.

Barlow, R. F. and F. Proschan (1975) *Statistical Theory of Reliability and Life Testing*. Holt, Rinehart and Winston, New York.

Basu, A. P. (1971) Bivariate Failure Rate. *J. Amer. Statist. Assoc.*, **60**: 103–104.

Bather, J. A. (1965) Invariant Conditional Distributions. *Ann. Math. Stat.*, **36**: 829–846.

Benkherouf, L. and J. A. Bather (1988) Oil Exploration: Sequential Decisions in the Face of Uncertainty. *J. Appl. Prob.*, **25**: 529–543.

Berger, J. O. (1985) *Statistical Decision Theory and Bayesian Analysis.* Second Edition, Springer-Verlag, New York.

Berger, J. O. and R. Wolpert (1984) *The Likelihood Principle.* Institute of Mathematical Statistics, Hayward, CA.

Bernardo, J. M. (1979) Reference Posterior Distributions for Bayesian Inference. *J. of the Roy. Statist. Soc.*, series B, **41**: 113–147.

Bernardo, J. M. (1997) Non-informative Priors Do Not Exist. A Dialogue with José M. Bernardo. *J. of Statist. Planning and Inference,* **65** 1: 159–189.

Bernardo, J. M. and A. F. M. Smith (1994) *Bayesian Theory.* Wiley, Chichester.

Bickel, P. J., V. N. Nair, and P. C. Wang (1992) Nonparametric Inference Under Biased Sampling from a Finite Population. *The Ann. Statist.*, **20**: 853–878.

Box, G. E. P. (1980) Sampling and Bayes Inference in Scientific Modeling and Robustness. *J. Royal Statist. Soc.*, Series A, **143**: 383–430.

Box, G. E. P. and G. M. Jenkins (1970) *Time Series Analysis: Forecasting and Control.* Revised Edition. Holden-Day, CA.

Box, G. E. P. and G. M. Jenkins (1976) *Time Series Analysis: Forecasting and Control.* Holden-Day, CA.

Brooks, S. P. (1998) Markov Chain Monté Carlo Method and Its Application. *The Statistician*, **47**, Part I: 69–100.

Brownlie, R., J. Prowse, and M. S. Phadke (1992) Robust Testing of AT&T PMX/StartMail Using OATS. *AT&T Tech. J.*, **71**: 41–47.

Campodónico, S. (1993) The Signature as a Covariate in Reliability and Biometry. PhD Thesis, School of Engineering and Applied Science, The George Washington University, Washington, DC.

Campodónico, S. and N. D. Singpurwalla (1994) A Bayesian Analysis of the Logarithmic-Poisson Execution Time Model Based on Expert Opinion and Failure Data. *IEEE Trans. Soft. Eng.*, **20**: 677–683.

Campodónico, S. and N. D. Singpurwalla (1995) Inference and Predictions from Poisson Point Processes Incorporating Expert Knowledge. *J. Amer. Statist. Assoc.*, **90**: 220–226.

Casella, G. and E. I. George (1992) Explaining the Gibbs Sampler. *Amer. Statist.*, **46** 3: 167–174.

Charette, R. N. (1989) *Software Engineering, Risk Analysis and Management*. McGraw-Hill, New York.

Chatfield, C. (1983) *Statistics for Engineering*. Third Edition, Chapman and Hall, London.

Chen, J. and N. D. Singpurwalla (1996) Composite Reliability and Its Hierarchical Bayes Estimation. *J. Amer. Statist. Assoc.*, **91** 436: 1474–1484.

Chen, Y. and N. D. Singpurwalla (1994) A Non-Gaussian Kalman Filter Model for Tracking Software Reliability. *Statistica Sinica*, **4** 2: 535–548.

Chen, Y. and N. D. Singpurwalla (1997) Unification of Software Reliability Models Via Self-Exciting Point Processes. *Advances in Applied Probability*, **29** 2: 337–352.

Chib, S. and E. Greenberg (1995) Understanding the Metropolis-Hastings Algorithm. *Amer. Statist.*, **49** 4: 327–335.

Cochran, W. G. (1977) *Sampling Techniques*. Wiley, New York.

Cohen, D. M., S. R. Dalal, A. Kajla, and G. C. Patton (1994) The Automatic Efficient Test Generator (AETG) System. In Proceedings of Fifth International Symposium on Software Reliability Engineering. *IEEE Computer Society Press*, Los Alamos, CA, 303–309.

Cox, D. R. and V. Isham (1980) *Point Processes*. Chapman and Hall, London.

Cox, D. R. and P. A. Lewis (1966) *Statistical Analysis of Series of Events*. Methuen, London.

Crosby, P. B. (1979) *Quality is Free*. McGraw Hill, New York.

Crow, L. H. and N. D. Singpurwalla (1984) An Empirically Developed Fourier Series Model for describing Software Failures. *IEEE Trans. Reliability*, R-**33**: 176–183.

Dalal, S. R. and C. L. Mallows (1988) When Should One Stop Testing Software? *J. of Amer. Statist. Assoc.*, **83**: 872–879.

Dalal, S. R. and C. L. Mallows (1990) Some Graphical Aids for Deciding When to Stop Testing Software. *IEEE J. on Selected Areas in Communications*, **8**: 169–175.

Dalal, S. R. and C. L. Mallows (1998) Factor-Covering Designs for Testing Software. *Technometrics*, **40** 3: 234–243.

Davis, A. M. (1990) *Software Requirements—Analysis and Specification*. Prentice-Hall, New York.

Dawid, A. P. (1984) The Prequential Approach. *J. of the Roy. Statist. Soc.* Series A, **147**, Part 2: 278–292.

Dawid, A. P. (1992) Prequential Analysis, Stochastic Complexity and Bayesian Inference. In *Bayesian Statistics 4* (J. M. Bernardo, J. O. Berger, A. P. Dawid, and A. F. M. Smith, Eds.), Oxford University Press, New York.

de Finetti, B. (1937) La Prévision: Ses Lois Logiques, Ses Sources Subjectives. *Ann. Inst. H. Poincaré* (Paris), **7**: 1–68 [see (1964) for English transl.], (cited pp. 5, 27, 29, 68, 78, 143, 149, 151, 186, 192–3, 213, 215, 229).

de Finetti, B. (1964) Foresight: Its Logical Laws, Its Subjective Sources. In H. E. Kyburg and H. E. Smokler (Eds.), *Studies in Subjective Probability*. Wiley, New York (English transl. of B. de Finetti, 1937).

de Finetti, B. (1972) *Probability, Induction and Statistics*. Wiley, New York.

de Finetti, B. (1974) *Theory of Probability, 1*. Wiley, New York.

DeGroot, M. H. (1970) *Optimal Statistical Decisions*. McGraw-Hill, New York.

Duane, J. T. (1964) Learning Curves Approach to Reliability Monitoring. *IEEE Trans. on Aerospace*, AS-**2**: 563–566.

Dwass, M. (1964) Extremal Processes. *Annals of Mathematical Statistics*, **35**: 1718–1725.

Efron, B. and R. Thisted (1976) Estimating the Number of Unseen Species: How Many Words Did Shakespeare Know? *Biometrika*, **63**: 435–447.

Fakhre-Zakeri, I. and E. Slud (1995) Mixture Models for Software Reliability with Imperfect Debugging Identifiability of Parameters. *IEEE Trans. Rel.*, **44** 1: 104–113.

Ferguson, T. S. and J. P. Hardwick (1989) Stopping Rules for Proofreading. *J. Appl. Prob.*, **26**: 304–313.

Fisher, R. A. and F. Yates (1953) *Statistical Tables for Biological, Agricultural and Medical Research,* (fourth edition.), Oliver & Boyd, Edinburg and London.

Forman, E. H. and N. D. Singpurwalla (1977) An Empirical Stopping Rule for Debugging and Testing Computer Software. *J. Amer. Statist. Assoc.*, **72** 360: 750–757.

French, S. (1980) Updating of Belief in the Light of Someone Else's Opinion. *J. of the Roy. Statist.* Soc. Series A, **143**: 43–48.

Freund, J. E. (1961) A Bivariate Extension of the Exponential Distribution. *J. Amer. Statist. Assoc.,* **56**: 971–977.

Gaffney, J. E., Jr. (1984) Estimating the Number of Faults in Code. *IEEE Trans. on Soft. Eng.*, SE-**10**: 459–464.

Geisser, S. (1984) On Prior Distributions for Binary Trials. *Amer. Statist.,* **38** 4: 244–247.

Gelfand, A. E., and Smith, A. F. M. (1990) Sampling-Based Approaches to Calculating Marginal Densities. *J. Amer. Statist. Assoc.*, **85**: 398–409.

Gill, R. D. (1984) Understanding Cox's Regression Model: A Martingale Approach. *J. Amer. Statist. Assoc.*, **79** 386: 441–447.

Goel, A. L. (1983) A Guide Book for Software Reliability Assessment. Technical Report, RADC-TR-83-176. Rome Air Development Center, Rome, New York.

Goel, A. L. (1985) Software Reliability Models: Assumptions, Limitations and Applicability. *IEEE Trans. on Soft. Eng.*, SE-**11**: 1411–1423.

Goel, A. L. and K. Okumoto (1978) An Analysis of Recurrent Software Failures on a Real-Time Control System. In *Proceedings of the ACM Annual Technical Conference*, 496–500.

Goel, A. L. and K. Okumoto (1979) Time-Dependent Error Detection Rate Model for Software Reliability and Other Performance Measures. *IEEE Trans. Rel.*, R-**28**: 206–211.

Gokhale, S. S., M. R. Lyu, and K. S. Trivedi (1998) Reliability Simulation of Component-Based Software Systems. In Proceedings of the Ninth International Symposium on Software Engineering (ISSRE-98): 192–201. *IEEE Computer Society*, Los Alamitos, CA.

Good, I. J. (1983) *Good Thinking: The Foundations of Probability and Its Applications*. University of Minnesota Press, Minneapolis.

Gordon, L. (1983) Successive Sampling in Large Finite Populations. *Ann. Statist.* **11**: 702–706.

Hicks, C. R. (1982) Fundamental Concepts in the Design of Experiments. Holt, Rinehart and Winston, Orlando, FL.

Hill, B. M. (1993) Dutch Books, the Jeffreys–Savage Theory of Hypothesis Testing and Bayesian Reliability. Chapter 3 in *Reliability and Decision Making* (R. E. Barlow, C. A. Clarotti, and F. Spizzichino, Eds.), Chapman and Hall, London, 31–85.

Hogg, R. V. and A. T. Craig. (1978) *Introduction to Mathematical Statistics*. Fourth edition, Macmillan, New York.

Howson, C. and P. Urbach (1989) *Scientific Reasoning: The Bayesian Approach*. Open Court, Il.

Hudson, A. (1967) Program Errors as a Birth and Death Process. Technical Report SP-3011, Systems Development Corp., Santa Monica, CA.

Humphrey, W. S. (1989) *Managing the Software Process*. SEI (The SEI Series in Software Engineering), Addison-Wesley, Reading, MA.

Humphrey, W. S. and N. D. Singpurwalla (1991) Predicting (Individual) Software Productivity. *IEEE Trans. Soft. Eng.*, **17** 2: 196–207.

Humphrey, W. S. and N. D. Singpurwalla (1998) A Bayesian Approach for Assessing Software Quality and Productivity. *Int. J. Reliability, Quality and Safety Eng.*, **5** 2: 195–209.

Humphrey, W. S. and W. L. Sweet (1987) A Method for Assessing the Software Engineering Capability of Contractors. SEI Technical Report SEI-87-TR-23.

Iannino, A., J. D. Musa, and K. Okumoto (1987) *Software Reliability: Measurement, Prediction, Application*. Wiley, New York.

Jalote, P. (1991) *An Integrated Approach to Software Engineering*. Springer-Verlag, New York.

Jaynes, E. T. (1968) Prior Probabilities. *IEEE Trans. System Science and Cybernetics*, SSC-**4**: 227–241.

Jaynes, E. T. (1983) *Papers on Probability, Statistics and Statistical Physics*. Reidal, Dordrecht.

Jeffreys, H. (1961) *Theory of Probability*. Third edition, OUP, Oxford.

Jelinski, Z. and P. Moranda (1972) Software Reliability Research. In W. Freiberger, Ed., *Statistical Computer Performance Evaluation,* Academic Press, New York.

Johnson, N. L. and S. Kotz (1970) *Continuous Univariate Distributions*, 2. Houghton Mifflin, New York.

Kass, R. E. and L. Wasserman (1996) The Selection for Prior Distribution by Formal Rules. *J. Amer. Statist. Assoc.*, **91** 435: 1343–1370.

Kaufman, G. M. (1996) Successive Sampling and Software Reliability. *J. Statistical Planning and Inference,* **49**: 343–369.

Knight, J. C. and N. G. Levenson (1986) An Experimental Evaluation of the Assumption of Independence in Multiversion Programming. *IEEE Trans. Soft. Eng.*, SE-**12** 1: 96–109.

Koch, G. and P. J. C. Spreij (1983) Software Reliability as an Application of Martingale and Filtering Theory. *IEEE Trans. Rel.,* R-**32**: 342–345.

Kolmogorov, A. N. (1950) *Foundations of the Theory of Probability*. Chelsea, New York.

Kuo, L. and T. Y. Yang (1995) Bayesian Computation of Software Reliability. *J. Comput. Graphical Stat.*, **4**: 65–82.

Kuo, L. and T. Y. Yang (1996) Bayesian Computation for Nonhomogeneous Poisson Processes in Software Reliability. *J. Amer. Statist. Assoc.*, **91** 434: 763–773.

Kurtz, T. G. (1983) Gaussian Approximations for Markov Chains and Counting Processes. *Bulletin of the International Statistical Institute*, **50**: 361–375.

Langberg, N. and N. D. Singpurwalla (1985) A Unification of Some Software Reliability Models. *SIAM J. Sci. Stat. Comput.* **6**: 781–790.

Lee, P. M. (1989) *Bayesian Statistics: An Introduction*. Oxford University Press, New York.

Lindley, D. V. (1972) Bayesian Statistics, A Review. Regional Conference Series in Applied Mathematics. *SIAM*, Philadelphia, PA.

Lindley, D. V. (1980) Approximate Bayesian Methods. *Trabajos Estadistica*, **31**: 223–237.

Lindley, D. V. (1982a) Scoring Rules and the Inevitability of Probability. *Inst. Statist. Rev.*, **50**: 1–26.

Lindley, D. V. (1982b) The Bayesian Approach to Statistics. In *Some Recent Advances in Statistics* (T. de Oliveira and B. Esptein Eds.). 65–87, London, Academic Press.

Lindley, D. V. (1983) Reconciliation of Probability Distributions. *Operations Research*, **31**: 866–880.

Lindley, D. V. and N. D. Singpurwalla (1986a) Reliability (and Fault Tree) Analysis Using Expert Opinion. *J. Amer. Statist. Assoc.*, **81**: 87–90.

Lindley, D. V. and N. D. Singpurwalla (1986b) Multivariate Distributions for the Lifelengths of Components of a System Sharing a Common Environment. *J. Appl. Prob.*, **23**: 418–431.

Littlewood, B. and J. L. Verall (1973) A Bayesian Reliability Growth Model for Computer Software. *Appl. Stat.*, **22**: 332–346.

Lynn, N. (1996) Software for a Shot-Noise Reliability Growth Model. The George Washington University Technical Memorandum TM-96-1.

Mandl, R. (1985) Orthogonal Latin Squares: An Application of Experimental Design to Compiler Testing. *Communications of the ACM*, **28** 10: 1054–1058.

Marshall, A. W. (1975) Some Comments on Hazard Gradients. *Stochastic Processes and Their Applications.*, **3**: 295–300.

Marshall, A. W. and I. Olkin (1967) A Multivariate Exponential Distribution. *J. Amer. Statist. Assoc.*, **62**: 30–44.

Martz, H. F. and R. A. Waller (1982) *Bayesian Reliability Analysis*. Wiley, New York.

Mazzuchi, T. A. and R. Soyer (1988) A Bayes Empirical-Bayes Model for Software Reliability. *IEEE Trans. Rel.*, R-**37**: 248–254.

McDaid, K. and S. P. Wilson (1999) Determining An Optimal Time to Test Software Assuming a Time-Dependent Error Detection Rate Model. Technical Report 96/02, Department of Statistics, Trinity College, Dublin.

Meinhold, R. J. and N. D. Singpurwalla (1983a) Bayesian Analysis of a Commonly Used Model for Describing Software Failures. *The Statistician*, **32** 2: 168–173.

Meinhold, R. J. and N. D. Singpurwalla (1983b) Understanding the Kalman Filter. *Amer. Statist.*, **37** 2: 123–127.

Miller, D. R. (1986) Exponential Order Statistic Models of Software Reliability Growth. *IEEE Trans. Soft. Eng.*, SE-**12**: 12–24.

Morali, N. and R. Soyer (1999) Optimal Stopping Rules for Software Testing. Under review.

Moranda, P. B. (1975) Prediction of Software Reliability and Its Applications. In *Proceedings of the Annual Reliability and Maintainability Symposium*, Washington, DC, 327–332.

Morris, P. A. (1974) Decision Analysis Expert Use. *Mgmt. Sci.*, **20**: 1233–1241.

Morris, P. A. (1977) Combining Expert Judgments: A Bayesian Approach. *Mgmt. Sci.*, **23**: 679–693.

Musa, J. D. (1975) A Theory of Software Reliability and Its Applications. *IEEE Trans. Soft. Eng.*, SE-**1**: 312–327.

Musa, J. D. (1979) Software Reliability Data. *IEEE Comput. Soc. Repository*, New York.

Musa, J. D. and K. Okumoto (1984) A Logarithmic Poisson Execution Time Model for Software Reliability Measurement. In *Proceedings of the seventh International Conference on Software Engineering.*, Orlando, FL, 230–237.

Musa, J. D., A. Iannino, and K. Okumoto (1987) *Software Reliability*. McGraw-Hill, New York.

Myers. G. J. (1978) *Composite/Structured Design*. Van Nostrand Reinhold, New York.

Nair, V. J. and P. C. C. Wang (1989) Maximum Likelihood Estimation Under a Successive Sampling Discovery Model. *Technometrics* **31**: 423–436.

Nair, V. N., D. A. James, W. K. Ehrlich, and J. Zevallos (1998) A Statistical Assessment of Some Software Testing Strategies and Application of Experimental Design Techniques. *Statistica Sinica*, **8** 1: 165–184.

Okumoto, K. and A. L. Goel (1980) Optimum Release Time for Software Systems, Based on Reliability and Cost Criteria. *J. Syst. Soft.*, **1**: 315–318.

Ozekici, S. and N. A. Catkan (1993) A Dynamic Software Release Model. *Computational Economics* **6**: 77–94.

Paulk, M. C., M. B. Chrissis, B. Curtis, and C. V. Weber (1993) Capability Maturity Model, Version 1.1. *IEEE Soft.*, 18–27.

Phadke, M. S. (1989) *Quality Engineering Using Robust Design*. Prentice- Hall, Englewood Cliffs, NJ.

Raftery, A. E. (1987) Inference and Prediction for a General Order Statistic Model With Unknown Population Size. *J. Amer. Statist. Assoc.*, **82**: 1163–1168.

Raftery, A. E. (1988) Analysis of a Simple Debugging Model. *Appl. Statist.*, **37** 1: 12–22.

Raftery, A. E. (1992) Discussion of Model Determination Using Predictive Distributions with Implementation via Sampling-Based Methods, by Gelfand et al. In *Bayesian Statistics 4* (J. M. Bernardo, J. O. Berger, A. P. Dawid and A. F. M. Smith, Eds.), Oxford University Press, New York.

Raiffa, H. and R. Schlaifer (1961) *Applied Statistical Decision Theory*. Division of Research, Harvard Business School, Boston.

Ramsey, F. P. (1964) Truth and Probability. In H. E. Kyburg Jr., and H. E. Smokler, Editors, *Studies in Subjective Probability*, Wiley, New York, 61–92.

Randolph, P. and M. Sahinoglu (1995) A Stopping Rule for a Compound Poisson Random Variable. *Applied Stochastic Models and Data Analysis*, **11**: 135–143.

Rao, C. R. (1987) Prediction of Future Observations in Growth Curve Models. *Statistical Science*, **2** 4: 434–471.

Roberts, H. V. (1965) Probabilistic Prediction. *J. Amer. Statist. Assoc.*, **60**: 50–61.

Ross, S. M. (1970) *Applied Probability Models with Optimization Applications*. Holden-Day, San Francisco.

Ross, S. M. (1985a) The Stopping Rule Problem. *IEEE Trans. Soft. Eng.*, SE-**11**: 1472–1476.

Ross, S. M. (1985b) Statistical Estimation of Software Reliability. *IEEE Trans. Soft. Eng.* SE-**11**: 479–483.

Sahinoglu, M. (1992) Compound-Poisson Software Reliability Model. *IEEE Trans. on Soft. Eng.* **18**: 624–630.

Savage, L. J. (1972) *The Foundations of Statistics*. Second edition, Dover, New York.

Schick, G. J. and R. W. Wolverton (1978) Assessment of Software Reliability. In *Proceedings in Operations Research*, Physica-Verlag, Vienna, 395–422.

Scholz, F. W. (1986) Software Reliability Modeling and Analysis. *IEEE Trans. Soft. Eng.*, SE-**12**: 25–31.

Singpurwalla, N. D. (1988a) Foundational Issues in Reliability and Risk Analysis. *SIAM Rev.*, **30**: 264–282.

Singpurwalla, N. D. (1988b) An Interactive PC-Based Procedure for Reliability Assessment Incorporating Expert Opinion and Survival Data. *J. Amer. Statist. Assoc.*, **83** 401: 43–51.

Singpurwalla, N. D. (1989a) A Unifying Perspective on Statistical Modeling. *SIAM Rev.*, **31** 4: 560–564.

Singpurwalla, N. D. (1989b) Preposterior Analysis in Software Testing. In *Statistical Data Analysis and Inference* (Y. Dodge, Ed.), Elsevier, North-Holland., Amsterdam, 581–595.

Singpurwalla, N. D. (1991) Determining an Optimal Time Interval for Testing and Debugging Software. *IEEE Trans Soft. Eng.*, **17** 4: 313–319.

Singpurwalla, N. D. (1992) A Bayesian Perspective on Taguchi's Approach to Quality Engineering and Tolerance Design. *Institute for Industrial Engineering Transactions*, **24** 5: 18–32.

Singpurwalla, N. D. (1993) Design by Decision Theory: A Unifying Perspective on Taguchi's Approach to Quality Engineering. Chapter 14 in *Reliability and Decision Making* (R. E. Barlow, C. A. Clarotti, and F. Spizzichino, Eds.), Chapman and Hall, London, 267–272.

Singpurwalla, N. D. (1995) The Failure Rate of Software: Does It Exist? *IEEE Trans. Rel.*, **44** 3: 463–469.

Singpurwalla, N. D. (1998a) A Paradigm for Modeling and Tracking Reliability Growth. In *Reliability Growth Modeling: Objectives, Expectations and Approaches* (Farquhar and Mosleh, Eds.), Center for Reliability Engineering, University of Maryland, College Park, MD.

Singpurwalla, N. D. (1998b) Software Reliability Modeling by Concatenating Failure Rates. In Proceedings of the ninth International Symposium on Software Reliability Engineering (ISSRE-98), *IEEE Computer Society*, Los Alamitos, CA, 106–110.

Singpurwalla, N. D. (1998c) The Stochastic Control of Process Capability Indices. *Sociedad Española de Estadística e Investigación Operativa. TEST*, **7** 4: 1–74.

Singpurwalla, N. D. (1999) A Probabilitistic Hierarchical Classification Model for Rating Suppliers. *J. Quality Tech.* To appear.

Singpurwalla, N. D. and M. S. Song (1988) Reliability Analysis Using Weibull Lifetime Data and Expert Opinion. *IEEE Trans. Rel.*, **37** 3: 340–347.

Singpurwalla, N. D. and R. Soyer (1985) Assessing (Software) Reliability Growth Using a Random Coefficient Autoregressive Process and Its Ramifications. *IEEE Trans. Soft. Eng.*, SE-**11** 12: 1456–1464.

Singpurwalla, N. D. and R. Soyer (1992) Nonhomogeneous Autoregressive Processes for Tracking (Software) Reliability Growth, and Their Bayesian Analysis. *J. Roy. Statist. Soc.* Series B, **54**: 145–156.

Singpurwalla, N. D. and R. Soyer (1996) Assessing the Reliability of Software: An Overview. In *Reliability and Maintenance of Complex Systems*, (S. Ozekici, Ed.), NATO ASI Series, Springer Verlag, New York. 345–367.

Singpurwalla, N. D. and S. Wilson (1994) Software Reliability Modeling. *International Statist. Rev.*, **62** 3: 289–317.

Singpurwalla, N. D. and S. P. Wilson (1995) The Exponentiation Formula of Reliability and Survival: Does It Always Hold? *Lifetime Data Analysis*, **1**: 187–194.

Slud, E. (1997) Testing for Imperfect Debugging in Software Reliability. *Scand. J. Statist.*, **24**: 527–555.

Smith, A. F. M. and A. E. Gelfand (1992) Bayesian Statistics Without Tears: A Sampling-Resampling Perspective. *Amer. Statist.*, **46** 2: 84–88.

Soyer, R. (1985) *Random Coefficient Autoregressive Processes and Their Ramifications: Applications to Reliability Growth Assessment.* PhD Thesis, School of Engineering and Applied Science, George Washington University, Washington, DC.

Soyer, R. (1992) Monitoring Software Reliability Using Non-Gaussian Dynamic Models. In *Proceedings of the Engineering Systems Design and Analysis Conference*, **1**: 419–423.

Tausworthe, R. C. and M. R. Lyu (1996) Software Reliability Simulation, in *Handbook of Software Reliability Engineering* (M. R. Lyu, Ed.), McGraw-Hill, New York, 661–698.

Tierney, L. and J. B. Kadane (1986) Accurate Approximations for Posterior Moments and Marginal Densities. *J. of Amer. Statist. Assoc.*, **81** 393: 82–86.

Tversky, A., D. V. Lindley and R. V. Brown (1979) On the Reconciliation of Probability Assessments (with discussion). *J. Roy. Statist. Soc.* series A, **142**: 146–180.

van Pul, M. C. J. (1993) *Statistical Analysis of Software Reliability Models.* Centre for Mathematics and Computer Science, Amsterdam.

von Mises, R. (1957) *Probability, Statistics and Truth.* Dover, New York.

West, M., P. J. Harrison, and H. S. Migon (1985) Dynamic Generalized Linear Models and Bayesian Forecasting (with Discussion). *J. Amer. Statist. Assoc.* **80**: 73–97.

Weyuker, E. J., and B. Jeng (1991) Analyzing Partition Testing Strategies. *IEEE Trans. Soft. Eng.*, **17**: 703–711.

Yamada, S. and S. Osaki (1984) Nonhomogeneous Error Detection Rate Models for Software Reliability Growth. In *Reliability Theory* (S. Osaki and Y. Hatoyama, Eds.), Springer-Verlag, Berlin, 120–143.

Yamada, S., H. Narihisa, and S. Osaki (1984) Optimum Release Policies for a Software System With a Scheduled Software Delivery Time. *Int. J. Syst. Sci.*, **15**: 905–914.

Zadeh, L. (1981) Possibility Theory and Soft Data Analysis. *Mathematical Frontiers of the Social and Policy Sciences* (L. Cobb and R. M. Thrall, Eds.), Westview Press, Boulder, CO: 69–129.

Zellner, A. (1971) *An Introduction to Bayesian Inference in Econometrics.* Wiley, New York.

Zellner, A. (1977) Maximal Data Information Prior Distributions. In *New Methods in the Applications of Bayesian Methods* (A. Aykac and C. Brumat, Eds.), North Holland, Amsterdam.

Author Index

Aalen, O. O. 224
Achcar, J. A. 90
Al-Mutairi, D. 59, 92, 156, 160
Anderson, P. K. 222
Andretta, G. 89, 202
Arjas, E. 70

Barlow, R. F. 74
Basu, A. P. 58
Bather, J. A. 81, 202
Benkherouf, L. 202
Berger, J. O. 25, 29, 105
Bernardo, J. M. 30, 105, 114–115, 146
Bickel, P. J. 89
Borgen, Ø. 222
Box, G. E. P. 91, 102–103, 172
Brooks, S. P. 249
Brown, R. V. 116, 150
Brownlie, R. 228, 232

Campodónico, S. 39, 44, 118, 120
Casella, G. 158, 249
Catkan, N. A. 214
Charette, R. N. 1
Chatfield, C. 106
Chen, J. 33, 69
Chen, Y. 59, 70, 72, 81, 84, 92, 143–144, 156, 160
Chib, S. 249, 259
Chrissis, M. B. 181
Cochran, W. G. 228
Cohen, D. M. 228, 232–234
Cox, D. R. 48, 91–92
Craig, A. T. 106
Crosby, P. B. 181
Crow, L. H. 85
Curtis, B. 181

Dalal, S. R. 86, 202, 204, 207, 228, 232–234, 238

Davis, A. M. 3
Dawid, A. P. 145–147
de Finetti, B. 9, 20, 26, 104, 194
DeGroot, M. H. 20
Dey, D. 90
Duane, J. T. 91
Dwass, M. 90

Efron, B. 202
Ehrlich, W. K. 228–229, 231, 234

Fakhre-Zaken, I. 70, 86
Ferguson, T. S. 202
Fisher, R. A. 236
Forman, E. H. 24, 107, 202
French, S. 116
Freund, J. E. 17

Gaffney, J. E., Jr. 120–121
Geisser, S. 33
Gelfand, A. E. 251, 253
George, E. I. 158, 249
Gill, R. D. 226
Goel, A. L. 70, 75, 77–79, 89–90, 110, 120, 202
Gokhale, S. S. 52
Good, I. J. 115
Gordon, L. 89
Greenberg, E. 249, 259

Haara, P. 70
Hardwick, J. P. 202
Harrison, P. J. 172
Hicks, C. R. 236–238
Hill, B. M. 194
Hogg, R. V. 106
Howson, C. 20
Hudson, A. 72
Humphrey, W. S. 174, 176, 180–181

Iannino, A. 72, 102

Isham, V. 48, 92

Jalote, P. 2
James, D. A. 228–229, 231, 234
Jaynes, E. T. 115
Jeffreys, H. 113, 147
Jelinski, Z. 59, 68, 72, 106, 109–110
Jeng, B. 228
Jenkins, G. M. 91, 102, 172
Johnson, N. L. 30

Kadane, J. B. 113
Kajla, A. 228, 232–234
Kass, R. E. 113
Kaufman, G. M. 88–89, 202
Knight, J. C. 17, 236
Koch, G. 86, 225
Kolmogorov, A. N. 6, 20
Kotz, S. 30
Kuo, L. 70, 72, 77, 86, 89
Kurtz, T. G. 224

Langberg, N. 70, 73, 76, 83, 89
Lee, P. M. 105, 159
Levenson, N. G. 17, 236
Lewis, P. A. 91
Lindley, D. V. 17, 20, 29, 116–117, 125, 141, 150, 195–196
Littlewood, B. 77, 83
Lynn, N. 158, 160
Lyu, M. R. 52, 82

Mallows, C. L. 86, 202, 204, 207, 238
Mandl, R. 228, 232, 235
Marshall, A. W. 17, 58
Martz, H. F. 113
Mazzuchi, T. A. 76, 83, 124–126, 153
McDaid, K. 112, 207, 215
Meinhold, R. J. 81, 107, 114
Migon, H. S. 172
Miller, D. R. 70, 88
Morali, N. 214, 216
Moranda, P. B. 59, 68, 72, 75, 77, 106, 109–110, 166
Morris, P. A. 116
Musa, J. D. 44, 70, 72, 78–79, 89, 91, 102–103, 106–107, 118, 120, 131, 209
Myers, G. J. 120

Nair, V. J. 89
Nair, V. N. 228–229, 231, 234
Narihisa, H. 202
Niverthy, M. 90

Okumoto, K. 44, 70, 72, 75, 77–79, 89, 91, 102–103, 110, 118, 120, 202, 209
Olkin, I. 17
Osaki, S. 90, 202
Ozekici, S. 214

Patton, G. C. 228, 232–234
Paulk, M. C. 181
Phadke, M. S. 228, 232–233
Proschan, F. 74
Prowse, J. 228, 232

Raftery, A. E. 88, 146, 151
Raiffa, H. 113
Ramsey, F. P. 194
Randolph, P. 202
Rao, C. R. 170
Roberts, H. V. 147
Ross, S. M. 88, 202, 215

Sahinoglu, M. 47, 85, 202
Savage, L. J. 194
Schick, G. J. 76
Schlaifer, R. 113
Scholz, F. W. 89
Singpurwalla, N. D. 17, 24, 29, 33, 38–39, 44, 54, 58–59, 68–73, 76, 80–81, 83–85, 89, 92, 107, 114, 116–118, 120, 131, 141, 143–144, 156, 160, 174–176, 181, 191, 202, 205–206, 215
Slud, E. 70, 86, 225

Smith, A. F. M. 30, 105, 146, 251, 253
Song, M. S. 116
Soyer, R. 38, 68–69, 76–77, 80–81, 83, 124–126, 139, 141, 153, 174–175, 214, 216
Spreij, P. J. C. 86, 225
Sweet, W. L. 181

Tausworthe, R. C. 82
Thisted, R. 202
Tierney, L. 113
Trivedi, K. S. 52
Tversky, A. 116, 150

Urbach, P. 20

van Pul, M. C. J. 86, 225
Verall, J. L. 77, 83
von Mises, R. 7

Waller, R. A. 113
Wang, P. C. C. 89
Wasserman, L. 113
Weber, C. V. 181
West, M. 172
Weyuker, E. J. 228
Wilson, S. P. 58, 69, 71, 112, 207, 215
Wolpert, R. 25
Wolverton, R. W. 76

Yamada, S. 90, 202
Yates, F. 236
Yang, T. Y. 70, 72, 77, 86, 89

Zadeh, L. 13
Zellner, A. 115
Zevallos, J. 228–229, 231, 234

Subject Index

aposteriori 7, 64, 109, 164
apriori 7, 109, 125, 151
absence of knowledge 115
absolute assessment 146
absolutely continuous 14, 40
accumulation of data 151, 174
adaptive
    concatenated failure rate model 91, 98, 99
    Gaussian Kalman filter (See Adaptive Kalman Filter) 81
    Kalman filter 129, 130, 141, 256
    model 142
adaptivity 92, 160, 188
AETG design 234, 238
allowable service time 68
autocorrelated 170
autoregressive 130, 223

band plots 161
baseline failure rate 226
Bayes factors 102, 149, 159
Bayes law 13, 60, 104, 185
Bayesian
    analysis 107, 114
    approach to analysis (See Bayesian inference)
    approach to prediction 104
    inference 7, 102, 122, 249
    paradigm 83, 102
    statistics 25
Bernoulli distribution (See Distribution)
binomial
    approximation to distribution 32
biostatistical 222
birth and death processes 72, 221
blanketing function 253
branch 196, 205

branch testing 228
bug counting model (See model)
bugs 3

classification of models 70, 95
CMM (See Capability Maturity Model)
coherent 20, 195
compensator 223
completely confounded 237
completely randomized design 237
component reliability theory 14
composite reliability 238
compound Poisson process (See Process)
concatenated failure rate function 58, 72, 84
concatenation 58
concatenation points 84
condition gate 239
conditional
    distribution 16 (See also Distribution)
    expectation 40
    independence 16, 54, 190
    mean 40
    orderliness 51, 84
    probabilities 16, 53
    probability density (See conditional probabilities)
    variance 40
confidence interval 106
consistency 106, 233
continuous random variables 5
control flow 239
control theory 129
convex combination 62
convexity 18
correlation 40
countable additivity 18
counting process (See Process)

covariance 40
covariates 57, 224
    internal 58
Cox regression model (See Model)
cumulative distribution 119
customer goodwill 205

data flow 239, 244
data flow diagram 2, 242
debugging 2, 47, 121, 145
debugging efficiency 48
decision
    making 191, 195, 224
    node 196, 198
    problem 198, 199, 202
    table 196
    theory 197, 217
    tree 196
        one-stage 199
        solving 196
        sequential 199
        two-stage 200
    -theoretic 146
de-eutrophication model (See Model)
defect classification 188
defective 89
degenerate 83, 105, 208
dependence structure 139, 190
dependent 17, 26, 78
design document 3
design of experiment 228, 232, 236
development time 169
discounting factor 204
distribution function 14
    Bernoulli 30
    beta 33, 61
    binomial 32, 41, 69
    DeMoivre's 81 (See also Gaussian distribution)
    exponential 34, 35, 36, 73, 88, 254
    failure rate 53, 54, 55
    gamma 35, 36, 37, 55, 63, 97
    Gaussian 38, 81, 129
    geometric 32, 33

distribution function (cont'd)
    GOS 88
    joint 58 (See also joint distribution)
    k-fold convolution 47
    lognormal 37, 38, 171
    marginal 58, 59, 115, 230
    marginal posterior 157, 159
    Normal (See Gaussian distribution)
    Pareto 77, 97, 141, 255
    Poisson 32
    posterior 61, 101, 114, 115
    prediction 105
    prior 24, 29, 101, 113, 209, 211
    Rayleigh 76
    Standard Normal 38, 119 (See also Gaussian distribution)
    Student's t 175
    truncated normal 38, 39
    of time to first failure 64
    uniform 31, 129
    Weibull 36, 37, 55
disutility 204
DOE (See design of experiment)
Doob decomposition 223
Doob-Meyer decomposition 222, 224, 227
doubly stochastic Poisson process (See process)
DSPP (See doubly stochastic Poisson process)
Dutch book 20, 195
dynamic 129, 130, 164, 188, 221, 222, 245
    linear model (See dynamic)
    modeling (See dynamic)
    statistical model 223 (See also dynamic)

efficiency 102, 106
empirical Baye's method 114
empirical formula
    length of code 120
    bugs per line of code 209

enhanced predictivity 129
EOS (See model, exponential order statistics)
error of prediction 222
estimation 106
    Bayes factors 158
    interval 102, 105, 106, 164
    point 102, 106
evolution
    of concatenated failure rate 84
    of decision tree 196
    of software reliability 82, 86, 221
exchangeability 14, 25, 81, 104, 186
exchangeable model (See model)
exhaustive testing 3, 198, 228
expected
    failure rate 229
    number of failures 77
    number of undetected failures 78
    partition failure rate 230
    utility 194
        single-stage 207, 210
        multi-stage 212
        principle of maximization 194, 217
    value 39, 225 (See also Mean)
experimental design 221, 237
expert opinion 29, 79, 266
exponential distribution (See distribution)
exponentail order statistics model (See model)
exponential smoothing 176
exponentiation formula 54

factorial design 234
    fractional 234
failure
    detection probability 231
    epochs 88, 89
    intensity 93
    model 29, 41, 88
    rate
        function 53, 58

failure rate (cont'd)
    models
        exponential 54
        gamma 55
        lognormal 56
        Weibull 56
    of marginal distribution-function 58
    of software 53, 58, 72
fault detection rate 78
fault tolerance 57, 236
fault-tolerant system 26
finite additivity 18
first moment 39 (See also mean, expected value)
    of a probability model 41
first-order nonhomogeneous-autoregressive process 172
fixed time lookahead 202, 212, 215
forecast density 155
fractional factorial design (See factorial design)
frequentist inference 102, 106
frequentist theory 7

gamma distribution (See distribution)
gamma function 34, 36
Gaussian distribution (See distribution)
Gaussian Kalman filter model (See model)
general order statistics model (See model)
generalization of the exponential
    to gamma 36, 37
    to Weibull 36, 37
generic model (See model)
geometric distribution (See distribution)
Gibbs
    sampling 110, 157, 249, 250
    sequence 158, 251
goodness of fit 25, 146
GOS (See model, general order statistics)

290 Subject Index

Greco-Latin 235
growth curve models (See model)

hierarchical
    Bayes 76, 83, 85, 97
    classification scheme 170
    model (See model)
    priors 113
    structure 60, 182, 187
history 5, 45, 50, 92, 222
homogeneous Poisson process (See process)
HPP (See homogeneous Poisson process)
hyperparameters 113, 172, 176
hypothesis testing 103

impossible events 18
improper prior 115
incoherent 20
incomplete block design 237
incomplete gamma integral 90
incomplete Latin square 236
independence 14, 17, 26, 40
    loss of 88
independent increment 45, 48, 49, 60
innovation 223
input domain 228
input specific reliability 30, 238
intensity function 45
intensity process (See process)
interarrival times 42, 45
interfailure times 58, 69, 70, 71
internal covariate (See covariate)
invariant conditional distributions 81
invertibility 91
iteration 239

joint
    distribution 45, 59, 120, 251
    distribution of the k-order statistics 88
    expectation 40
    k out of n 88

joint (cont'd)
    posterior distribution 108, 112, 157
    predictive density 147
    prior density 112
    probability 21
judgment of indifference 26
jump discontinuities 45

Kalman filter model (See model)
Key Process Areas 181, 261
k-fold convolution 47
KPA (See Key Process Areas)
kth moment 39
kth order autoregressive process 91

lack of memory 55
Latin hypercube 235
Latin square 235
Law of
    inverse probability 21
    the extension of conversation 20
    total probability 13, 21, 29, 30, 184
learning
    environment 162
    phenomenon 171, 176
    process 170, 173
    trend 179
lifelength 17, 29, 55, 57
lifetimes 79
likelihood 14, 24, 25, 157
    prequential 146, 147, 151
    prequential ratio 147, 151
    principle 25
logic engine 68
lognormal distribution (See distribution)
loss of consumer confidence 193, 217

marginal 21
    density 21, 251
    distribution (See distribution)

marginal (cont'd)
    posterior distribution (See distribution)
marked point process 70
Markov
    chain 190, 215
    Chain Monte Carlo simulation 142, 157, 249
    dependence 186
    property 45
martingale 223, 224, 225, 226
    Central Limit Theorem 224
    difference 223
    Law of Large Numbers 224
    semi- 224
    theory 86, 222
maturity levels 181, 183, 266
maturity questionairre 181, 261, 265
maximum entropy priors 115
maximum likelihood estimate 105
MCMC (See Markov Chain Monte Carlo simulation)
mean 39
    conditional 40
    of Bernoulli 41
    of binomial 41
    of gamma 41
    of Gaussian 38
    of Poisson 41
    residual life 95, 155
    square errors 122, 167
    time between failures 39, 54
    time to failure 39, 41
    value function 45, 83, 89, 90
median 118
memory 52, 84, 85, 180
    lack of- 55
    of the self-exciting Poisson process 52
method of maximum likelihood 25, 102, 105, 120
Metropolis–Hastings algorithm 249, 259
MEU (See expected utility - maximization)

minutes per line of code 171
mission time 53, 67
mixture model (See model)
mode 118, 126, 140
model
    adaptive Gaussian Kalman filter 81
    adaptive Kalman filter 130, 139, 145, 256
    averaging 148, 149, 164
    bug counting 74, 76, 227
    complexity 150
    Cox regression 226
    de-eutrophication 73, 77, 166
    EOS 88 (Also exponential order statistics model)
    exchangeable 130, 141
    exponential order statistics 88
    failure rate 53
    Gaussian Kalman filter 81
    general order statistics 88, 98
    generic 154
    GOS 88, 98
    growth-curve 170, 188
    hierarchical 69, 113, 126, 255
    Kalman filter 129
    logarithmic Poisson execution time 80, 103
    mixture 86
    non-Gaussian Kalman filter 81, 130, 141, 145, 258
    nonbug counting 73
    of Goel and Okumoto 70, 75, 83, 206, 211, 218, 227
    of Jelinski and Moranda 72, 82, 95, 202, 218
    of Langberg and Singpurwalla 73, 83
    of Littlewood and Verall 77, 83
    of Mazzuchi and Soyer 76, 83
    of Musa–Okumoto 79, 91, 96, 209
    of Ohba and Yamada 90
    of Schick and Wolverton 75, 76, 85, 91

model (cont'd)
  record value statistics 89
  selection 102, 146, 149, 150
  shock 74
  software reliability 29, 67, 69, 86
    Type I 71, 82, 86, 214
    Type II 71, 77
  time dependent error detection 77, 86, 110
  uncertainty 172
modular decomposition 245
modules 238
Monte Carlo simulation (See MCMC)
MRL (See mean residual life)
MSE (See mean square error)
MTBF (See mean time between failures)
MTTF (See mean time to failure)
multinomial 188
multiplication law (See multiplicativity)
multiplicative 18
multistage hierarchy 130
mutually exclusive 18

natural conjugate priors 113
Naval Tactical Data Systems 109, 126, 127, 153, 216 (Also NTDS)
NHPP (See nonhomogeneous Poisson process)
nonbug counting 73
noninformative priors 114
nonunique estimators 121
normalizing constant 39
normative approach 170, 230
NTDS (See Naval Tactical Data Systems)

observation equation 81, 130
Occam's razor 150
omnibus prior 156
one-bug lookahead 202, 214
one-stage lookaheed 199, 202, 206, 211, 215

one-stage testing 198, 202, 206
one-step-ahead prediction 121, 165
operational profile 3, 15, 50, 67, 221, 227
opportunity cost 204, 210
optimal testing time 192, 193
order statistics 86, 87, 98
  ith 87
  largest 87
  process 226
  smallest 87
ordering 86, 88
orthogonal array 233
orthogonal increments 223
orthogonal Latin square (See Latin square)

parallel redundant 86
parameters 14
partition testing 228, 229, 231
penalty for late release 204
permutation invariance 28
point mass 14
point of saturation 176
point process (See process)
Poisson process (See process)
possibility theory 4
posterior
  density 117
  distribution (See distribution)
  inference 109
  mean 62
  mode 140
  odds 149, 151
  probability 24
  weight 149
power law 80, 204
predictable process 223
prediction
  error 222
  interval 174, 178
  limits 179
predictive
  ability 145, 150, 151
  density 110, 111, 254, 256

Subject Index 293

predictive (cont'd)
    distribution (See distribution)
    failure rate function (See failure rate function)
    mean 94, 160, 161
    variance 94
predictivity 92, 129, 139, 141
preposterior analysis 196, 232
prequential likelihood (See likelihood)
prequential prediction 102, 146, 147
principle of indifference 6
prior
    distribution (See distribution)
    mean 62, 207
    odds 149
    probability (See also prior distribution)
probability density function 14 (See also prior probability)
probability models 14, 29, 39 (See also model)
probability specification process 101
probability theory 4, 191, 223
process
    counting 14, 42, 84, 224, 226
    intensity 225, 227
    intensity function of 43
    management 180, 263
    point 14, 41, 45, 46, 83, 86
    Poisson 43, 71
        compound 46, 47, 60
        doubly stochastic 48, 50, 52, 60
        homogeneous 45, 77
        nonhomogeneous 44, 45, 78, 164
        self-exciting 51, 52, 60, 84, 85, 222
    shot 92
    shot noise 92, 93
    stochastic 42, 44, 48, 222, 227
    stochastic counting (See process - counting)
    stress 92

product moment 40 (See also joint expectation)
product obsolescence 192, 193, 204
productivity data 179, 188
productivity rates 175
projection 10, 174, 175, 179

random
    coefficient autoregressive process 71, 80, 188
    coefficient exchangeable model (See model - exchangeable)
    events 5
    node 196
    quantities 5
    sampling 228
    testing 228, 231
    variables 5, 68, 69
        continuous 5
        discrete 5
        exchangeable 26
        mixed 14
randomized complete block design 236
RCAP (See random coefficient autoregressive process)
record
    pairs 90
    times 90
    value statistics model (See model)
    values 86, 90
recursive probabilities classifications scheme 186
recursive relationship 185
reference
    priors 113
    time 5
rejection kernel 253, 256
rejection sampling 253
relative frequency 7, 102
relative growth in reliability 95
reliability 10
    assessment 17
    decay 82, 157, 159

294 Subject Index

reliability (cont'd)
    function 52, 97
    growth 76, 77, 95, 154, 159
    modeling 82, 88
risk
    aversion 195
    neutral 196
    proneness 196
    set 226
robust 180, 232

sample path 42
scale parameter 35
scoring rules 20
second moment 39
self-exciting Poisson process (See process - Poisson)
SEI (See Software Engineering Institute)
semimartingale (See martingale)
sensitivity 113, 114, 122
SEPP (See self-exciting Poisson process)
sequential testing 192, 198, 217
series system 19, 87
several steps ahead predictions 122
shape parameter 37
shifted exponential density 155
shifted gamma 77
shock model (See model)
significance testing 146
simulation (See MCMC)
single-stage testing (See one-stage testing)
software 2
    credibility (See software reliability)
    development 2
    development cycle 228
    downtime 47
    failure 3
    productivity 170, 176
    reliability (See reliability)
    reliability model (See model)

Software Engineering Institute 10, 170
specialist knowledge 115
stage-by-stage growth 80, 180
stages of testing 131, 199, 218
standard deviation 40, 120, 209
Standard Normal distribution (See distribution)
states of nature 193
stationarity 157
statistical decision theory (See decision theory)
statistical inference 4, 67, 228
step function 14, 42, 225
stochastic counting process (See counting process)
strata 228, 230
stress process 92
structure function 244
subjective interpretation of probability 8, 11, 25
subjectivistic Bayesian inference 7
    (See also Bayesian inference)
survival analysis 222, 225
survival function 52, 98
System 40 data 131
system equation 81, 130, 214

test cases 228, 231
testing phases 3, 192
testing strategies 228, 232
time dependent error detection model (See model)
time sequence 239
time series
    models 80, 91
    processes 43
total debugging time 47
truncated normal distribution (See distribution)
tuning coefficients 116
two-stage testing 198, 206

unbiasedness 102, 106

unconditional 46
unconditional orderliness 51
uniform distribution (See
  distribution)
uniqueness 106
unit testing 176
universal model 82
unreliability 19, 92
utilities 193, 195, 206, 215
   assigning 193
   expected 194, 205, 210,
      principle of maximization
         194

utilities (cont'd)
   function 203, 209, 214
      concave 195
      convex 196
   of money 195
   theory 194, 217

variance 38, 40
   conditional 40

waiting times 45

Youden square 236

Springer Series in Statistics

(continued from p. ii)

Kotz/Johnson (Eds.): Breakthroughs in Statistics Volume II.
Kotz/Johnson (Eds.): Breakthroughs in Statistics Volume III.
Kres: Statistical Tables for Multivariate Analysis.
Küchler/Sørensen: Exponential Families of Stochastic Processes.
Le Cam: Asymptotic Methods in Statistical Decision Theory.
Le Cam/Yang: Asymptotics in Statistics: Some Basic Concepts.
Longford: Models for Uncertainty in Educational Testing.
Manoukian: Modern Concepts and Theorems of Mathematical Statistics.
Miller, Jr.: Simultaneous Statistical Inference, 2nd edition.
Mosteller/Wallace: Applied Bayesian and Classical Inference: The Case of the Federalist Papers.
Parzen/Tanabe/Kitagawa: Selected Papers of Hirotugu Akaike.
Politis/Romano/Wolf: Subsampling.
Pollard: Convergence of Stochastic Processes.
Pratt/Gibbons: Concepts of Nonparametric Theory.
Ramsay/Silverman: Functional Data Analysis.
Rao/Toutenburg: Linear Models: Least Squares and Alternatives.
Read/Cressie: Goodness-of-Fit Statistics for Discrete Multivariate Data.
Reinsel: Elements of Multivariate Time Series Analysis, 2nd edition.
Reiss: A Course on Point Processes.
Reiss: Approximate Distributions of Order Statistics: With Applications to Non-parametric Statistics.
Rieder: Robust Asymptotic Statistics.
Rosenbaum: Observational Studies.
Ross: Nonlinear Estimation.
Sachs: Applied Statistics: A Handbook of Techniques, 2nd edition.
Särndal/Swensson/Wretman: Model Assisted Survey Sampling.
Schervish: Theory of Statistics.
Seneta: Non-Negative Matrices and Markov Chains, 2nd edition.
Shao/Tu: The Jackknife and Bootstrap.
Siegmund: Sequential Analysis: Tests and Confidence Intervals.
Simonoff: Smoothing Methods in Statistics.
Singpurwalla and Wilson: Statistical Methods in Software Engineering: Reliability and Risk.
Small: The Statistical Theory of Shape.
Stein: Interpolation of Spatial Data: Some Theory for Kriging
Tanner: Tools for Statistical Inference: Methods for the Exploration of Posterior Distributions and Likelihood Functions, 3rd edition.
Tong: The Multivariate Normal Distribution.
van der Vaart/Wellner: Weak Convergence and Empirical Processes: With Applications to Statistics.
Vapnik: Estimation of Dependences Based on Empirical Data.
Weerahandi: Exact Statistical Methods for Data Analysis.
West/Harrison: Bayesian Forecasting and Dynamic Models, 2nd edition.
Wolter: Introduction to Variance Estimation.
Yaglom: Correlation Theory of Stationary and Related Random Functions I: Basic Results.
Yaglom: Correlation Theory of Stationary and Related Random Functions II: Supplementary Notes and References.